Photochemical and Photobiological Reviews

Volume 2

Photochemical and Photobiological Reviews

Volume 2

Edited by

Kendric C. Smith
Stanford University School of Medicine

PLENUM PRESS · NEW YORK AND LONDON

Library of Congress Cataloging in Publication Data

Main entry under title:
  Photochemical and photobiological reviews.

  Includes bibliographies and index.
  1. Photobiology—Collected works. 2. Photochemistry—Collected works. I.
Smith, Kendric C., 1926-   [DNLM: 1. Radiobiology—Periodicals. 2. Photo-
chemistry—Periodicals.   W1 PH653]
QH515.P48   574.1'9153   75-43689
ISBN 0-306-33802-5 (v. 2)

©1977 Plenum Press, New York
A Division of Plenum Publishing Corporation
227 West 17th Street, New York, N.Y. 10011

Printed in the United States of America

Preface

The science of photobiology is a dynamic multidisciplinary field whose relevance to the needs of man is growing more apparent each day. Publicity about supersonic transports (the SSTs) and spray cans, their possible deleterious effects on the stratospheric ozone layer, and the possible resultant consequences of enhanced solar ultraviolet radiation on man and his environment have helped to focus attention on both the beneficial and the detrimental effects of light. In addition, considerable activity is currently being directed toward harnessing solar energy as one solution to the world energy crisis. Some mechanisms for accomplishing this involve photobiological systems or photochemical models based on these systems. It would thus seem that modern man has rediscovered the sun and is now actively considering new uses of light rather than thinking of light only as an aid to vision. Photobiology has become a major new scientific field.

The American Society for Photobiology has divided the science of photobiology into 14 subspecialty groups:

Phototechnology	Chronobiology
Photochemistry	Photoreception
Spectroscopy	Vision
Photosensitization	Photomorphogenesis
Ultraviolet and Visible Radiation Effects	Photomovement
Environmental Photobiology	Photosynthesis
Medicine	Bioluminescence

The goals of the science of photobiology have been divided into four categories*: "(1) The development of ways to protect organisms, including man, from the detrimental effects of light; (2) the development of ways to control the beneficial effects of light upon our environment; (3) the continued development of photochemical tools for use in studies of life processes; and (4) the development of photochemical therapies in medicine."

* K. C. Smith, *BioScience* 4:45–48 (1974).

In every scientific field it is important that the leaders review the field periodically, as a service to the younger scientists in the field and to senior scientists in related fields. Such reviews provide a ready access to the recent literature in the field, and, more importantly, they often provide a critical evaluation of the direction that the field is taking, and frequently suggest a redirection when appropriate. To achieve these objectives, the present series, *Photochemical and Photobiological Reviews,* was inaugurated.

Kendric C. Smith
Editor

Contents

Biological, Photochemical, and Spectroscopic Applications of Lasers

Michael W. Berns

*Department of Developmental and Cell Biology, University of California, Irvine,
Irvine, California 92717*

1. INTRODUCTION

Not more than 5 years ago, it was common to refer to the laser as an
"instrument in search of a problem." Here was this marvelous device that
could generate electromagnetic radiation that was naturally monochromatic
($+0.05$ nm) anywhere from 250 nm through the visible and infrared regions

of the spectrum, and both the peak intensities and the average intensities were thousands of orders of magnitude higher than obtainable with the older classical sources. In addition, it was possible to generate ultrashort pulses (below 10^{-12} s) of this intense, monochromatic radiation. Furthermore, because of the physical mechanism involved in stimulated emission of radiation, and the design of laser cavities, the output beam was always plane polarized and virtually nondivergent.

It was felt by many people that a radiation source with the above features would literally revolutionize the fields of photochemistry and photobiology. Whether or not these lofty expectations will ever be realized is certainly debatable. However, even now, it is clear that the special features of the laser have led to more sophisticated and sensitive studies on photochemical reactions and in the area of spectroscopy (Rentzepis, 1968.)

The spectroscopic processes examined by classical means have generally been one-photon absorption processes involving singlet–singlet transitions. However, these are not the only energy levels that excited molecules can occupy. The extremely high photon density attainable with lasers ($>10^{24}$ photons cm^{-2} s^{-1}) finally permits the study of the "forbidden" processes that can occur. Such processes as multiphoton absorption (Pao and Rentzepis, 1965) and spin-forbidden transitions (Novak and Windsor, 1967) can now be studied. Other nonradiative transitions (Jortner and Berry, 1968) such as vibrational relaxation of large molecules, intersystem crossing, and energy transfer can also be studied, because these transitions, which occur very rapidly, can be separated from the stimulatory pulse by using ultrafast bursts of intense laser light as the stimulatory radiation. In addition, entire new areas, such as laser-induced isotope separation and laser-induced emission flash spectroscopy, are being developed.

In biological areas, the laser has become a very useful radiation source in the older techniques of microbeam irradiation and microfluorimetry, greatly expanding the types of studies performed and the amount of information obtained from the material. In addition, several unique biological systems have been developed that employ the coherent properties of the laser in light scattering and diffraction studies.

It is the purpose of this chapter to examine those areas in spectroscopy, photochemistry, and biology where the laser is having its greatest impact. In addition, the more promising areas in which the laser is just beginning to be used will be discussed. For the sake of simplicity, the studies discussed herein will be divided into the areas of spectroscopy, photochemistry, and biology. However, it must certainly be recognized that the photochemistry of biological molecules (some of which will be treated in the section on photochemistry) falls under the purview of the three major subdivisions.

2. SPECTROSCOPY

There are at least four different types of spectroscopy in which the laser is used. Two of these, stimulated Raman spectroscopy and flash photolysis, are quite widely employed in diverse kinds of problems. The others, emission microspectroscopy and fluorescence (and luminescence), are more limited in their applications, and are still in the developmental stages.

2.1. Raman Spectroscopy

Raman spectroscopy is an old technique that provides spectra that contain very precise information about the vibrational electronic structure of molecules. However, it has enjoyed only limited success, principally because it has been difficult to obtain spectra that are clearly resolvable from background noise. This problem is further complicated by the fact that the intensity of the emitted Raman wavelengths is often extremely weak, and therefore not easily detected. The highly intense, coherent, monochromatic beam from the laser stimulates the emission of clear and strong Raman spectra. In addition, in certain molecules it is possible to obtain "resonance enhancement" in Raman spectra when the excitation frequency is close to the absorption peak of the scattering molecules (Behringer, 1967; Heyde *et al.,* 1972). The result of this resonance enhancement is the production of greater Raman intensities of certain fundamental vibrations for the specific molecules.

Laser Raman spectroscopy has been employed to study a large variety of biologically important molecules. These have generally been pigments and proteins associated with vision, respiratory proteins, nucleic acids, and numerous other complex molecules such as antibiotics, hormones, and neurotoxins.

2.1.1. Pigments and Proteins Associated with Vision

Extensive studies involving the visual system have been performed by Rimai *et al.* (1970, 1971*a,b*), Gill *et al.* (1971), and Lewis and Spoonhower (1974). By generating resonance-enhanced Raman spectra of the visual pigments in intact bovine retinas at low temperatures, it was suggested that the pigment in the retina is bound as a protonated Schiff base. It was concluded that the Raman spectra observed were contributed only by the visual pigments, and, in particular, the strongest line was contributed by the ethylenic C=C stretching mode of retinal (Rimai *et al.*, 1970). In addition, Raman emissions were used in order to study inaccessible electronic levels in molecules of retinal, retinol, and naphthalene (Rimai *et al.*, 1971*b*). Raman spectra of dilute solutions of vitamin A demonstrated that the technique of laser Raman spectroscopy was very sensitive, not only to the nature of the terminal group on the vitamin A, but also to the particular isomeric configuration. The authors felt that the Raman spectra identifying the differences between the various vitamin A isomers were much more apparent than for the other commonly employed techniques of absorption, fluorescence, and infrared spectroscopy (Rimai *et al.*, 1971*a*). In another study on the 11-*cis* isomer of retinaldehyde (Gill *et al.*, 1971), the uniqueness of the Raman spectrum of each of the mono-*cis*-retinal isomers was confirmed. It was concluded that the identification of the stereoisomers could be obtained by rapid scanning in the limited range of 1100–1300 cm^{-1} even with far-red laser excitation (674 nm). The possibility was suggested that Raman studies could be performed on rhodopsin and its less stable bleaching intermediates.

The visual system has also been studied extensively by Lewis *et al.* (1974). In particular, they have studied both mammalian rhodopsin (Lewis and Spoonhower, 1974) and "bacteriorhodopsin," the rhodopsinlike compound from the bacterium *Halobacterium halobium*. One of the major methods of study employed laser resonance Raman spectroscopy. In conjunction with biochemical and other spectroscopic data, the Raman spectroscopic data suggested that the Schiff-base linkage of retinal to opsin is protonated. This result differed from the case of bacteriorhodopsin, where it had been suggested that retinal was linked to opsin by an unprotonated Schiff base (Mendelsohn, 1973). Lewis *et al.* (1974) reexamined this question using laser resonance Raman spectroscopy. They concluded that the retinylidene–lysine linkage in bacteriorhodopsin is protonated. In addition, their results suggested that the proton is removed when the bacteriorhodopsin undergoes its photochemical cycle, and is restored when the pigment returns to its original form.

In quite a different Raman study on the visual system, Yu and East (1975) investigated isolated protein fractions of the ocular lens. They performed Raman spectral analysis on water-soluble proteins designated α-, β_1-, β_2-, β_3-, and γ-crystallin. The laser-generated Raman spectra indicated that the proteins contained a predominantly antiparallel pleated sheet structure in the main chains, and that sulfhydryl groups were highly localized in the γ-crystallin. They were further able to make specific conclusions about the homogeneity of the protein structure of the intact lens and the distribution of certain amino acids throughout the lens. Also, by examining the Raman spectra in the 2582 cm^{-1} and amide I and III regions, they demonstrated that the sulfhydryl groups and the β conformation of the lens proteins were unaffected when the transparent lens was converted to an opaque lens by heat denaturation. From this result, it was concluded the lens opacification does not necessarily involve oxidation of sulfhydryl groups or conformation changes.

2.1.2. Respiratory Chain Molecules

Raman spectroscopy has been used to study several of the molecules in animals and plants involved in respiration. Salmeen *et al.* (1973) have performed resonance Raman spectroscopy of cytochrome oxidase, both solubilized and in electron transport particles. The excitation wavelengths employed were near the Soret band and apparently provided much greater sensitivity in the resonance Raman spectra than when excitation wavelengths near the visible α and β bands were used, as in earlier studies on cytochrome c and hemoglobin (Strekas and Spiro, 1972; Spiro and Strekas, 1972; Brunner *et al.*, 1972). As in the spectra of other hemoproteins, such as cytochrome c, the intensity and shape of several of the resonant bands changed with changes in the oxidation state. However, it appeared that in one of the hemes of cytochrome c oxidase the redox behavior was anomalous. In addition, it was demonstrated that the spectra of the electron transport particles were dominated by cytochrome c oxidase.

In other studies on hemoproteins, Yamamoto *et al.* (1973) performed resonant Raman spectra using the excitation wavelengths in the Soret region. The main purpose of the studies was to characterize the features of the valence and spin states of iron. Their spectral results strongly supported the assignment of a low-spin ferric structure to the iron ion in oxyhemoglobin.

Laser Raman spectroscopy has also been used as a mechanistic probe of the phosphate transfer from ATP (Lewis *et al.*, 1975). The spectra obtained in these studies support a mechanism involving Mg^{2+} binding of the α and β phosphates of ATP, leaving the third phosphate free for the transfer reaction. It is also suggested that a relatively stable intermediate is formed and facilitated by the presence of dimethyl sulfoxide and maleate. This intermediate has a Raman spectrum similar to that of the end-product ADP. Since the *in vitro* model appeared to have many features in common with the transfer reactions catalyzed by coupling factors from spinach chloroplasts, it was suggested that laser Raman spectroscopy could prove to be a useful tool in the elucidation of biological energy transfer reactions.

2.1.3. Nucleic Acids

Laser Raman spectroscopy also has been used to probe the intramolecular organization and secondary structure of nucleic acids. Laser-excited Raman spectra of ribosomal RNA of *Escherichia coli* in H_2O and ^2H_2O have been studied by Thomas (1970). Five major conclusions were made: (1) the Raman lines caused by the phosphodiester group were clearly distinguishable from those of sugar or base residues, (2) the symmetrical stretching vibration of the PO_2 group at 814 cm^{-1} was particularly sensitive to ionic strength while the PO_2 vibration at 1100 cm^{-1} was not, (3) Raman lines due to bases involved in purine–pyrimidine hydrogen bonds occur prominently in the double-bond region, (4) characteristic ring vibrations allow each of the four heterocyclic bases to be distinguished from each other in one single spectrum, and (5) information about the effects of isotopic exchange may be obtained from comparison of H_2O and ^2H_2O spectra. The overall conclusions from the above experiment were that Raman spectra of aqueous nucleic acids and polynucleotides will be useful for the determination of base compositions, for the characterization of secondary structure in terms of paired and unpaired bases, and for the study of secondary structure changes produced as a result of ionic concentration changes, temperature changes, etc. In addition, it was suggested that Raman spectroscopy will be useful for the study of molecular interactions between nucleic acids and other biologically important molecules and ions. In a study on transfer RNA, Tsuboi *et al.* (1971) have investigated the intramolecular environments of the base residues. They were able to demonstrate the extent to which the vibrational states of each base residue are perturbed in the secondary structure of the tRNA molecule.

DNA also has been studied by laser Raman spectroscopy (Erfurth and Peticolas, 1975). These authors studied melting and premelting conformational changes in DNA. The Raman bands increased gradually in intensity for each of the four DNA bases prior to the melting region, and abruptly increased at the melting temperature for DNA. The conclusion was made that changes in the bases relative to each other begin to occur at around 50°C, well below the melting region of 70–85°C. From the various spectra obtained, it appears that DNA remains in the B conformational state until melting is achieved, at which time the DNA progresses into a disordered random coil form. No A-form conformation was found in either the premelting or the melting region.

2.1.4. Miscellaneous Compounds

Raman spectroscopy also has been applied to study the molecular organization of insulin and proinsulin (Yu *et al.,* 1972), antibiotics gramicidin A′ (Rothschild and Stanley, 1974) and valinomycin (Asher *et al.*, 1974), and sea snake venoms (Yu *et al.*, 1975). And as a final application of Raman spectroscopy, Inaba and Kobayashi (1969) have employed laser–Raman radar to analyze the chemicals in polluted air.

2.2. Flash Photolysis

The other major spectroscopic area in which the laser is being widely applied is in flash photolysis. In this technique, the material is exposed to a short, bright flash of radiation, producing photoinduced products and/or reactions which can be studied either spectroscopically or chemically. An important aspect of this approach is that if the flash of stimulating radiation is short enough, it is then possible to detect and study species and reactions with very short lifetimes. There are several advantages of laser flash photolysis over conventional xenon flash photolysis: (1) laser flashes are much shorter, with comparable or greater quanta (i.e., it is now possible to obtain subnanosecond pulses, Andreoni *et al.*, 1975*a*, and subpicosecond pulses, Ippen and Shank, 1975); (2) there is greater ease in spectrally separating the exciting and measuring light; (3) highly specific excitation is obtainable with monochromatic light; and (4) there is a minimum of electrical and acoustic interference because the high degree of collimation of laser radiation permits the operation of the laser energy generator some distance from the detector.

Over the past 10 years, there have been hundreds of studies employing laser photolysis. However, as in the previous section on Raman spectroscopy, a large number of the biologically relevant investigations have dealt with respiratory molecules in plants and animals and with the molecules involved in the visual process.

One of the first applications of lasers to biological problems was the use of a pulsed ruby laser for the rapid activation of cytochrome oxidation in the photosynthetically pale green mutant of *Chlamydomonas reinhardii* (Schleyer and Chance, 1962; Chance *et al.*, 1963). In a later study, the same technique was used to study *Chromatium* (Chance and Schoener, 1964). The results suggested the laser flashes contained enough power and shortness (of time) to detect a turnover of about one cytochrome *b* per ten chlorophyll *a* molecules per second. Following these studies (DeVault and Chance, 1966), it was possible to measure very rapid cytochrome oxidation (half-time 2×10^{-6} s). These studies employed laser flash photolysis in which the laser flash duration was 20×10^{-9} s.

In more recent studies (Chance *et al.*, 1970), the fast changes in hydrogen ion concentration following photoactivation of bacteriochlorophyll have been measured by observing changes of bromcresol purple. A dye laser pumped by a ruby laser was employed because it was possible to generate an output pulse corresponding to an absorption peak of the chromatophore. The system has been used to obtain the time-resolved difference spectrum for the binding of H^+ to the chromatophore membrane following a 20×10^{-9} s pulse. The results were interpreted to indicate a very rapid light-induced change.

Rentzepis and his colleagues (Netzel *et al.*, 1973) have studied the picosecond kinetics of the bleaching of the 865-nm absorption band. A pheophytin band with a maximum at 535 nm was excited with the second harmonic (530 nm) of a mode-locked neodymium glass laser, and the kinetics of the bleaching of the bacteriochlorophyll band at 865 nm were measured in the picosecond range. According to the authors, this work permitted study of the very first steps in photosynthesis. The authors conclude that

> when the picosecond range is monitored, great care must be taken to ensure that long-lived intermediates do not contribute artificial results. This work has enabled us to observe the very first steps in photosynthesis. What was found was that photooxidation of the reaction center is not an instantaneous process when it is excited with green light. Rather an incubation period of a few picoseconds is needed to transfer the energy to P865. If the bleaching of P865 is concomitant with the electron ejection, then this work is a direct measure of the rate of photooxidation. However, the strong interactions between the bacteriopheophytins and the bacteriochlorophyll (P865) are clearly demonstrated. This confirms a general belief that could not be directly substantiated with previous experimental techniques.

In addition, the state of the membrane was characterized by an increased hydrogen ion binding, approximately equal to one H^+ per hundred molecules of chlorophyll. This entire reaction had a half-time of 400×10^{-6} s and was completed by 800×10^{-6} s.

Numerous additional studies on the photosynthetic, respiratory, and visual system reactions have involved even shorter stimulating pulses. For example, picosecond pulses (10^{-12}–10^{-9} s) have been used to study ultrafast kinetics of chlorophyll reaction centers (Netzel *et al.*, 1973). The picosecond kinetics leading to bacteriochlorophyll oxidation were studied (Kaufmann *et al.*, 1975). Mathis *et al.* (1975) have studied the primary reactions in photosystem-2 of spinach chloroplasts at low temperature. Windsor *et al.* (1975) have used 8×10^{-12} s flashes to study the photosynthetic reaction centers of *Rhodopseudomonas*. They were able to demonstrate a transient state immediately following excitation that was characterized by new absorption bands near 500 and 600 nm. This transient state had a lifetime of about 246×10^{-12} s following the flash. As this transient state decayed, the radical cation of the reaction center bacteriochlorophyll complex appeared.

This result indicated that the transient state is an intermediate in the photooxidation of bacteriochlorophyll. Further spectral analysis permitted conclusions concerning electron transfer within the reaction center.

Another molecule involved in photosynthesis (and also mitochondrial energy transport) that has been studied by nanosecond laser flash photolysis is ubiquinone (coenzyme Q-6). Bensasson and Land (1971) and Bensasson *et al.* (1972) have studied the heretofore undetected triplet state of this coenzyme and its derivatives. As pointed out by these authors, the excited states of the naturally occurring quinones are of interest with respect to the degradative action of light on biological materials, photophosphorylation, and possibly in the initiation of electron transport in mitochondria and chloroplasts by light absorption. The characterization of the physico-chemical properties of the excited states of the quinones is an important aspect in better understanding the function of these molecules.

In studies on several derivatives of coenzyme Q, Bensasson and Land (1971) were able to determine the triplet absorption spectrum, extinction coefficient, lifetime, energy level, and quantum efficiency of formation. Furthermore, by comparison with previous studies on coenzyme Q, it was determined that the low triplet energy and quantum efficiency of the formation of triplet ubiquinone-30 are caused by two adjacent methoxy substitutes. The low quantum efficiency of triplet formation suggested that little of the ubisemiquinone observed in bacterial photosynthesis is formed via excited ubiquinone.

Laser photolysis studies have been conducted on other important respiratory molecules such as hemoglobin (Alpert *et al.*, 1974) and cytochrome a_3 (Chance and Erecińska, 1971). In the hemoglobin study, the kinetics of conformational changes were studied. Photolysis of carbon monoxide and oxygen derivatives of hemoglobin produced transient species that rapidly decayed to normal deoxyhemoglobin. This effect was interpreted as corresponding to structural changes in the heme pocket on ligand dissociation. The decay of the transient species followed first-order kinetics. The kinetic constants were pH dependent, although they remained first order or pseudo first order at all wavelengths. The authors interpreted this as demonstrating the close linkage of tertiary and quaternary structure changes in normal hemoglobin.

In their studies on cytochrome a_3, Chance and Erecińska (1971) developed a new dual wavelength spectrophotometric technique called "cycle selection," which permitted the recording of absorbancy changes from 150×10^{-6} s onward. This technique was combined with laser flash photolysis employing 0.4×10^{-6} s pulses of light from a newly developed liquid dye laser. The system was employed to study the kinetic properties of cytochrome oxidase in intact mitochondrial membranes, the associated crossover responses of cytochromes a_3 and c, and the energy coupling between cytochrome c and cytochrome oxidase. The experimental data permitted explanations for the existence of crossover points between cytochrome a_3 and oxygen and between oxygen and cytochromes c and a. It was concluded that electron flow operates on both oxidizing and reducing reactions of cytochrome oxidase.

The symmetry properties and rotational mobility of cytochrome oxidase in inner mitochondrial membranes have been studied by Junge and DeVault (1975). They used a linearly polarized laser flash to cause the photodissociation of the complex between cytochrome oxidase and CO. The resulting absorption changes of the cyt-a_3-heme were evaluated for transient linear dichroism. Their general results indicated that cytochrome oxidase carries out only anisotropic rotational diffusion in the mitochondrial membrane, and the rotational axis coincides with the symmetry axis of the cyt-a_3-heme.

As mentioned in Section 2.1.1, laser photolysis also has been used to study molecules of the visual system. Fisher and Weiss (1974) have performed laser photolysis studies on retinal and its protonated and unprotonated Schiff base. They have employed the second harmonic wavelength (347 nm) of the ruby laser to measure the triplet–triplet absorption spectra, and triplet lifetimes of *trans*-retinal, *N-trans*-retinylidene-*n*-butylamine (NRBA), and protonated NRBA (NRBAH$^+$) at room temperature. Evidence was presented for a significant solvent effect in the isomerization of retinal via the triplet state and for the occurrence of $cis \rightarrow trans$ isomerization from the triplet state of NRBAH$^+$. All of these results were discussed in light of rhodopsin photochemistry.

Rhodopsin itself has been studied using nanosecond flash photolysis at 337 nm emitted from a nitrogen gas laser (Rosenfeld *et al.*, 1972). Application of fast flash photolysis makes possible the study of the intermediate products and reactions in the bleaching of rhodopsin $\rightarrow$ all-*trans* retinal + opsin at physiological temperatures. All previous studies required low temperatures. With a time resolution of 10×10^{-9} s, it was possible to study the initial stages of rhodopsin photochemistry at physiological temperatures.

Laser photolysis studies have also been conducted on several of the flavins (Visser *et al.*, 1974; Katan *et al.*, 1971) to determine the effect of pH on the appearance of transient absorptions after the flavins had been flashed in aqueous solutions. Three transient species were identified and their decay curves analyzed. Two of the species spectra apparently were due to excited singlet and triplet states, and were not affected by pH. The third spectrum was dependent on pH, and the nature of this species and the pH changes were not fully resolved.

2.3. Microprobe-Emission Spectroscopy

Laser microprobe-emission spectroscopy is a type of spectroscopy that has been developed and employed in only a couple of laboratories. The most productive and initiating laboratory has been Glick's group at Stanford. The basic approach of this technique is to use a microscope to focus a high-power laser into a biological sample. At the focal point of the microscope objective, the material is vaporized. The vaporized material subsequently sets off the discharge of an electric spark (Fig. 1). The light from the spark passes through a spectrograph and is analyzed quantitatively. Current technology permits detection of elements such as Mg^{2+}, Ca^{2+}, and Ag at levels of 10^{-15}–10^{-19} g (10^{-10}–10^{-9} mol) in sample areas as small as 50–100 μm (Glick, 1966, 1969; Glick and Rosan, 1966). Comparative elemental determinations have been made between cells in culture, human skin, transplanted hearts, and human cancer tissues (Glick and Marich, 1975). Considerable differences were found in the elements in cancer vs. normal tissue, and several similarities were found in different cancer tissues. Explanations for the observations were not presented; however, the potential diagnostic value of the technique was pointed out.

In a different kind of study, Hillenkamp *et al.* (1975) have combined mass spectrometry with the laser microprobe technique. The laser system employed a Q-switched frequency-doubled ruby laser that was passed into a microscope and focused down to a spot approximately 0.5 μm in diameter. At the focal point of the laser, some of the material was ionized and ejected directly into a mass spectrometer, which analyzed its constituents according to atomic or molecular mass. The overall sensitivity of the technique in analyzing lithium was a mass detection of 1.4×10^{-19} g or 1.4×10^{-4} atoms.

2.4. Fluorescence and Luminescence

Lasers have been used as the excitatory source in fluorescent and luminescent studies. These studies can be divided into two general categories: (1) those that are aimed mainly at studying the photochemistry of atoms and molecules in nonliving systems and (2) studies in which the fluorescent signal is generated in cellular systems with the purpose of elucidating biological function and structure. This latter group of investigations will be discussed in the third section of this chapter.

Laser-induced fluorescence has developed as a powerful method for elucidating complex molecular spectra and determining molecular constants, potential curves, and dissociation energies. This is possible

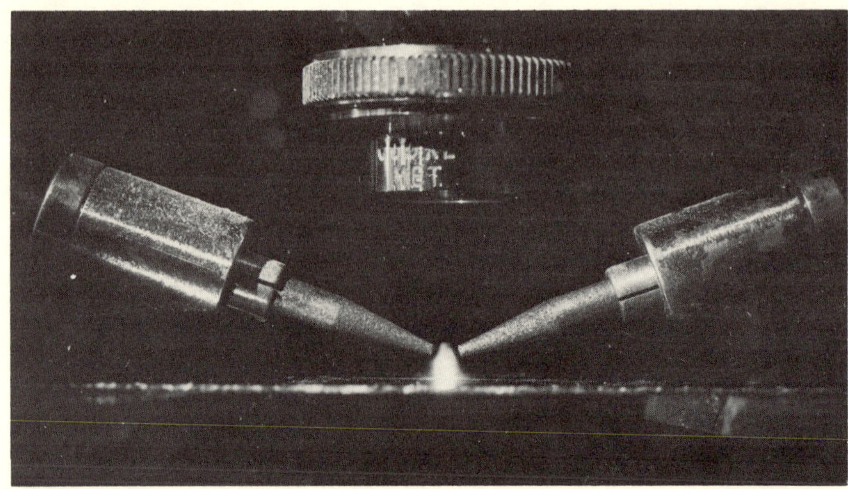

Fig. 1. Microprobe-emission spectroscopy. Top: Plume from a tissue sample, 45 μm diameter, without cross-excitation, 0.1 J laser energy output. Bottom: Plume from similar sample with 50 J cross-excitation energy. Note tracks of incandescent material leaving sample zone and large contact area between spark and sample. From Beatrice *et al.* (1969).

because the high intensities attainable with lasers permit excitation of a significant number of molecules into definite rotational-vibrational levels of electronically excited states. Through collisions with other atoms or molecules, the excited molecules may undergo energy transfer processes that can be studied by observing the fluorescence "from the collisionally populated neighboring levels" (Demtroder and Stock, 1975).

The laser-induced fluorescence technique permits the determination of absolute collision cross sections for rotational (Bergmann and Demtroder, 1972), vibrational, and dissociative transitions (Bergmann *et al.*, 1974). Investigations have demonstrated that precise information about the interaction potential between excited molecules and other collision partners can be obtained (Klar, 1973). In a detailed study on Na_2, using a single-mode CW argon ion laser, Demtroder and Stock (1975) were able to identify all the lines in the Na_2-fluorescence spectrum that originated from the B state. This was possible even for transitions with very high vibrational or rotational quantum numbers. Similar studies by Ault *et al.* (1975) have used the CW argon ion laser to generate the fluorescence spectra of halogen molecules in inert matrices. This has permitted the determination of spectroscopic constants and the determination of the degree of aggregation of the species in the matrix.

In a rather unique application of laser-induced fluorescence, Berman and Zare (1975) have employed a pulsed nitrogen laser (337.1 nm at 14×10^{-9} s/pulse) to stimulate fluorescence in chromatograms containing aflatoxins. These are carcinogenic mold metabolites whose accurate detection in various foodstuffs and compounds would be extremely useful. With the intense (100 kW/pulse) ultraviolet radiation of the nitrogen laser, it was possible to detect subnanogram quantities of these compounds.

Laser-fluorescence studies have been conducted on both chlorophyll *a* and chlorophyll *b*. Gee and Truscott (1968) measured the fluorescence of chlorophyll *b* as a result of CW irradiation with a helium–neon laser (632.8 nm). They found radiative transitions between 550 and 750 nm in the emission spectrum, and explained the short-wavelength emissions by transitions from different vibronic excited levels of the first excited chlorophyll *b* singlet level. In their study on chlorophyll *a*, Vacek *et al.* (1973) used a He–Ne laser to stimulate fluorescence. Their spectra were similar to those obtained with conventional light sources, and they were unable to show radiative transitions from nonzero vibronic states of the lowest singlet level of chlorophyll *a*.

One of the best reviews on the use of laser-excited fluorescence is the one by Moore and Zittel (1973). They present numerous examples of how laser-excited fluorescent techniques can be used to study energy transfer and chemical reaction rates for individual molecular energy levels. One of the most well-studied molecules in HCl.

In addition to vibrational relaxation, laser-fluorescence has been used to study excited electronic states—especially electronic energy transfer, vibrational and rotational energy transfer among the levels of the excited state, radiationless transitions of isolated molecules and of molecules undergoing collisions, and chemical reactions of excited states. For example, Leone and Wodarczyk (1974) have studied electronic-to-vibrational energy transfer of the type $Br^* + HCl \rightarrow Br + HCl(v = 1)$. The excited bromine ($Br^*$) atoms were produced by laser photodissociation of Br_2, and the infrared fluorescence of HCl was subsequently observed.

Steady-state laser-fluorescence spectroscopy has been used to study electronically excited alkali dimer molecules and I_2 (Moore and Zittel, 1973). In addition, the lifetimes of fluorescence decay from single vibrational levels have been measured, using pulsed tunable lasers, for molecules such as IC1, NO_2, formaldehyde, and glyoxal (Moore and Zittel, 1973). It should be pointed out that considerably higher resolution and sensitivity are possible with lasers as the excitation sources, as compared with conventional flashlamps and monochromators. Furthermore, as more sophisticated and versatile laser instruments are designed, the resolution and sensitivity of the measurements will improve even more. For example, the construction of a cavity-dumped argon ion laser as an excitation source for time-resolved fluorimetry has permitted the generation of shorter pulses and greater wavelength selectivity than was previously available (Lytle and Kelsey, 1974).

Lasers are also being used to study luminescence. Anderson and Ricchio (1973) used the 266-nm line of a neodymium YAG laser—the primary output of which was 1.06 μm. By passing this beam through two frequency-doubling crystals, it was possible to generate the 266-nm wavelength with good efficiency. The peak power pulses in the UV region were about 1 kW, with a pulse duration of 75×10^{-9} s. This system was used to measure the delay time between excitation and luminescence of inorganic phosphor compounds of varying types. The delays were found to be much shorter than anticipated on the basis of low-intensity pulsed luminescence. In addition, the delays, as well as the rise and decay times of luminescence, were found to be intensity dependent.

3. PHOTOCHEMISTRY

A major application of lasers in photochemistry is to control selective chemical reactions. The principle of the approach is to use the differences in absorption spectra of substances to selectively influence them to change their composition and properties. A rather extensive review of this field was written by Letokhov in 1973. Several areas were discussed in which the laser was used to drive specific chemical reactions. For example, selective catalysis of chemical reactions with infrared radiation became possible with the use of powerful molecular lasers, such as the 2.7-μm hydrogen fluoride laser. This system was used to cause selective reactivity of vibrationally excited CH_3OH molecules (Mayer *et al.*, 1970).

Another area of laser application is in the dissociation of excited molecules. Selective two-step photodissociation of HCl has been accomplished by Ambartzumian and Letokhov (1972; Letokhov and Ambartzumian, 1971). These experiments employed a pulsed neodymium laser with a very narrow frequency-tuned emission (line width less than 0.002 nm). A 20-ns pulse at 1060 nm was frequency shifted by stimulated Raman scattering to 1.18 μm, the appropriate wavelength for excitation of the third vibrational level of HCl. This pulse was directed into the HCl simultaneously along with another pulse at 265 nm, which was obtained by twice doubling the 1.06-μm line of the neodymium laser. The infrared line served to stimulate the HCl into an excited state. The UV radiation was at the appropriate frequency so that it did not fall within the absorption range of normal unexcited HCl, but it did coincide with the absorption band of excited molecules. The result was the selective photodissociation of the excited molecule. Furthermore, by careful tuning of the infrared radiation, it was possible to selectively photodissociate HCl with either isotope, ^{35}Cl or ^{37}Cl. In further applications of the two-step photodissociation, it was possible to separate out isotopes of nitrogen from NH_3 (either $^{14}NH_3$ or $^{15}NH_3$) with a very high selectivity coefficient (degree of isotope enrichment).

Several methods of isotope enrichment have been developed that do not require two-step photodissociation. For example, Lamotte *et al.* (1975*a,b*) have employed selective irradiation of specific isotopic transitions of $CsCl_2$ with a tunable dye laser. Following this, they reacted the excited species with diethoxyethylene. This resulted in a significant isotopic reaction in products and starting materials. Milligram amounts of chlorine were enriched by several hours of irradiation with a 7-mW laser. In another study, Liu *et al.* (1975) reported the photochemical separation of ^{35}Cl and ^{37}Cl where a mixture of ICl, *trans*-ClHC=CHCl, and 1,2-dibromoethylene was irradiated with a CW tunable dye laser that selectively excited $I^{37}Cl$.

There are other laser-driven chemical reactions besides those that result in isotope enrichment. For example, a CO_2 infrared laser has been used to convert CF_2Cl_2 (Freon 12) into $C_2F_4Cl_2$ (Freon 114) and Cl_2. No other products were detected by either gas chromatography or mass spectrometry. The authors presented evidence that the laser energy dissipation was not by simple thermal conversion. Rather, it was suggested that the vibrational mode of CF_2Cl_2 excited in the experiments represented a rocking motion of the CF_2 group against the Cl_2 group or *vice versa* (Zitter *et al.*, 1975).

Several studies designed to test the usefulness of the CW argon ion laser in selective photochemistry have been conducted by Wilson and his colleagues (Wilson and Wunderly, 1974*a,b;* Wilson *et al.*, 1974). In all of their studies, a 6-W CW argon ion laser was employed set at the visible argon wavelengths (514.5, 501.7, 496.5, 488, 476, and 457.9 nm), and radiation times upward of 20 min were used. In one study (Wilson *et al.,* 1974), the laser radiation was used to initiate oxidative photoaddition of *p*-benzoquinone to cyclooctatetraene. It was suggested that the formation of the photoproducts was dependent on subtle stereoelectronic factors that influenced the lifetime of the oxetane precursor biradical and stabilized it long enough to permit trapping by oxygen. In another experiment (Wilson and Wunderly, 1974*b*), the sulfur dioxide trapping of photochemically generated biradicals was demonstrated. The general conclusion from these experiments was that biological trapping was not necessarily confined to oxygen but in fact might be a more generally occurring phenomenon in synthetic schemes. Finally, Wilson and Wunderly (1974*a*) used the argon ion laser to induce the formation of 1,2,4-trioxans from *p*-benzoquinone and olefins under aerobic conditions.

4. BIOLOGY

It should already be evident from the previous section of this chapter that the laser has developed into a useful tool with which to study numerous biologically important reactions. Those reactions associated with photosynthesis, aerobic respiration, and vision have received the most attention.

4.1. Laser Effects (Photosensitivity)

There are several other areas in biology where the laser has been effectively applied. One such area would simply have to be termed "laser photosensitivity." Historically, when any new radiation source is developed, much effort is devoted to the straightforward description of its varied effects on organisms, tissues, cells, and, ultimately, molecules. These studies may be purely descriptive, aimed at defining safe radiation levels, or their purpose may be to describe particular effects as well as to employ the radiation as a tool to better understand structure and function. There are two excellent volumes reviewing laser applications in medicine and biology (Wolbarsht, 1971, 1974).

One way to deal with laser effects on biological systems is to start with the whole organism and work down in levels of complexity to organs, tissues, cells, and molecules. There have been numerous studies in which either whole organisms or rather large areas of an organism were exposed to laser radiation of varying wavelengths and intensities. These early studies are contained in two major symposia volumes (Whipple, 1965; Litwin and Earle, 1965). A few other studies have dealt with whole-organism irradiation. For example, both the morphological (Wilde, 1965, 1967) and physiological (Feir and Lough, 1969) effects of laser irradiation have been studied in insects. In the latter study, several physiological parameters were examined following whole-body exposure to 0.4–0.6 J/pulse (10–12 $\times$ 10^{-9} s) of a Q-switched ruby laser (694.3 nm). The radiation caused a significant decrease in oxygen uptake on the first 2 days of the fifth stadium. In addition, a decrease in hemolymph protein concentration was found on the fourth day of the stadium. There was a decrease in whole-body ATP content on the first day of the stadium, and there was an increase over control values on days 2, 5, and 9. The authors could not provide an explanation of the observed findings other than general organism "shock" and recovery.

Edlow *et al.* (1965) studied the effects of a 20–40 J ruby laser on rat embryos and fetuses *in utero*. The laser was focused on the developing embryo or fetus while *in utero* following the cutting open of the mother. A whole spectrum of effects was observed, ranging from embryonic death and resorption to a slight discoloration of the fetal surface. The authors suggested that the technique might be useful for studies on organ formation and on fetal reparative processes to local injury. However, very few studies along these lines have been reported since that time.

In a rather novel study, Paleg and Aspinall (1970) used a He–Ne laser to irradiate lettuce seedlings more than a quarter of a mile away from the laser source. They were able to cause activation of the phytochrome system, and thus control plant growth and development. The authors suggested that this approach might have potential application for crop growth control.

Laser irradiation of organs and tissues has mainly been for the purpose of clinical surgery or for the purpose of defining safe levels of exposure (Wolbarsht, 1971, 1974). In the latter case, much concern has been expressed over the safe threshold levels of exposure for ocular tissue. This has become particularly important with the development of ultrashort picosecond pulses with high power densities. As examined at the ultrastructural level with the electron microscope, threshold levels of injury were produced with an energy density of 13 μJ. The damage was characterized by highly localized damage on the photoreceptor and pigmented epithelial cells in the outer retina (Ham *et al.,* 1974). Other studies have demonstrated the potential usefulness of using the laser as a surgical "knife" (Wolbarsht, 1971, 1974; Goldman *et al.,* 1970; Hall *et al.,* 1971).

In a rather detailed investigation, Délèze (1970) studied the electrophysiological responses and recovery phenomena in sheep heart that had been injured by laser "cutting." These studies were useful in assessing the effects of laser surgery on heart tissue, and, more importantly, they were useful in elucidating the passive electrical properties of heart fibers. By producing precisely controlled lesions on the heart tissues, and then studying the recovery and healing process, the author was able to make specific conclusions about the electrical properties of heart tissue.

Another organ irradiated was the liver (Nicholls *et al.*, 1974). In this study, the whole liver of anesthetized rats was irradiated with an unfocused beam of the ruby laser. Each rat was exposed to 150 J of laser energy in a total of 600 pulses. Animals were sacrificed 48 h following irradiation, and the livers were removed and assayed with respect to protein synthesis. A significant increase in liver elongation factor 1, but not of factor 2, was detected in the experimental animals. The authors point out that this response is similar to the response of liver during regeneration, rather than to the response following exposure to ionizing radiation. It was also suggested that the response to laser irradiation could be a secondary response merely reflecting a general increase in protein synthesis.

Because of the developing interest in using lasers in clinical practice, it was felt that the effects of lasers on bones should be investigated (Kolar *et al.*, 1969). The infrared wavelength (1.06 μm) of a pulsed CO_2 laser was used. The unfocused 4-nm laser beam was applied directly to the skin above the knee. The bones in the area of irradiation and elsewhere in the animal were examined 48 h following irradiation. The bones were examined for total calcium content and for ^{45}Ca. In general, the results indicated that laser radiation affected the calcium metabolism not only in the irradiated area but also elsewhere in the animal. The authors suggested that the effects on the bones may not be negligible when lasers are used clinically.

Lasers have also been widely applied in studies at the cellular level. In succeeding sections of this chapter, two such areas will be discussed: cytofluorimetry and partial cell irradiation. However, both of these applications involve focusing the laser beam within the cell. Numerous studies on the irradiation of whole cells and/or cell populations have been conducted. Much of the work has been done by D. E. Rounds and his co-workers. Early in 1965, this group published a paper describing the potential of the laser in cell research (Rounds *et al.*, 1965a). It was pointed out that specific molecular components within a cell may be selectively altered by laser irradiation with an appropriate wavelength and a suitable power density. These authors were particularly interested in the selective interruption of cellular respiration by "knocking out" the various cytochromes involved in cellular respiration. Indeed, the wavelength specificity of laser-induced biological damage was shown (Rounds *et al.*, 1967b) and the ability to block respiration at various points was demonstrated (Rounds *et al.*, 1965b, 1967a).

Other studies involving natural chromophores and their use in selective laser damage to cells have been conducted in plant systems where the chlorophylls are bound to the chloroplast membranes. Studies by Keyhani *et al.* (1971) and Floyd *et al.* (1971) were undertaken to describe the effects of laser irradiation on cellular membranes, and also to elucidate the structural and functional organization of the chloroplast membranes.

Electron microscope analysis (Keyhani *et al.*, 1971) revealed that ruby laser irradiation (2.3 J/pulse) caused a decrease in granularity and an increase in vacuolization of the membranes of isolated chloroplasts. However, it was demonstrated that the chlorophyll itself was not damaged. It was suggested that the laser radiation blocked energy coupling without affecting the photosystems. In the subsequent study, Floyd *et al.* (1971) correlated the previous structural changes with biochemical changes. They were able to demonstrate a considerable decrease in photoinducible absorption changes of cytochromes b_{559}, b_{563}, and P_{520}, and a slight reduction in cytochrome *f*. The authors suggest that membrane parameters may be the most labile components in highly absorbing biological systems.

Several studies have been conducted on generally nonpigmented whole-cell cultures and a variety of effects have been reported. For example, ruby laser irradiation of salamander lung epithelial cells (Okigaki and Rounds, 1967), human leukocyte cultures (Gordon *et al.*, 1968), and rabbit endothelial cells in culture (Rounds *et al.*, 1965c) has resulted in various types of chromosome anomalies and mitotic blockage. It is not known whether these effects were due to a thermal effect or to some other effect of the radiation (such as two-photon effects or dielectric breakdown). In other studies in tissue culture systems (Rounds *et al.*, 1965c), laser effects were also demonstrated on the contractile rate of chick heart muscle cells, isolated skeletal muscle cells, and involuntary smooth muscle of the intestine. The mechanism of contractility inhibition in all three muscle types is not known. However, it was suggested that enzymes such as lactic dehydrogenase which are essential to muscular function might be affected by the radiation. Other studies by Rounds (1965) had demonstrated that the conversion of pyruvate to lactate in the presence of DPNH and lactic dehydrogenase was inhibited by irradiation of either the DPNH or the lactic dehydrogenase with the 347.1-nm line of a frequency-doubled ruby laser.

In addition to the previous studies where the laser effects were due to absorption by some natural chromophore (such as cytochromes, hemogloin, or chlorophyll), several studies have been conducted where selective photosensitizing agents have been added that bind to specific structures or regions in the cell (see review by Cameron *et al.* 1972). In one of these studies, the mitochondria of HeLa cells in culture had their sensitivity to laser light increased by treatment with Janus Green B (Baba, 1970). Previous studies by Rounds *et al.* (1965*c*) had demonstrated the same phenomenon with Janus Green B and acridine orange (Rounds *et al.*, 1967*b*). Selective photosensitization also has been applied to prokaryotic organisms. Herczegh *et al.* (1971) treated T7 phages with methylene blue (0.4×10^{-6} M) and then irradiated them with a ruby laser. The killing results followed a typical multihit curve.

In a rather unusual experiment, Jamieson *et al.* (1969) irradiated mouse and human melanoma cells in culture with sublethal doses of ruby laser radiation. This treatment resulted in growth enhancement. Although this effect could not be explained by the authors, it led to the suggestion of extreme caution in the use of the laser for cancer therapy. Furthermore, earlier studies (Hoye *et al.*, 1967) had indicated that tumor destruction by laser irradiation actually resulted in the spreading of tumor cells, because cells were ejected out of the irradiated area by the explosive force of the focused laser beam. These early studies have resulted in the general nonapplication of lasers to cancer therapy, although some limited applications have developed (Wolbarsht, 1971, 1974).

Taking the laser down to the molecular and subcellular level, a few studies on isolated nucleic acids and enzymes have been conducted. In a novel experiment, Rounds *et al.* (1966) demonstrated two-photon absorption in samples of NADH irradiated with the 532-nm beam of a doubled Q-switched laser. They were able to detect fluorescence at wavelengths shorter than the visible stimulating wavelength. The fluoresence spectrum was identical to the spectrum obtained in conventional studies where UV radiation was used as the excitatory stimulus. More recently, Berns (unpublished) has derived a UV action spectrum for chromosome paling when only focused visible laser light is used. (These studies will be discussed in Section 4.3 on partial cell irradiation.) A two-photon effect was suggested in these experiments.

Ultraviolet laser radiation has been employed in the irradiation of DNA (Matsui *et al.*, 1971; Berns, unpublished) and proteins (Berns *et al.*, 1970*a*). In the case of DNA, Matsui *et al.* (1971) studied the template activity of the DNA by incubating the irradiated DNA with RNA polymerase and then assaying the amount of RNA synthesized by determining the amount of [^{14}C]UMP incorporated. The results demonstrated that 265-nm light from a frequency-quadrupled neodymium YAG laser affected the template activity of DNA. The estimated energy at 265 nm was 150–300 MW in a 15×10^{-9} s pulse. Furthermore, the data suggested a lack of reciprocity using the high peak power of the UV laser. In a somewhat different experiment (Berns and Sutherland, unpublished), highly radioactive DNA was exposed to a varied number of pulses from a frequency-quadrupled neodymium laser (266 nm) and was subsequently assayed for the production of thymine dimers. The estimated peak power was 1 kW in an 85×10^{-9} s exposure. The laser system employed is diagrammed in Fig. 2. It is interesting to note that the amount of dimerization produced in a total of 850×10^{-9} s of laser exposure would have required about 40 s exposure to a classical UV source.

4.2. Cytofluorimetry

One of the major developments in cellular research has been the use of laser radiation to stimulate fluorescence in single cells. By combining the laser stimulation with rapid cell flow systems and sophisticated electronics, it has been possible to record and determine numerous parameters from large populations of cells. Recently, Crissman and Steinkamp (1973) have demonstrated the ability to rapidly perform simultaneous measurements of DNA, protein, and cell volume in single cells from large populations. By treating the cells with specific fluorescent and other staining compounds, it is possible to detect fluorescent signals for several parameters. The principle of the apparatus and typical curves are presented in Figs. 3 and 4. In a more sophisticated experiment, it has been possible to determine the quantitative amount of DNA per cell and then plot the population distribution (Kraemer *et al.*, 1974). In this way, the proportion of a cell population that is in various stages of the cell cycle may be determined. It is anticipated that this kind of instrumentation will greatly facilitate analysis of cell population parameters, and, as a result, open the way for numerous new experimental plans.

The other major application of the laser to cytofluorimetry is in performing specific fluorimetric measurements on subcellular organelles (Andreoni *et al.*, 1975*b*; Sacchi *et al.*, 1974). The apparatus employs a pulsed tunable organic dye laser with 2×10^{-9} s pulses and power outputs up to 100 kW. The laser beam is passed through a microscope and into the biological specimen, where a fluorescent signal is generated and subsequently recorded by a photomultiplier tube. The system permits, for the first time, the measurement of the fluorescence decay time within a living cell. In addition, when compared to conventional methods, the laser-stimulated fluorescence technique has the advantages of not resulting in photodecomposition and of having a potential increase in sensitivity by several orders of magnitude over conventional methods.

One of the first biological questions approached with this technique has been study of the fluorescent pattern in the bands of individual chromosomes stained with quinacrine mustard (Andreoni *et al.*, 1975*b*). The apparatus has permitted measurement of the main parameters of the

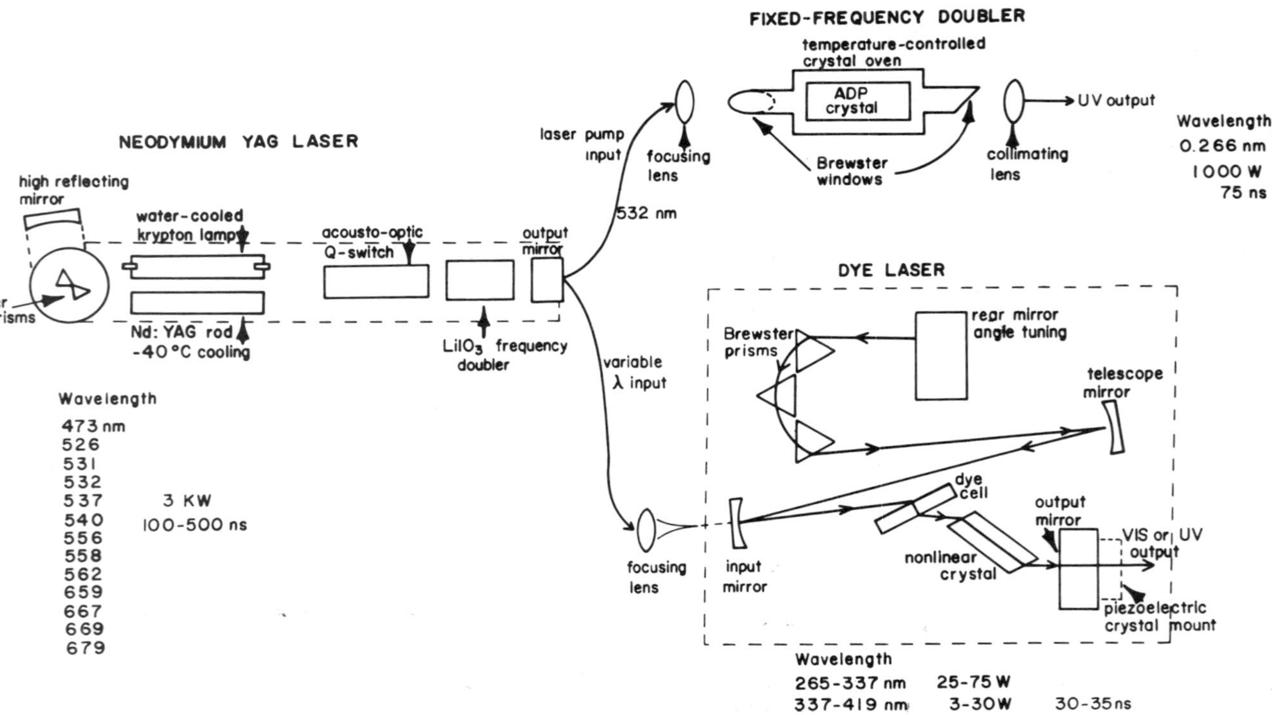

Fig. 2. Tunable neodymium YAG-dye laser system. The three major components in this system are (1) the YAG laser, (2) the dye laser, and (3) the external fixed-frequency second harmonic doubling unit. Discrete wavelengths in the range of 473–679 nm can be obtained directly from the YAG laser. High power in the ultraviolet at 266 nm can be obtained by passing the YAG 532-nm wavelength into the external fixed-frequency ADP doubling unit. A wide spectrum of ultraviolet and visible wavelengths (265–670 nm) can be obtained by using the wavelengths of the YAG laser to stimulate the dye laser.

fluorescent transition (fluorescence saturation intensity and quantum yield) in the *single fluorescent* chromosome bands. Preliminary measurements were made in the band and interband regions of quinacrine mustard-stained metaphase chromosomes of *Vicia faba*. Using this technique, it was possible to determine that within the bands the adenine–thymine (A-T) concentration was 60% of the total DNA. Furthermore, it was determined that the fluorescent waveforms contained a rapid and a slow decay component. It was suggested that the rapid decay component was due to the intercalation of the quinacrine binding to guanine residues, and the slower decay component was due to the intercalation of the quinacrine between the A–T base pairs. Presently, the apparatus cannot give accurate measurements for out-of-band (interband) regions, and the authors are in the process of improving the resolution of the instrumentation to permit these kinds of determinations. The general application of this technique to chromosome studies (using quinacrine as well as other selectively binding fluorescent agents) and to studies on other organelles is great and could greatly expand the already existing analytical techniques of microscopy.

4.3. Partial Cell Irradiation

The previous fluorimetric studies involved focusing the laser into cells for the purpose of extracting analytical information. Another application is to focus the laser into living cells with the purpose of selectively altering the subcellular organelles and molecules, while keeping the cells alive for further analysis. In this way, the laser is used to selectively alter cell organelles in order to study their function and organization. The first studies of this nature were reported by Bessis *et al.* (1962), employing a ruby laser focused to a small spot inside living cells. From that time on, numerous studies involving virtually every kind of laser have been undertaken. Two major reviews (Berns and Salet, 1972; Berns, 1974*a*) and a book (Berns, 1974*b*) have been devoted to this subject. Therefore, a complete review of the field is not justifiable here.

The studies can be broadly divided into two general categories: (1) those that simply describe the structural and biochemical effects of laser microbeam irradiation and (2) those that attempt to approach a specific functional question. Often the distinction is not so clear-cut because it may be necessary to describe the nature of the alteration, and, in the same study, to probe a particular cellular function. For example, in our most recent studies (Berns and Rattner, 1975) on the role of the centriolar region in the organization of the mitotic apparatus, it was first necessary to describe the kinds of ultrastructural alterations obtained by varying such parameters as

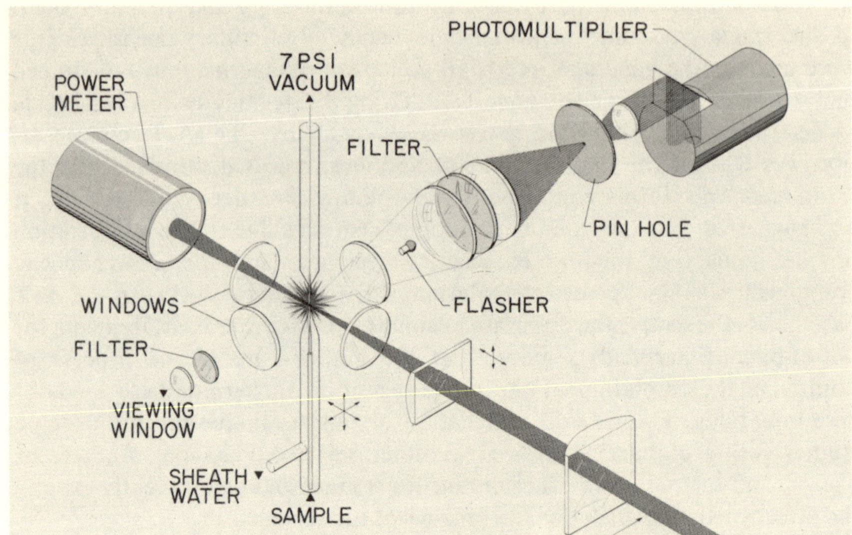

Fig. 3a. Cytofluorimetric flow system. Schematic showing cells flowing through the water jet system. The laser beam comes in from the right and stimulates fluorescence in the stained cells. The power meter on the left monitors the laser power. The photomultiplier records the amount of fluorescence given off by the cells.

laser energy, vital dye, and time of fixation of tissue. Only after these parameters were established was it possible to perform consistent experiments on the centrioles. The results indicate that following treatment with the vital stain, acridine orange, it is possible to selectively alter the pericentriolar material by microirradiation with the green wavelengths of a pulsed argon ion laser. The effect on mitosis is the inability of the irradiated zone to organize microtubules. The chromosomes can proceed into metaphase, but they do not undergo anaphase movements. Despite this disruption of mitosis, the cell does undergo cytokinesis with all of the chromosomes ending up in one of the daughter cells.

The preceding study illustrates how the laser microbeam can be used to study functional biological questions. By and large, this technique has been used to study the organization and function of cellular organelles in both major cell compartments, the nucleus and the cytoplasm. In the nucleus, the two prominent kinds of structures are the nucleoli and the chromosomes. Several studies on the organization and biochemistry of the nucleolus have been undertaken (Berns *et al.*, 1969, 1970*b*; Brinkley and Berns, 1974). The chromosomes have been studied from the structural point of view (Rattner and Berns, 1974), for selective inactivation of ribosomal genes (Berns and Cheng, 1971; Ohnuki *et al.*, 1972), and for the selective removal of whole chromosomes followed by cell cloning (Berns, 1974*c,d*).

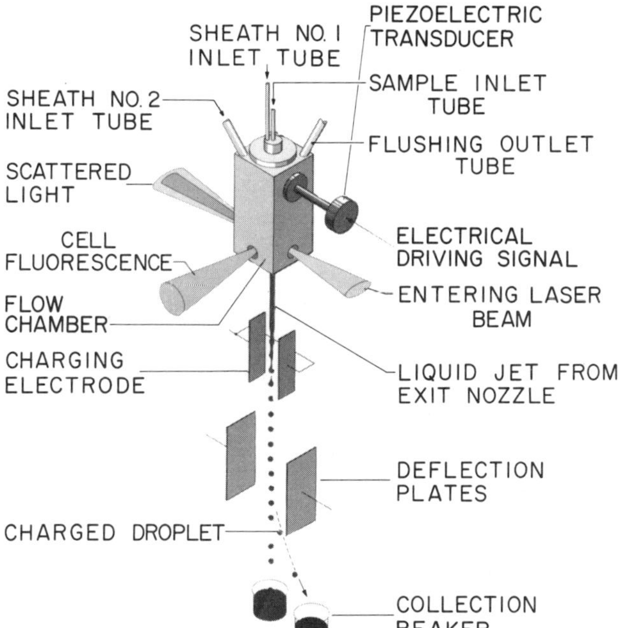

Fig. 3b. Cytofluorimetric flow system. Schematic showing a more sophisticated system than in 3a; here the fluorescent signal is fed into a computer which in turn controls electrodes that charge the individual cells. The charged cells are then selectively separated into different receptacles. This system permits the isolation of cells based on their DNA content. From Kraemer *et al.* (1974).

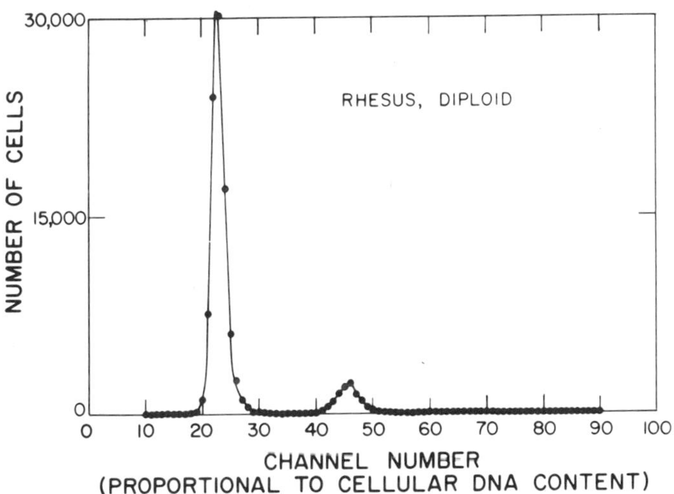

Fig. 4. Typical curve showing the cytofluorimetric determination of the cell cycle position of a population of cells. The large peak represents cells prior to S phase, and the small peak represents post-S-phase cells. From Kraemer *et al.* (1974).

The results of these studies suggest the feasibility of using laser micro-irradiation to (1) selectively alter molecular species (DNA and/or histone) in a 0.25-μm (or larger) region of a preselected chromosome (Fig. 5), (2) inactivate a chromosome region as small as 0.25 μm, and (3) remove desired whole chromosomes or chromosome regions from cells (Fig. 6) and then follow the fate of the cell and its descendants.

Cytoplasmic organelles such as mitochondria in contracting heart cells, chloroplasts in developing algae, and centrioles have been probed with the laser microbeam. All of these studies (except for the centriolar investigations) are adequately reviewed by Berns (1974a).

In addition to the above studies, earlier studies by Amy *et al.* (1967) on cytoplasmic mitochondrial structure and function, Bessis and Ter-Pogossian (1965) on numerous cell types, Goldstein *et al.* (1970) on sperm flagellar activity, and McKinnel *et al.* (1969) on amphibian egg enucleation have all contributed to a better understanding of biological structure and function.

Microbeam studies have been conducted using ruby, argon, helium-neon, neodymium YAG, and organic dye lasers. As a result, the available wavelengths span the entire visible spectrum, and with frequency-shifting techniques extend into the ultraviolet. The materials irradiated may be in their natural state, or they may be stained with selectively binding, nontoxic vital dyes (e.g., Berns and Salet, 1972). Of particular interest is the production of distinct molecular alterations with visible laser radiation when no known chromophore is present. This was the situation in the production of distinct histone lesions with the focused blue-green radiation of a 35-W pulsed (50 μs) argon ion laser (Rattner and Berns, 1974; Berns *et al.*, 1971). Similar effects were observed with a wider range of visible wavelengths when a completely tunable dye laser was developed. In the tunable system, the frequency-doubled line (532 nm) of a neodymium YAG laser was used to pump Rhodamine 6G and fluorescein in an organic dye laser. Such a system has provided complete tunability throughout the visible spectrum and with a second set of frequency-doubling crystals provides complete tunability down to 240 nm (Berns, unpublished). With this system, it has been possible to generate the visible action spectrum for the histone lesion response. When this action spectrum is plotted as $\lambda/2$ instead of λ, the curve is a near perfect match to the UV absorption spectrum of histones (Fig. 7). This result strongly suggests a two-photon effect because a UV action spectrum would be expected for such an effect. Furthermore, preliminary reciprocity studies indicate that reciprocity does not hold in these experiments (all unpublished data). Calculations and measurements indicate that the photon density in the irradiated zone is on the order of 10^{24}–10^{27} photons s^{-1} cm^{-2}. This is clearly within the realm of two-photon

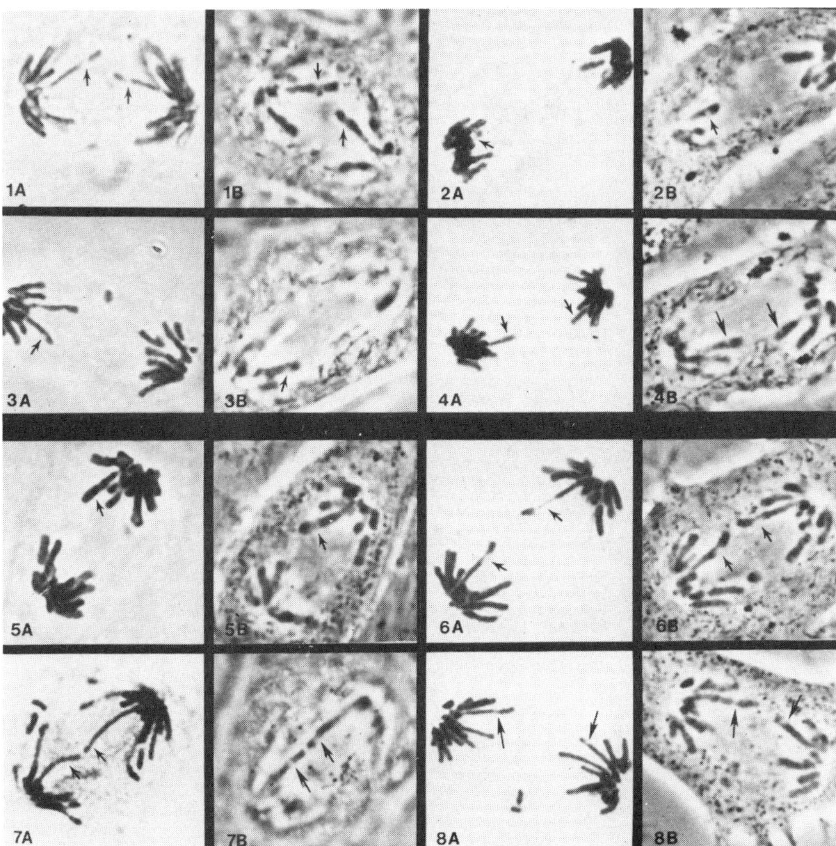

Fig. 5. Cytochemically stained chromosomes and corresponding live light micrographs immediately preceding irradiation. Each pair is a different cell. Arrows indicate lesion areas: 1A, alkaline fast green (histone) negative, no acridine orange, severe lesion, high energy (1000 μJ/μm²); 1B, corresponding live cell; 2A, Feulgen (DNA) negative, no acridine orange, severe lesion, high energy; 2B, corresponding live cell; 3A, alkaline fast green negative, no acridine orange, threshold lesion, moderate energy (500 μJ/μm²); 3B, corresponding live cell; 4A, Feulgen positive, no acridine orange, threshold lesion, moderate energy; 4B, corresponding live cell; 5A, alkaline fast green positive, acridine orange used, severe lesion, moderate energy; 5B, corresponding live cell; 6A, Feulgen negative, acridine orange used, severe lesion, moderate energy; 6B, corresponding live cell; 7A, alkaline fast green positive, acridine orange used, threshold lesion, low energy (50 μJ/μm²); 7B, corresponding live cell; 8A, Feulgen negative, acridine orange used, threshold lesion, low energy; 8B, corresponding live cell. From Berns (unpublished).

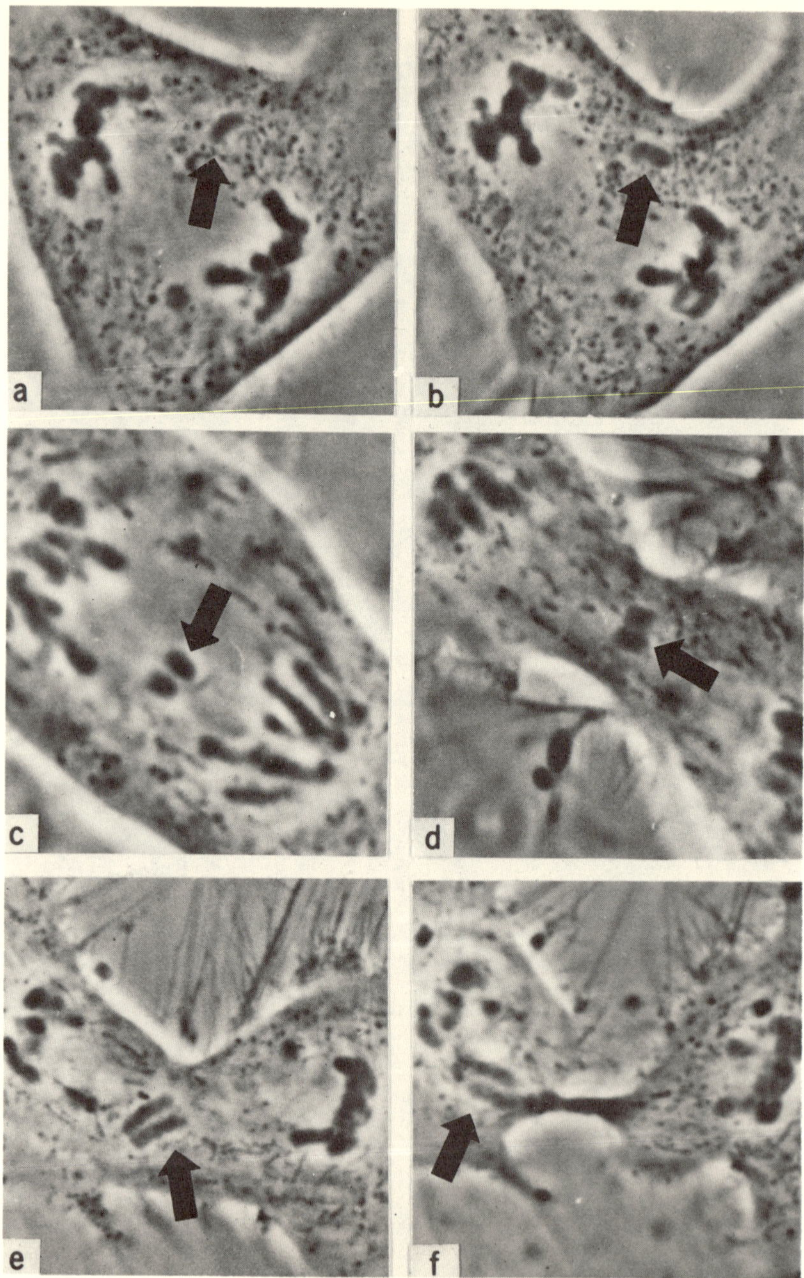

Fig. 6. (a,b) Anaphase cell demonstrating the loss of an irradiated chromosome from the nucleus. The irradiated chromosome can be seen in the cytoplasm; the two light micrographs

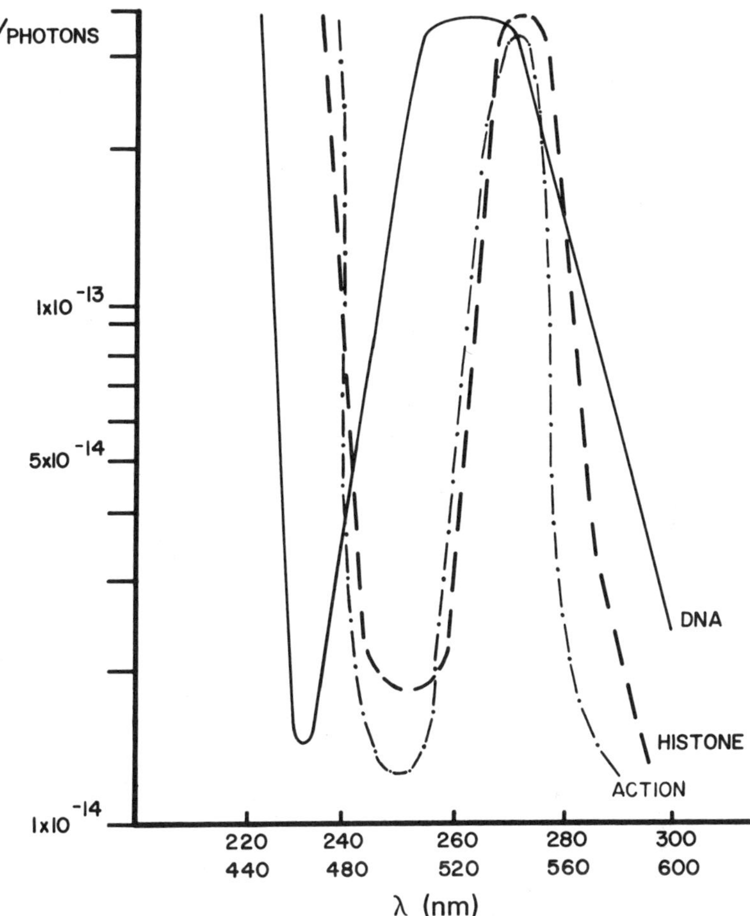

Fig. 7. Ultraviolet action spectrum derived using *visible* laser light. Note that the curve obtained by plotting half of each visible wavelength exactly matches a histone UV absorption spectrum. From Berns, 1976, *Biophys J.* **16**:973–977.

show slightly different focal planes in the cell; the two chromatids are indicated by the arrows. (c,d) Another cell in which the irradiated chromosomes are lost during mitosis as a result of being caught in the stem body: (c) anaphase cell showing the two chromosomes remaining behind at the metaphase plate; (d) the same cell in telophase showing the two chromosomes being caught in the stem body; neither daughter cell received one of these chromosomes. (e,f) A third cell in which both microirradiated chromosomes are incorporated into the nucleus of one daughter cell: (e) early telophase showing both chromosomes being incorporated in one daughter cell; (f) late telophase showing both chromosomes being incorporated into one nucleus. From Berns (1974*d*).

effects as well as of other uncommon physical effects such as pressure, dielectric breakdown, and acoustic phenomena.

Finally, the conversion of the optical systems to quartz will now permit the complete spectrum of laser microbeam experiments from the infrared through the visible and down to the far UV. In combination with tissue culture, electron microscopy, electrophysiology, cytochemistry, and biophysical techniques, it will be possible to apply microbeam irradiation to even a more diverse set of questions and problems.

4.4. Laser Light Scattering

The laser is an ideal tool for light-scattering studies because of its high degree of monochromaticity and its well-defined degree of collimation. The far-field fringe patterns produced by passing a laser into a particular structure or solution will be highly sensitive to the orientation and position of the structure and its component molecules with respect to the incident laser light. It has been pointed out (Dubin, 1972; Dubin *et al.*, 1967) that classical light-scattering measurements can provide information on the molecular weight of macromolecules in solution on the basis of the intensity of scattered light. "Quasi-elastic laser light scattering" considers the fluctuations of light. The intensity of the light scattered by a suspension of particles or a solution of macromolecules at a particular time is determined by the superposition of the phases of scattered waves. The intensity of the scattered light fluctuates because the phase of the light scattered by each particle is randomly changing as the particle undergoes a random motion. The scattered intensity must reflect a concentration fluctuation that obeys the standard translational diffusion equation (Arrio *et al.*, 1974). As a result, it is possible to make size determinations under the same conditions as activity measurements. This can be done rapidly, and can therefore be a useful tool in the description of biological material. This technique has been applied to studies on preparations of sarcoplasmic reticulum vesicles (Arrio *et al.*, 1974), bacterial flagella (Fujime *et al.*, 1972), and ribosomes (N. C. Ford, personal communication).

Although the technique of quasi-elastic laser light scattering is just beginning to be used, it could develop into a more generally applied and useful approach to complex biological problems.

5. CONCLUSIONS

In this chapter, I have attempted to present a comprehensive picture of how the laser is being applied in the areas of spectroscopy, photochemistry, and biology. It is clear that in some areas of spectroscopy, such as Raman spectroscopy and flash photolysis, the laser is permitting far more sensitive and diverse studies than ever before. In other spectroscopic areas, such as emission spectroscopy and fluorescence and luminescence, the laser is leading to the development of new and innovative approches. In the area of photochemistry, the laser is contributing to studies on light-driven reactions that heretofore have been virtually impossible to perform. The entire area of isotope enrichment could prove to be one of the most important developments in future applications of lasers.

In the areas of biology and medicine, the laser is also finding its niche. While many of the early studies simply dealt with the effects of lasers and instrument development, the more recent studies have actually employed the laser as a tool to investigate biological phenomena. This is certainly the case in partial cell irradiation and cytofluorimetry.

In summary, it appears that the laser is a truly unique device that has enhanced many of the already existing approaches in spectroscopy, photochemistry, and biology, and in addition has led to numerous new approaches that were not feasible with the classical radiation sources.

ACKNOWLEDGMENTS

Some of the research reported in this chapter has been supported by the following grants: NIH HL 15740, NIH GM 22754, and NSF GB 43527.

6. REFERENCES

Alpert, B., Banerjee, R., and Lindqvist, L., 1974, The kinetics of conformational changes in hemoglobin, studied by laser photolysis, *Proc. Natl. Acad. Sci. USA* **71**:558–562.

Ambartzumian, R. V., and Letokhov, V. S., 1972, Selective two-step (STS) photoionization of atoms and photodissociation of molecules by laser radiation, *Appl. Opt.* **11**:354–358.

Amy, R. J., Storb, R., Fauconnier, B., and Wertz, R. K., 1967, Ruby laser microirradiation of single tissue cells vitally stained with Janus Green B. I. Effects observed with the phase contrast microscope, *Exp. Cell Res.* **45**:361–373.

Anderson, R. J., and Ricchio, S. G., 1973, Luminescent rise times of inorganic phosphors excited by high intensity ultraviolet light, *Appl. Opt.* **12**:2751–2758.

Andreoni, A., Benetti, P., and Sacchi, C. A., 1975a, Subnanosecond pulses from a single-cavity dye laser, *Appl. Phys.* **7**:61–64.

Andreoni, A., Sacchi, C. A., Cova, S., Bottiroli, G., and Prenna, G., 1975b, Pulsed tunable laser in cytofluorometry: A study of the fluorescence pattern of chromosomes, in: *Lasers in Physical Chemistry and Biophysics* (J. Joussot-Dubien, ed.), pp. 413–423, Elsevier Scientific, Amsterdam.

Arrio, B., Chevallier, J., Jullien, M., Yon, J., and Calvayrac, R., 1974, Description by quasi elastic laser light scattering of a biological preparation: Sarcoplasmic reticulum vesicles, *J. Membr. Biol.* **18**:95–112.

Asher, I. M., Rothschild, K. J., and Stanley, H. E., 1974, Raman spectroscopic study of the Valinomycin-KSCN complex, *J. Mol. Biol.* **89**:205–222.

Ault, B. S., Howard, W. F., Jr., and Andrews, L., 1975, Laser-induced fluorescence and Raman spectra of chlorine and bromine molecules isolated in inert matrices, *J. Mol. Spectrosc.* **55**:217–228.

Baba, K., 1970, Selective injury of mitochondria with Janus Green B and ruby laser light: Enzyme morphological and ultrastructural study, *Acta Pathol. Jpn.* **20**(1):59–78.

Beatrice, E. S., Harding-Barlow, I., and Glick, D., 1969. Electric spark cross-excitation in laser microprobe-emission spectroscopy for samples of 10–25 μ diameter, *Appl. Spectrosc.* **23**:257–259.

Behringer, J., 1967, Observed resonance Raman spectra, in: *Raman Spectroscopy,* Vol. 1 (H. A. Szymanski, ed.), pp. 168–223, Plenum Press, New York.

Bensasson, R., and Land, E. J., 1971, Triplet–triplet extinction coefficients via energy transfer, *Trans. Faraday Soc.* **67**:1904–1915.

Bensasson, R., Chachaty, C., Land, E. J., and Salet, C., 1972, Nanosecond irradiation studies of biological molecules. I. Coenzyme Q6 (ubiquinone-30), *Photochem. Photobiol.* **16**:27–37.

Bergmann, K., and Demtroder, W., 1972, Inelastic collision cross section of excited molecules. II. Asymmetries in the cross section for rotational transitions in the $Na_2(B^1II_u)$ state, *J. Phys. B* **5**:1386–1395.

Bergmann, K., Demtroder, W., Stock, M., and Vogl, G., 1974, Inelastic collision cross section of excited molecules. IV. Rotational transitions for very high rotational quantum numbers and temperature effects in $Na_2(B^1II_u)$, *J. Phys. B* **7**:2036–2046.

Berman, M. R., and Zare, R. N., 1975, Laser fluorescence analysis of chromatograms: Subnanogram detection of aflatoxins, *Anal. Chem.* **47**:1200–1201.

Berns, M. W., 1974a, Recent progress with laser microbeams, *Int. Rev. Cytol.* **39**:383–411.

Berns, M. W., 1974b, *Microbeams and Partial Cell Irradiation,* Prentice-Hall, Englewood Cliffs, N.J.

Berns, M. W., 1974c, Laser microirradiation of chromosomes, *Cold Spring Harbor Symp. Quant. Biol.* **38**:165–174.

Berns, M. W., 1974d, Directed chromosome loss by laser microirradiation, *Science* **186**:700–705.

Berns, M. W., and Cheng, W. K., 1971, Are chromosome secondary constrictions nucleolar organizers: A re-evaluation using a laser microbeam, *Exp. Cell Res.* **69**:185–192.

Berns, M. W., and Rattner, J. B., 1975, Irradiation of the centriolar region in mitotic *Potorous* cells with a laser microbeam, *J. Cell Biol.* **67**:30a.

Berns, M. W., and Salet, C., 1972, Laser microbeams for partial cell irradiation, *Int. Rev. Cytol.* **33**:131–156.

Berns, M. W., Olson, R. S., and Rounds, D. E., 1969, Argon laser microirradiation of nucleoli, *J. Cell Biol.* **43**:621–626.

Berns, M. W., Matsui, S., Olson, R. S., and Rounds, D. E., 1970a, Enzyme inactivation with ultraviolet laser energy (2650 Angstroms), *Science* **169**:1215–1217.

Berns, M. W., El-Kadi, S., Olson, R. S., and Rounds, D. E., 1970b, Laser photosensitization and metabolic inhibition of tissue culture cells treated with quinacrine hydrochloride, *Life Sci.* **9**:1061–1069.

Berns, M. W., Cheng, W. K., Floyd, A. D., and Ohnuki, Y., 1971, Chromosome lesions produced with an argon laser microbeam without dye sensitization, *Science* **171**:903–905.

Bessis, M., and Ter-Pogossian, M., 1965, Micropuncture of cells by means of a laser beam, *Ann. N.Y. Acad. Sci.* **122**:689–694.

Bessis, M., Gires, F., and Mayer, G., 1962, Irradiation des organites à l'aide d'un laser à rubis, *C.R. Acad. Sci.* **255**:1010–1012.

Brinkley, L., and Berns, M. W., 1974, Laser microdissection of actinomycin D segregated nucleoli, *Exp. Cell Res.* **87**:417–422.

Brunner, H., Mayer, A., and Sussner, H., 1972, Resonance Raman scattering on the haem group of oxy- and deoxyhaemoglobin, *J. Mol. Biol.* **70**:153–156.

Cameron, L., Burton, A. L., and Hiatt, C. W., 1972, Photodynamic action of laser light on cells, in: *Concepts in Radiation Cell Biology* (G. L. Whitson, ed.), pp. 245–258, Academic Press, New York.

Chance, B., and Erecińska, M., 1971, Flow flash kinetics of the cytochrome a_3-oxygen reaction in coupled and uncoupled mitochondria using the liquid dye laser, *Arch. Biochem. Biophys.* **143**:675–687.

Chance, B., and Schoener, B., 1964, *Abst. 8th Ann. Mtg. Biophys. Soc. FD9.*

Chance, B., Schleyer, H., and Legallais, V., 1963, Activation of electron transfer in a *Chlamydomonas* mutant by light impulses from an optical maser, in: *Microalgae and Photosynthetic Bacteria* (Japan Soc. Plant Physiol., ed.), pp. 337–346, University of Tokyo Press, Tokyo, Japan.

Chance, B., McCray, J. A., and Bunkenburg, J., 1970, Fast spectrophotometric measurement of H^+ changes in *Chromatium* chromatophores activated by a liquid dye laser, *Nature (London)* **225**:705–708.

Crissman, H. A., and Steinkamp, J. A., 1973, Rapid, simultaneous measurement of DNA, protein, and cell volume in single cells from large mammalian cell populations, *J. Cell Biol.* **59**:766–771.

Délèze, J., 1970, The recovery of resting potential and input resistance in sheep heart injured by knife or laser, *J. Physiol.* **208**:547–562.

Demtroder, W., and Stock, M., 1975, Molecular constants and potential curves of Na_2 from laser-induced fluorescence, *J. Mol. Spectrosc.* **55**:476–486.

DeVault, D., and Chance, B., 1966, Studies of photosynthesis using a pulsed laser. I. Temperature dependence of cytochrome oxidation rate in *Chromatium*. Evidence for tunneling, *Biophys. J.* **6**:825–847.

Dubin, S. B., 1972, Measurement of translational and rotational diffusion coefficients by laser light scattering, in: *Methods in Enzymology,* Vol. 26 (C. M. W. Hirs and S. N. Timasheff, eds.), Part C, pp. 119–174, Academic Press, New York.

Dubin, S. B., Lunacek, J. H., and Benedek, G. B., 1967, Observation of the spectrum of light scattered by solution of biological macromolecules, *Proc. Natl. Acad. Sci. USA* **57**:1164–1171.

Edlow, J., Fine, S., Vawter, G. F., Jockin, H., and Klein, E., 1965, Laser irradiation: Effect on rat embryo and fetus *in utero, Life Sci.* **4**:615–623.

Erfurth, S., and Peticolas, W. L., 1975, Melting and premelting phenomenon in DNA by laser Raman scattering, *Biopolymers* **14**:247–264.

Feir, D., and Lough, J. W., Jr., 1969, Physiology of the large milkweed bug after laser irradiation, *Comp. Biochem. Physiol.* **28**:759–764.

Fisher, M. M., and Weiss, K., 1974, Laser photolysis of retinal and its protonated and unprotonated n-butylamine Schiff base, *Photochem. Photobiol.* **20**:423–432.

Floyd, R. A., Keyhani, E., and Chance, B., 1971, Membrane structure and function. II. Alterations in the photo-induced absorption changes after treatment of isolated chloroplasts with large pulses of the ruby laser, *Arch. Biochem. Biophys.* **146**:627–634.

Fujime, S., Maruyama, M., and Asakura, S., 1972, Flexural rigidity of bacterial flagella studied by quasielastic scattering of laser light, *J. Mol. Biol.* **68**:347–359.

Gee, R. A., and Truscott, T. G., 1968, Fluorescence spectra of chlorophyll excited by a continuous gas laser, *Chem. Commun.* **15**:839–841.

Gill, D., Heyde, M. E., and Rimai, L., 1971, Raman spectrum of the 11-*cis* isomer of retinaldehyde, *J. Am. Chem. Soc.* **93**:6288–6289.

Glick, D., 1966, The laser microprobe. Its use for elemental analysis in histochemistry, *J. Histochem. Cytochem.* **14**:862–868.

Glick, D., 1969, Cytochemical analysis by laser microprobe-emission spectroscopy, *Ann. N.Y. Acad. Sci.* **157**:265–274.

Glick, D., and Marich, K. W., 1975, Potential for clinical use of the analytical laser microprobe for element measurement, *Clin. Chem.* **21**:1238–1244.

Glick, D., and Rosan, R. C., 1966, Laser microprobe for elemental microanalysis, application in histochemistry, *Microchem. J.* **10**:393–401.

Goldman, L., Rockwell, R. J., Jr., Naprstek, Z., Siler, V. E., Hoefer, R., Hobeika, C., Hishimoto, C., Polanyi, T., and Bredmeier, H. C., 1970, Some parameters of high output CO_2 laser experimental surgery, *Nature (London)* **228**:1344–1345.

Goldstein, S. F., Holwill, M. E. J., and Silvester, N. R., 1970, The effects of laser microbeam irradiation on the flagellum of *Crithidia (Strigomonas) oncopelti, J. Exp. Biol.* **53**:401–409.

Gordon, T. E., Bishop, K., Carter, C. H., and Connolly, M. J., 1968, Laser blockage or delay of cell division at prophase in human leukocyte cultures, *J. Dent. Res.* **47**:171.

Hall, R. R., Beach, A. D., Baker, E., and Morison, P. C. A., 1971, Incision of tissue by carbon dioxide laser, *Nature (London)* **232**:131–132.

Ham, W. T., Jr., Mueller, H. A., Goldman, A. I., Newnam, B. E., Holland, L. M., and Kuwabara, T., 1974, Ocular hazard from picosecond pulses of Nd:YAG laser radiation, *Science* **185**:362–363.

Herczegh, M., Mester, E., and Ronto, G., 1971, Examination of laser-inactivation on T7 phages, *Acta Biochim. Biophys. Acad. Sci. Hung.* **6**(1):41–44.

Heyde, M. E., Rimai, L., Kilponen, R. G., and Gill, D., 1972, Resonance-enhanced Raman spectra of iodine complexes with amylose and polyvinyl alcohol, and of some iodine-containing trihalides, *J. Am. Chem. Soc.* **94**:5222–5227.

Hillenkamp, F., Unsold, E., Kaufmann, R., and Nitsche, R., 1975, Laser microprobe mass analysis of organic materials, *Nature (London)* **256**:119–120.

Hoye, R. C., Ketcham, A. S., and Riggle, G. C., 1967, The air-borne dissemination of viable tumor by high-energy neodymium laser, *Life Sci.* **6**:119–125.

Inaba, H., and Kobayashi, T., 1969, Laser-Raman radar for chemical analysis of polluted air, *Nature (London)* **224**:170–172.

Ippen, E. P., and Shank, C. V., 1975, Sub-picosecond spectroscopy with a mode-locked CW dye laser, in: *Lasers in Physical Chemistry and Biophysics* (J. Joussot-Dubien, ed.), pp. 293–302, Elsevier Scientific, Amsterdam.

Jamieson, C. W., Litwin, M. S., Longo, S. E., and Krementz, E. T., 1969, Enhancement of melanoma cell culture growth rate by ruby laser radiation, *Life Sci.* **8**:101–106.

Jortner, J., and Berry, R. S., 1968, Radiationless transitions and molecular quantum beats, *J. Chem. Phys.* **48**:2757–2766.

Junge, W., and DeVault, D., 1975, Symmetry, orientation and rotational mobility of heme A_3 of cytochrome-*c*-oxidase in the inner membrane of mitochondria, in: *Lasers in Physical Chemistry and Biophysics* (J. Joussot-Dubien, ed.), pp. 439–447, Elsevier Scientific, Amsterdam.

Katan, M. B., Giling, L. J., and van Voorst, J. D. W., 1971, pH dependence of the transient absorptions on the flash photolysis of 3-methyllumiflavin, *Biochim. Biophys. Acta* **234**:242–248.

Kaufmann, K. J., Dutton, P. L., Netzel, T. L., Leigh, J. S., and Rentzepis, P. M., 1975, Picosecond kinetics of events leading to reaction center bacteriochlorophyll oxidation, *Science* **188**:1301–1304.

Keyhani, E., Floyd, R. A., and Chance, B., 1971, Membrane structure and function. I. An electron microscope study of the alteration induced by laser irradiation on the chloroplast lamellar membranes, *Arch. Biochem. Biophys.* **146**:618–626.

Klar, H., 1973, Theory of collision induced rotational energy transfer in the π state of diatomic molecules, *J. Phys. B* **6**:2139–2149.

Kolar, J., Babicky, A., and Blabla, J., 1969, Some effects of laser upon the bones, *Experientia* **25**:365–366.

Kraemer, P. M., Deaven, L. L., Crissman, H. A., Steinkamp, J. A., and Petersen, D. F., 1974, On the nature of heteroploidy, *Cold Spring Harbor Symp. Quant. Biol.* **38**:133–144.

Lamotte, M., Dewey, H. J., Keller, R. A., and Ritter, J. J., 1975a, Laser induced photochemical enrichment of chlorine isotopes, *Chem. Phys. Lett.* **30**:165–170.

Lamotte, M., Dewey, H. J., Ritter, J. J., and Keller, R. A., 1975b, Laser induced photochemical enrichment of chlorine isotopes, in: *Lasers in Physical Chemistry and Biophysics* (J. Joussot-Dubien, ed.), pp. 153–162, Elsevier Scientific, Amsterdam.

Leone, S. R., and Wodarczyk, F. J., 1974, Laser-excited electronic-to-vibrational energy transfer from bromine ($4^2P_{1/2}$) to hydrogen chloride and hydrogen bromide, *J. Chem. Phys.* **60**:314–315.

Letokhov, V. S., 1973, Use of lasers to control selective chemical reactions, *Science* **180**:451–458.

Letokhov, V. S., and Ambartzumian, R. V., 1971, Selective two-step (STS) photoionization of atoms and photodissociation of molecules by laser radiation, *IEEE J. Quantum Electron.* **7**:305–306.

Lewis, A., and Spoonhower, J., 1974, Tunable laser resonance Raman spectroscopy in biology, in: *Spectroscopy in Biology and Chemistry* (S. Yip and S. Chen, eds.), pp. 347–376, Academic Press, New York.

Lewis, A., Spoonhower, J., Bogomolni, R., Lozier, R., and Stoeckenius, W., 1974, Tunable laser resonance Raman spectroscopy of bacteriorhodopsin, *Proc. Natl. Acad. Sci. USA* **71**:4462–4466.

Lewis, A., Nelson, N., and Racker, E., 1975, Laser Raman spectroscopy as a mechanistic probe of the phosphate transfer from adenosine triphosphate in a model system, *Biochemistry* **14**:1532–1535.

Litwin, M. S., and Earle, K. M. (eds.), 1965, *Proceedings of the First Annual Conference on Biologic Effects of Laser Radiation, Fed. Proc.,* Suppl. 14, Vol. 24, No. 1, Part III.

Liu, D. D.-S., Datta, S., and Zare, R. N., 1975, Laser separation of chlorine isotopes. The photochemical reaction of electronically excited iodine monochloride with halogenated olefins. *J. Am. Chem. Soc.* **97**:2557–2558.

Lytle, F. E., and Kelsey, M. S., 1974, Cavity-dumped argon-ion laser as an excitation source in time-resolved fluorimetry, *Anal. Chem.* **46**:855–860.

Mathis, P., Vermeglio, A., and Haveman, J., 1975, Primary reactions of photosynthesis in green plants. A study of photosystem-2 at low temperature, in: *Lasers in Physical Chemistry and Biophysics* (J. Joussot-Dubien, ed.), pp. 465–474, Elsevier Scientific, Amsterdam.

Matsui, S., Rounds, D. E., and Olson, R. S., 1971, The effect of laser power at 2650 Å on deoxyribonucleic acid, *Life Sci.* **10**:217–221.

Mayer, S. W., Kwok, M. A., Gross, R. W., and Spencer, D. J., 1970, Isotope separation with the CW hydrogen fluoride laser, *Appl. Phys. Lett.* **17**:516–519.

McKinnel, R. G., Mims, M. F., and Reed, L. A., 1969, Laser ablation of maternal chromosomes in eggs of *Rana pipiens, Z. Zellforsch. Mikrosk. Anat.* **93**:30–35.

Mendelsohn, R., 1973, Resonance Raman spectroscopy of the photoreceptor-like pigment of *Halobacterium halobium, Nature (London)* **243**:22–24.

Moore, C. B., and Zittel, P. F., 1973, State-selected kinetics from laser-excited fluorescence, *Science* **182**:541–546.

Netzel, T. L., Rentzepis, P. M., and Leigh, J., 1973, Picosecond kinetics of reaction centers containing bacteriochlorophyll, *Science* **182**:238–241.

Nicholls, D. M., Petryshyn, R., and Warner, L., 1974, Laser irradiation induces increased activity of liver elongation factor 1, *Radiat. Res.* **60**:98–107.

Novak, J. R., and Windsor, M. W., 1967, Laser photolysis and spectroscopy in the nanosecond time range: Excited singlet state absorption in coronene, *J. Chem. Phys.* **47**:3075–3076.

Ohnuki, Y., Olson, R. S., Rounds, D. E., and Berns, M. W., 1972, Laser microbeam irradiation of the juxtanucleolar region of prophase nucleolar chromosomes, *Exp. Cell Res.* **71**:132–144.

Okigaki, T., and Rounds, D. E., 1967, Effect of laser radiation on mitosis, *Chromosome Info. Serv. No. 8,* pp. 16–19.

Paleg, L. G., and Aspinall, D., 1970, Field control of plant growth and development through the laser activation of phytochrome, *Nature (London)* **228**:970–973.

Pao, Y.-H., and Rentzepis, P. M., 1965, Multiphoton absorption and optical-harmonic generation in highly absorbing molecular crystals, *J. Chem. Phys.* **43**:1281–1286.

Rattner, J. B., and Berns, M. W., 1974, Light and electron microscopy of laser microirradiated chromosomes, *J. Cell Biol.* **62**:526–533.

Rentzepis, P. M., 1968, Lasers in chemistry, *Photochem. Photobiol.* **8**:579–588.

Rimai, L., Kilponen, R. G., and Gill, D., 1970, Resonance-enhanced Raman spectra of visual pigments in intact bovine retinas at low temperatures, *Biochem. Biophys. Res. Commun.* **41**:492–497.

Rimai, L., Gill, D., and Parsons, J. L., 1971*a*, Raman spectra of dilute solutions of some stereoisomes of vitamin A type molecules. *J. Am. Chem. Soc.* **93**:1353–1357.

Rimai, L., Heyde, M. E., Heller, H. C., and Gill, D., 1971*b*, Raman excitation profiles as probes for inaccessible electronic levels in molecules: Retinal, retinol and naphthalene, *Chem. Phys. Lett.* **10**:207–211.

Rosenfeld, T., Alchalal, A., and Ottolenghi, M., 1972, Nanosecond laser photolysis of rhodopsin in solution, *Nature (London)* **240**:482–483.

Rothschild, K. J., and Stanley, H. E., 1974, Raman spectroscopic investigation of Gramicidin A′ conformations, *Science* **185**:616–618.

Rounds, D. E., 1965, Effects of laser radiation on cell cultures, *Fed. Proc. Suppl. 14,* **24**(1):S-116–S-121.

Rounds, D. E., Olson, R. S., and Johnson, F. M., 1965*a*, The laser as a potential tool for cell research, *J. Cell Biol.* **27**:191–197.

Rounds, D. E., Olson, R. S., and Johnson, F. M., 1965*b*, The effect of the laser on cellular respiration, *IEEE/NEREM Rec.* **7**:106–108.

Rounds, D. E., Chamberlain, E. C., and Okigaki, T., 1965*c*, Laser radiation of tissue cultures, *Ann. N.Y. Acad. Sci.* **122**:713–727.

Rounds, D. E., Olson, R. S., and Johnson, F. M., 1966, Two photon absorption in reduced nicotinamide-adenine denucleotide (NADH), *NEREM Rec.* **8**:158–159.

Rounds, D. E., Olson, R. S., and Johnson, F. M., 1967*a*, The effect of the laser on cellular respiration, *Z. Zellforsch. Mikrosk. Anat.* **87**:193–198.

Rounds, D. E., Olson, R. S., and Johnson, F. M., 1967*b*, Wavelength specificity of laser-induced biological damage, *IEEE 9th Ann. Symp. Electron, Ion, Laser Beam Technology,* pp. 363–370.

Sacchi, C. A., Svelto, O., and Prenna, G., 1974, Pulsed tunable lasers in cytofluorometry, *Histochem. J.* **6**:251–258.

Salmeen, I., Rimai, L., Gill, D., Yamamoto, T., Palmer, G., Hartzell, C. R., and Beinert, H., 1973, Resonance Raman spectroscopy of cytochrome *c* oxidase and electron transport particles with excitation near the Soret band, *Biochem. Biophys. Res. Commun.* **52**(3):1100–1107.

Schleyer, H., and Chance, B., 1962, Abst. 6th Ann. Mtg. Biophys. Soc., FC9.

Spiro, T. G., and Strekas, T. C., 1972, Resonance Raman spectra of hemoglobin and cytochrome *c*: Inverse polarization and vibronic scattering, *Proc. Natl. Acad. Sci. USA* **69**:2622–2626.

Strekas, T. C., and Spiro, T. G., 1972, Hemoglobin: Resonance Raman spectra, *Biochim. Biophys. Acta* **263**:830–833.

Thomas, G. J., Jr., 1970, Raman spectral studies of nucleic acids in laser-excited spectra of ribosomal RNA, *Biochim. Biophys. Acta* **213**:417–423.

Tsuboi, M., Takahashi, S., Muraishi, S., Kajiura, T., and Nishimura, S., 1971, Raman spectrum of a transfer RNA, *Science* **174**:1142–1144.

Vacek, K., Vavrinec, E., and Kalousek, I., 1973, Fluorescence of chlorophyll a excited by a He-Ne laser, *Photochem. Photobiol.* **17**:63–64.

Visser, A. J. W. G., van Ommen, G. J., van Ark, G., Muller, F., and van Voorst, J. D. W., 1974, Laser photolysis of 3-methyllumiflavin, *Photochem. Photobiol.* **20**:227–232.

Whipple, H. E. (ed.), 1965, *The Laser, Ann. N.Y. Acad. Sci.,* Vol. 122.

Wilde, W. H. A., 1965, Laser effects on two insects, *Can. Entomol.* **97**:88–92.

Wilde, W. H. A., 1967, Laser effects on some phytophagous arthropods and their hosts, *Ann. Entomol. Soc. Am.* **60**:204–207.

Wilson, R. M., and Wunderly, S. W., 1974*a*, Laser-induced formation of 1,2,4-trioxans: The trapping oxetan precursors with molecular oxygen, *J. Chem. Soc. Chem. Comm.* **12**:461–462.

Wilson, R. M., and Wunderly, S. W., 1974*b*, Sulfur dioxide trapping of photochemically generated 1,4-biradicals, *J. Am. Chem. Soc.* **96**:7350–7351.

Wilson, R. M., Gardner, E. J., Elder, R. C., Squire, R. H., and Florian, L. R., 1974, The laser initiated oxidative photoaddition of *p*-benzoquinone to cyclooctatetraene, *J. Am. Chem. Soc.* **96**:2955–2963.

Windsor, M. W., Rockley, M. G., Cogdell, R. J., and Parson, W. W., 1975, Picosecond flash photolysis and spectroscopy and kinetics of intermediates in bacterial photosynthesis, in: *Lasers in Physical Chemistry and Biophysics* (J. Joussot-Dubien, ed.), pp. 369–376, Elsevier Scientific, Amsterdam.

Wolbarsht, M. L. (ed.), 1971, *Laser Applications in Medicine and Biology,* Vol. 1, Plenum Press, New York.

Wolbarsht, M. L. (ed.), 1974, *Laser Applications in Medicine and Biology,* Vol. 2, Plenum Press, New York.

Yamamoto, T., Palmer, G., Gill D., Salmeen, I. T., and Rimai, L., 1973, The valence and spin state of iron in oxyhemoglobin as inferred from resonance Raman spectroscopy, *J. Biol. Chem.* **248**:5211–5213.

Yu, N.-T., and East, E. J., 1975, Laser Raman spectroscopic studies of ocular lens and its isolated protein fractions, *J. Biol. Chem.* **250**:2196–2202.

Yu, N.-T., Liu, C. S., and O'Shea, D. C., 1972, Laser Raman spectroscopy and the conformation of insulin and proinsulin, *J. Mol. Biol.* **70**:117–132.

Yu, N.-T., Lin, T.-S., and Tu, A. T., 1975, Laser Raman scattering of neurotoxins isolated from the venoms of sea snakes *Lapemis hardwickii* and *Enhydrina schistosa, J. Biol. Chem.* **250**:1782–1785.

Zitter, R. N., Lau, R. A., and Wills, K. S., 1975, Infrared laser induced reaction of CF_2Cl_2, *Am. Chem. Soc.* **97**:2578.

2

Photochemistry of the Nucleic Acids

Leonhard Kittler and Günter Löber

Akademie der Wissenschaften der DDR, Forschungszentrum für Molekularbiologie und Medizin, Zentralinstitut für Mikrobiologie und experimentelle Therapie Jena, Abteilung Biophysikochemie, DDR-69 Jena, Beuthenbergstrasse 11

1. INTRODUCTION

Deoxyribonucleic acid (DNA) and ribonucleic acid (RNA) contain four major bases—adenine, guanine, cytosine, and thymine (uracil instead of thymine in RNA)—each of which strongly absorbs ultraviolet (UV) radiation with a maximum at about 260 nm. Therefore, UV irradiation of nucleic acids and related model compounds produces electronically excited bases that are able to undergo various photochemical reactions involving neighboring bases, sugar residues, surrounding water molecules, proteins, and bound organic ligands. The photochemical properties of minor bases and base analogues usually differ from those of the major normal bases. In view of this, the study of photochemical processes may well lead to important advances in the recognition of their biochemical functions.

Nucleic acids and nucleic acid constituents do not absorb visible light. When a fluorescent sensitizer is in close association with the nucleic acids, the fluorescence properties of the sensitizer are influenced by the nucleic acid structure. This statement is discussed as an important determinant for the occurrence of fluorescence banding patterns on eukaryotic chromosomes. Moreover, through the association of sensitizer molecules with the nucleic acids, one can photooxidize guanine residues under visible light irradiation in the presence of molecular oxygen (photodynamic action). Dye-sensitized autooxidation can proceed by different reaction pathways. Photodynamic effects on nucleic acids differ greatly from those produced by UV radiation.

The photochemistry of nucleic acids and the nature and biological significance of the damage produced have been the subject of several review articles (McLaren and Shugar, 1964; Johns, 1966; Burr, 1968; Smith and Hanawalt, 1969; Fahr, 1970; Varghese, 1972a; Löber and Kittler, 1973, 1977; Berg, 1975). Our attention will be directed toward recent results on UV-induced and dye-sensitized photoreactions of nucleic acids, to investigate how such studies help in understanding the influence that nucleic acid structure has on the mechanism of photochemical reactions. In addition, photochemical experiments with isolated nucleic acids help to interpret photobiological effects on a molecular level.

2. PHOTOREACTIONS OF NUCLEIC ACIDS AND THEIR CONSTITUENTS

2.1. Natural Bases

2.1.1. Dimer Photoproducts of the Cyclobutane Type

The formation of a cyclobutane photodimer of thymine was first reported by Beukers *et al.* (1958, 1959) and Wang (1958, 1961). Experiments were carried out in an ice matrix. Uracil dimers were subsequently found by the same technique (Smith, 1963). Since that time, research in this field has progressed rapidly and many of the well-known facts have been summarized in numerous monographs and review articles (McLaren and Shugar, 1964; Smith, 1964*b*; Wacker *et al.*, 1964*a*; Wang, 1965; J. K. Setlow, 1966; Johns, 1966, 1971; Burr, 1968; Smith and Hanawalt, 1969; Fahr, 1969, 1970; Kochetkov *et al.*, 1970; Varghese, 1972*a*; Rahn, 1973; Löber and Kittler, 1973; 1977). Therefore, we want to confine this chapter to the major results, and to a few of the recent papers.

The formation of cyclobutane dimers is observed only under certain experimental conditions. Pyrimidine bases dimerized if UV irradiation was performed in an ice matrix (Wang, 1964, 1965; Varghese, 1970*a*), in KBr pellets (Lisewski and Wierzchowski, 1970), or in solid films (Varghese, 1971*b*), while generally no dimerization has been detected after the UV irradiation of solutions. However, as mentioned by different authors, photodimers are formed in solutions of oligomers, polymers, apurinc acid, and nucleic acids (Johns *et al.*, 1964; J. K. Setlow, 1966; Hariharan and Johns, 1968*b*; Rahn and Landry, 1971; Stepien *et al.*, 1973*b*; Rahn and Schleich, 1974). Aside from the thymine dimer, the homodimers of uracil (Smith, 1963; Johns, 1968; Brown and Johns, 1968; Kondo and Witkop, 1969), cytosine (Varghese, 1971*a*; Varghese and Rupert, 1971), orotic acid (Sztumpf and Shugar, 1965), and different heterodimers (Smith, 1963; Weinblum, 1967; Varghese, 1971*a*; Fahr *et al.*, 1974; Rahn and Schleich, 1974) can be generated in significant yield. Pyrimidine dimers exhibit short-wavelength reversal. At 280 nm the yield of dimers is greater than that at 240 nm, reflecting the large cross section for dimer photoreversal at the latter wavelength. The absorption spectra indicate that dimers do not absorb at 280 nm, while molar extinction coefficients of dimers and monomers are comparable at 240 nm.

Pyrimidine bases form dimers in solutions containing sensitizers (ketones) when irradiated at energies below the first excited singlet state of the pyrimidines. Therefore, the immediate precursor of the dimer is the first excited triplet state (Ben-Hur *et al.,* 1967; Charlier and Hélénè, 1967; Lamola and Yamane, 1967; Greenstock and Johns, 1968; Jennings *et al.,* 1970, 1972; Varghese, 1972*b*). A photosensitized splitting of dimers has been found when quinone derivatives are used as sensitizers (Ben-Hur and Rosenthal, 1970). Moreover, dimers are split by a photorepair mechanism (Harm, 1970; Böhme and Adler, 1972; Hanawalt, 1972). Section 4.2.6 of this chapter is devoted to the sensitized splitting of pyrimidine dimers.

A cyclobutane dimer could theoretically exist in 12 isomeric forms; however, only four of them with the configuration *cis-syn, cis-anti, trans-syn,* and *trans-anti* are produced in excellent yield (Khattak and Wang, 1972). Irradiation of pyrimidines in an ice matrix yields dimers in the *cis* configuration, while dimers in the *trans* configuration are formed by ketone sensitization (Table 1) (Weinblum and Johns, 1966; Varghese and Wang, 1967*a*; Lamola, 1970; Khattak and Wang, 1972). The different isomers of a particular dimer are usually identified on the basis of infrared absorption spectra, nuclear magnetic resonance spectra (Wulff and Fraenkel, 1961; Weinblum and Johns, 1966; Weinblum *et al.,* 1968; Varghese, 1971*a*; Fahr *et al.,* 1972*a,b,* 1974), mass spectra (Fenselau and Wang, 1969), chromatographic mobilities (Smith, 1966*b*; Varghese and Wang, 1967*a*; Weinblum, 1967; Weinblum and Johns, 1966), and X-ray crystallography (Camerman and Camerman, 1968; Camerman *et al.,* 1969; Adman and Jensen, 1970). The structures of the different pyrimidine dimers have been confirmed by their chemical synthesis (Blackburn and Davies, 1966*a,b*; Dönges and Fahr, 1966; Dörhöfer and Fahr, 1966; Fahr, 1969). In contrast

TABLE 1.   Relative Amounts of Isomeric Thymine Dimers[a]

	Irradiation condition	*cis-syn*	*cis-anti*	*trans-syn*	*trans-anti*
Thymine	Frozen	100	—	—	—
	83% acetone	20	18	22	40
	Acetophenone (saturated)	24	33	20	23
Thymidine	Frozen	38	48	—	14
	Acetone	26	13	6	55
	Solid	26	55	—	19

[a] From Varghese (1972*a*).

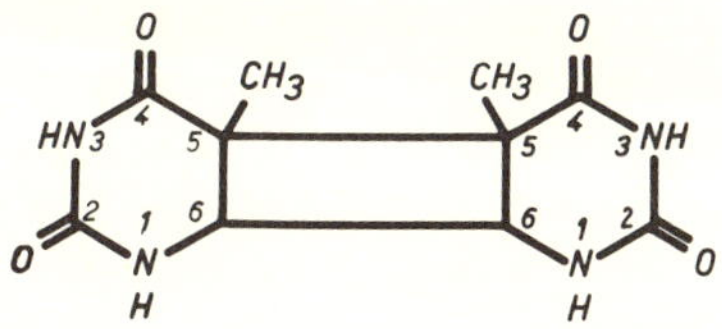

Fig. 1. Formula of the *cis-syn* thymine dimer.

to the behavior of dimers possessing a *cis* configuration, the influence of alkali or acid gives rise to structural alterations of dimers in the *trans* configuration (Herbert *et al.,* 1969; Khattak and Wang, 1972). There is evidence that the *cis-syn* pyrimidine dimers (Fig. 1) play a dominant role in most biological systems. In support of this suggestion, it was found that the photorepairing enzyme preferentially eliminates this type of dimer (Ben-Hur and Ben-Ishai, 1968).

2.1.1a. Mechanisms for Dimerization

Originally, Beukers and Berends (1961) demonstrated a triplet-state intermediate for thymine dimerization in an ice matrix through quenching studies using paramagnetic ions which should be effective in quenching the triplet but not the singlet state of thymine. The triplet state as an effective precursor was also confirmed for orotic acid dimerization. Results were obtained from quenching studies, electron paramagnetic resonance (EPR) investigations, flash photolysis, and kinetic experiments (Haug and Douzou, 1965; Sztumpf and Shugar, 1965; Herbert *et al.,* 1968; Yip *et al.,* 1970; Herbert and Johns, 1971). Although conditions for thymine dimerization and uracil dimerization in fluid aqueous solution are far from ideal (Alcantara and Wang, 1965a,b), the particular role of triplet precursors is entirely expected. Taking into consideration that the lifetime of the thymine singlet state in aqueous solution at room temperature is on the order of 10^{-11} s, and assuming that the dimerization is diffusion controlled, the short lifetime of the monomer singlet state makes it an unlikely precursor for the dimer because of the much longer time necessary for the monomers to diffuse together. For this reason, the relatively long-lived triplet states, having lifetimes on the order of 10^{-6}–10^{-5} s, are much more probable as active intermediates. Various experimental (Greenstock *et al.,* 1967; Johns, 1968; Brown and Johns, 1968; Szabo *et al.,* 1970; Whillans and Johns, 1971, 1972b) and theoretical (Pullman, 1968; Danilov *et al.,* 1969; Snyder *et al.,* 1970) considerations are in agreement with a triplet mechanism. On the other hand, indications are available that under proper conditions dimerization proceeds via the first excited singlet state of pyrimidine monomers. This holds true for the high efficiency of photodimerization of thymine in ice, where, in freezing, microcystals of thymine are formed in which neighboring thymines are parallelly stacked and suitably placed for dimerization (Davis and Tinoco, 1966). In this case, dimerization reaction is faster than the process of fluorescence emission (Lamola and Eisinger, 1968). At higher thymine concentration, i.e., $> 5 \times 10^{-4}$ M, self-association occurs. The existence of thymine associates favors dimer formation by a singlet mechanism (Stepien *et al.*, 1973*b*). Thymine residues that are covalently linked, e.g., in thymidylyl (3′-5′)-thymidine (TpT) and thymine-containing oligonucleotides, are able to dimerize via a singlet mechanism (Eisinger and Lamola, 1967).

There is evidence for the decreased formation of photodimers as a result of addition of organic solvents. It has been discussed that destacking of the bases induced by organic solvent molecules is the main effect of the organic environment in the cases of N_1,N_3-dimethylthymine (Lisewski and Wierzchowski, 1969) and DNA (Dellweg and Wacker, 1965). Otherwise, the transition from double-stranded native state to the single-stranded denatured state of DNA is accompanied by an enhanced ability of pyrimidines to dimerize (Dellweg and Wacker, 1965). This fact is probably caused by an untwisting process of the helical arrangement of DNA bases in going from the native to the denatured state of DNA. The first systematic investigation of solvent effects on the formation of dimers was reported by Wacker and Lodemann (1965). In conclusion, the effect of solvent environment on dimerization rate of TpT is characterized by a significant decrease (Table 2). It has been pointed out that both the excited singlet state and the excited

TABLE 2.   Dimer Formation of TpT in Various Solvents[a]

Solvent	F^b (kcal/mol)	Dimerization rate (%)
t-Butanol	−2.9	5
Ethanol	−4.2	5.5
Methanol	−4.5	6
n-Butanol	−4.7	7.5
Glycol	−11.3	19
Formamide	−16.3	21
Glycerol	−16.5	33
Water	−22.3	35

[a] From Wacker and Lodemann (1965).
[b] Free energy of dimerization reaction.

triplet state are involved in the photoreaction (Lamola, 1968; Morrison *et al.*, 1968; Wagner and Bucheck, 1968; Murcia *et al.*, 1972). Kleopfer and Morrison (1972), in a discussion of environmental effects on the excited singlet and triplet levels of thymine, correlated solvent effects with the distribution of different isomers. Dimerization in stacked thymine molecules could take place very rapidly, i.e., within the lifetime of a singlet or singlet excimer (Morrison *et al.*, 1968; Wagner and Bucheck, 1968). Nonetheless, triplet sensitization experiments ensure population of the triplet state of the thymine by energy transfer from the triplet state of a suitable sensitizer and confirm that the photodimerization of thymine can proceed by way of a triplet-state precursor. The triplet-state mechanism preferentially yields *trans* isomers, while the singlet-state mechanism prefers the formation of *cis* isomers (Khattak and Wang, 1972).

Results on the temperature dependence of dimerization reaction in DNA were described (Rahn and Hosszu, 1968*b*; Rahn, 1970). It was found that the dimerization rate remained constant when temperature was raised up to the melting point. Above the melting point, the yield of dimer formation cooperatively decreased. This is probably due to both the breakdown of the double helical structure of DNA and the reduced tendency of the bases to stack at a temperature above the melting point (Hosszu and Rahn, 1967). A greater yield of dimers is formed as compared with DNA in its native state, when, before irradiation, the DNA sample was heated above the melting point and subsequently cooled to room temperature. This treatment obviously induces a changed spatial arrangement of the stacked bases, which is in favor of the dimerization reaction. Fenster and Johns (1973) have discussed the temperature effect on dimer formation of thymine and orotic acid within the temperature range from 15°C to 90°C. They involved in their consideration both the decrease of the triplet lifetime of an excited base and the increasing rate of diffusion-controlled collisions at elevated temperatures. Füchtbauer and Mazur (1966) measured the quantum yield for disappearance of the thymine monomers in frozen aqueous solution to be between one and two. The yield is the same at 77°K and 268°K. On the contrary, Rahn and Hosszu (1968*b*) described a lowered efficiency of photodimerization of thymine in ice at 77°K. They discussed their result on the basis of a restricted mobility of thymine molecules in solid aqueous solution.

The influence of relative humidity of DNA films on dimer formation has been interpreted in terms of conformational effects. When irradiation was performed within a range of 65% up to 100% relative humidity, no alterations in the dimer yield could be found (Rahn, 1970). DNA is thought to exist between values of 75% and 85% relative humidity in the A-type conformation and above 85% in the B-type conformation (Pilet and Brahms,

1973). This conformational transition is obviously not reflected in the dimer yield. Below a value of 65% relative humidity, B-type conformation disappears in favor of a more disordered structure, which has less ability to dimerize (Rahn, 1970). Recently, Fritzsche *et al.* (1976) have reported on linear infrared dichroic studies done with oriented DNA films at about 80% relative humidity. These investigators suggested that UV irradiation could produce a change from the A-type to the C-type structure of DNA.

The formation of pyrimidine dimers has been studied as a function of pH of the sample during UV irradiation. The rate of dimerization is lowered in alkaline as well as in acidic medium. Repulsion forces acting on anionic and cationic forms of pyrimidines at high and low pH, respectively, induce a spatial rearrangement of pyrimidine residues in DNA which decreases dimerization to a great extent (Sztumpf and Shugar, 1965; Whillans and Johns, 1972*a*).

Dyes bound to nucleic acid cause a drastic decrease of the dimer yield (Beukers, 1965; Setlow and Carrier, 1967; Sutherland and Sutherland, 1969*a*, 1970; Roth, 1973; Klimek and Sevčikova, 1973). This fact is probably caused by a transfer of excitation energy from nucleic acid bases to bound dye molecules (see Section 4.2.5). Metal ions, e.g., mercury and copper, when bound to nucleic acids also reduce the dimer yield (Rahn *et al.*, 1970; Rahn and Landry, 1973). Such behavior has been interpreted by means of the "heavy atom effect." A shortening of the triplet lifetime of thymine as ensured in the presence of metal ions demonstrates again that the photodimerization of thymine can proceed by way of a triplet-state precursor (physicochemical factors influencing pyrimidine dimerization are summarized in Table 3 on pages 48–49).

2.1.1b. *Molecular Biological Effects of Pyrimidine Dimers*

A major part of the lethal effect of UV irradiation on biological systems is attributed to photochemical alterations of pyrimidine residues. With respect to thymine dimers, their existence in UV-irradiated bacteria was first described by Wacker *et al.* (1960, 1961). There are numerous phenotypic effects that are caused primarily by pyrimidine dimers and that could be explained at the genetic level (R. B. Setlow, 1968). Such genetic effects are (1) interference of dimer-containing DNAs with replication and transcription processes (Bollum and Setlow, 1963; Erikson and Szybalski, 1963; Setlow *et al.*, 1963; Cleaver, 1971; Wacker *et al.*, 1964*a;* Sinclair and Morton, 1965; Rauth and Whitmore, 1966; J. K. Setlow, 1964; Swenson and Setlow, 1966; Phillips *et al.*, 1967; Scholes *et al.*, 1967; Modak and Setlow, 1969; Rahn *et al.*, 1969; Horii and Suzuki, 1968, 1970; R. B. Setlow and J. K.

Setlow, 1970; Ramenda and Sinsheimer, 1971; Kahn, 1974); (2) inactivation of bacteria (Wacker *et al.*, 1960; Dellweg and Wacker, 1962; R. B. Setlow and J. K. Setlow, 1962; Jagger and Stafford, 1965; Yasuda and Fukutome, 1970), viruses Ono and Shimazu, 1967; Bishop *et al.*, 1967; McLaren and Takahashi, 1970), and mammalian cells (Klimek, 1966; Cleaver and Trosko, 1970; Klimek and Vasuček, 1970; Trosko *et al.*, 1970); (3) dimers as a lethal hit for a cell (R. B. Setlow and J. K. Setlow, 1962; J. K. Setlow, 1966; Giese, 1968); and (4) mutations induced by the dimers (R. B. Setlow, 1964; Träger *et al.*, 1964; Hill, 1965; Kondo and Kato, 1966; Ottensmeyer and Whitmore, 1968; Witkin, 1969; Bridges and Munson, 1970; Grahn, 1972).

Various biological systems exist under nearly continuous UV irradiation conditions. Thus, if DNA is accessible to UV light, DNA defects will occur. In view of this fact, it seems reasonable to assume that at least part of the DNA damage can be eliminated by enzymatic repair processes. Indeed, various types of dark repair and photorepair mechanisms are known at present. Such repair mechanisms might regulate the balance between variability and selectivity of the genetic material during evolution (comprehensive literature: R. B. Setlow, 1964; Smith and Hanawalt, 1969; Witkin, 1969; Harm, 1970; Böhme and Adler, 1972; Grahn, 1972; Hanawalt, 1972; Cleaver, 1973, 1974; Löber and Kittler, 1973; Strauss, 1975; Cerutti, 1975; Friedberg, 1975).

2.1.2. Solvent-Addition Photoproducts of the Pyrimidines

2.1.2a. Water-Addition Photoproducts

Another type of photochemical reaction of a pyrimidine base is the addition of a molecule of water across the 5,6 double bond to form a 5,6-dihydro-6-hydroxy derivative. This type of photoreaction was first described by Sinsheimer and Hastings (1949). These hydrates easily revert to the parent compounds. The reversal proceeds rapidly on heating and is favored in acidic medium (Sinsheimer, 1954; Moore and Thomson, 1955, 1957; Moore, 1958). It was possible to prepare 5,6-dihydro-6-hydroxyuracil by chemical synthesis and to ensure in this way the proposed structure of the photohydrate (Gattner and Fahr, 1963).

5,6-Dihydro-6-hydroxycytosine, which is formed on UV irradiation of cytosine in aqueous solution, was found to be less stable than the corresponding photohydrate of uracil (Wierzchowski and Shugar, 1957, 1961*b*). Photochemical experiments were carried out for cytosine and uracil, both dissolved in normal and heavy water. The ratio of the quantum

TABLE 3. Physicochemical Factors Influencing Pyrimidine Dimerization

Physicochemical factor	Pyrimidine-containing substrate	Effect on dimer formation	Photochemical precursor	References
Solvents				
Glycol	DNA	Decrease by base unstacking		Dellweg and Wacker (1965), Wacker and Lodemann (1965)
Alcohols, Formamide, water	TpT			
Water	N_1,N_3-Dimethylthymine (0.1 M)	Increase by base stacking in the ground state	Singlet state	Lisewski and Wierzchowski (1969)
Various nonpolar, polar, polar protonic	N_1,N_3-Dimethylthymine		Singlet and triplet state	Morrison *et al.* (1968), Murcia *et al.* (1972), Kleopfer and Morrison (1972)
Acetonitrile	Thymine, uracil, and their derivatives		Singlet and triplet states	Lamola and Mittal (1966)
Temperature				
Above T_m	DNA	Decrease by base unstacking		Rahn and Hosszu (1968*a*)
15–90°C	Thymine	Decrease by deactivation of excited states	Triplet state	Fenster and Johns (1973)
Ice matrix	Thymine	Increase by base stacking	Triplet state Singlet state	Beukers and Berends (1961), Eisinger and Shulman (1967)
Frozen solution at liquid nitrogen temperature	DNA Poly(dT)	Decrease by reduced movability of thymine residues	Singlet and triplet states	Rahn and Hosszu (1968*a,b*), Lamola (1968)

Metal ions			
Ag^+	DNA	Decrease by heavy atom effect	Rahn *et al.*, (1970)
Hg^+	DNA	Decrease by heavy atom effect	Rahn and Landry (1973)
Dyes			
Proflavine	Polynucleotides	Decrease by energy transfer from bases to dyes	Beukers (1965), Setlow and Carrier (1967), Sutherland and Sutherland (1969*a*, 1970), Klimek and Sevčikova (1973)
Acridine orange	DNA		
Ethidium bromide	DNA		
Various acridine derivatives	DNA		
pH			
Alkaline	Thymine, orotic acid	Decrease by electrostatic repulsion forces	Sztumpf and Shugar (1965), Whillans and Johns (1972*a*)
Acidic	Orotic acid		
Relative humidity (r.h.)			
65–100% r.h.	DNA films	B conformation, no effect	Rahn (1970)
Less than 65% r.h.	DNA films	Decrease by collapse of B conformation to dis-ordered structures	Rahn and Hosszu (1969*a*)
85–60% r.h.		A–C transition	Fritzsche *et al.* (1976)

yields $\phi H_2O/\phi D_2O$ equals 2.2 in both cases. This result suggests by analogy the formation of a water-addition product of cytosine (Wierzchowski and Shugar, 1959), while the proposed phototautomerism is less probable (Wang, 1959). Kleber *et al.* (1965) and Fahr *et al.* (1966) gave evidence for the suggested structure of the photohydrate of cytosine by means of thin-layer chromatographic studies. Later on, its isolation was described using electrophoresis as a tool (Freeman *et al.*, 1965; Johns *et al.*, 1965; Becker *et al.*, 1967). Miller and Cerutti (1968) confirmed the structure of uracil and cytosine photohydrates by means of a specific reaction of these products with sodium borohydride. De Boer and Johns (1970) found further evidence for the proposed chemical structure on the basis of results derived from isotopic exchange experiments. The corresponding photohydrate of thymine was proposed by Fahr *et al.* (1966) and, although rather unstable, was identified by Fisher and Johns (1973).

The photochemical generation of water-addition products was shown also for oligonucleotides (Freeman *et al.*, 1965; Brown *et al.*, 1966), polynucleotides (Wierzchowski and Shugar, 1959; Ono *et al.*, 1965a), apurinic acid (Wierzchowski and Shugar, 1959), RNA (Schuster, 1964; Mattern *et al.*, 1972), and DNA in its denatured or temporarily single-stranded state (Pearson and Johns, 1966a; Setlow and Carrier, 1966). However, photohydrate formation is suppressed in double-stranded DNA and double-stranded copolymers (Wierzchowski and Shugar, 1961a; Setlow *et al.*, 1965; Pearson *et al.*, 1966). In oligonucleotides and polynucleotides, the observed quantum yield of the photohydration reaction has a lower rate than that for single bases (Wierzchowski and Shugar, 1959, 1961a). In addition, the number of photoproducts that are not identical with water-addition products increases with increasing chain length of a polynucleotide (Freeman *et al.*, 1965). Elucidation of the mechanisms of photohydrate formation is complicated, since uracil and cytosine are known for their ability to form photohydrates as well as photodimers (Helleiner *et al.*, 1963; Hariharan and Johns, 1967). A mathematical model which considers the generation of a special photoproduct in the presence of possible other ones has been suggested by Johns *et al.* (1966) and Brown and Johns (1967).

According to Wacker *et al.* (1965), the water-addition reactions are favored at high UV fluences, while at low UV fluences (or with decreasing temperature) dimerization reactions are preferred. The yields of uracil, uracil dimer, and uracil hydrate were 58%, 36%, and 6%, respectively, after 1.2×10^6 ergs/mm^2; and 22%, 24%, and 54%, respectively, after 5.0×10^6 ergs/mm^2. In the presence of triplet quenchers, dimerization processes occurring in various polynucleotides can be selectively suppressed; on the other hand, photohydration processes remain unaffected (Burr *et al.*, 1968).

2.1.2b.　*Reversal of Water-Addition Photoproducts*

It has been shown that the photohydrates of uracil and cytosine are stable against further influence of UV light; however, they undergo chemical reactions in the dark which revert them to the parent compounds. Such chemical reactions were observed at elevated temperature or under extreme pH conditions (Hariharan and Johns, 1968*a*). The stability found for the hydrates of various pyrimidine derivatives decreases in the following order:

$$\text{dimethyluracil} \cdot H_2O > \text{uracil} \cdot H_2O > \text{cytosine} \cdot H_2O > \text{thymine} \cdot H_2O$$

The photohydrate of uracil is stable in the solid state but decomposes in aqueous solution (50°C) following a first-order decay with a rate constant of 3.2×10^{-4} min^{-1} (Gattner and Fahr, 1963). Moreover, the stability of respective uracil and cytosine photohydrates increases in the order

$$\text{base} \cdot H_2O < \text{nucleoside} \cdot H_2O < \text{nucleotide} \cdot H_2O$$

Hydrogen-bond formation between the hydroxy group at the C-6 atom and the sugar phosphate residue has been postulated to explain the stabilizing effect (Fahr *et al.*, 1966). When the pH is raised, a ring rupture process takes place that can involve about 35% of the uracil residues (Schuster, 1964; Fikus and Shugar, 1966).

The rate of reversal of uracil photohydrate decreases with increasing molecular weight of the copolymer (Logan and Whitmore, 1966). The situation becomes more complicated for the cytosine photohydrate, since, besides the reversal, a deamination of cytosine takes place which changes cytosine to uracil (Fig. 2). Thus it could be stated unequivocally that the occurrence of the uracil hydrate after UV irradiation of cytosine must be noted as the result of a dark reaction rather than of a photoreaction. Johns *et al.* (1965) investigated the effect of various parameters, i.e., temperature, pH, and ionic strength, on the deamination reaction. The deamination of cytosine hydrate and concomitant appearance of uracil hydrate equaled about 9% when physiological conditions were chosen. The reversal of cytosine hydrate to cytosine as a function of pH was studied by Fahr *et al.* (1966). The temperature dependence of reversal reaction yielded an activation energy of 16 kcal/mol. This value is in good agreement with that of earlier reports (Sinsheimer, 1957; Wierzchowski and Shugar, 1961*b*). The dehydration of photoproducts is an acid–base catalyzed first-order reaction (McLaren and Shugar, 1964). It follows the monomolecular elimination mechanism.

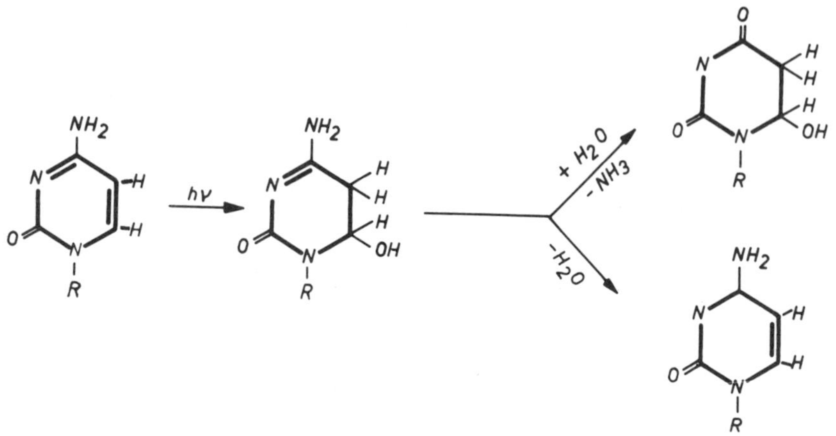

Fig. 2. Photohydration of cytosine in aqueous solution on UV irradiation at 254 nm and subsequent dark reactions, i.e., deamination reaction yielding the photohydrate of uracil (right side, top) and dark reversal yielding the starting compound, uracil (right side, bottom). R, Ribose. According to Johns *et al.* (1965).

2.1.2c. *Mechanisms of Nucleophilic Solvent-Addition Reactions*

The photochemical addition of the nucleophilic agents across the 5,6 double bond of pyrimidines proceeds according to the $S_N 1$ mechanism. Hydrazines, amines, water, alcohols, BH_4^- ions, and HSO_3^- ions can act as nucleophilic agents. Generally, the quantum yield increases with the nucleophilic character, provided that no steric hindrances occur (Wang and Nnadi, 1968; Summers *et al.*, 1973). As a result of solvent addition, the heteroaromatic character of the pyrimidine rings, and with it their ring planarity, is lost (Cerutti *et al.*, 1966; Summers *et al.*, 1973). Since the photohydration process is not quenched by oxygen (Greenstock *et al.*, 1967) and cannot be sensitized by triplet-state sensitizers (Eisinger and Lamola, 1967; Brown and Johns, 1968), it has been stated that the reaction involves singlet excited molecules. They are known to possess strongly polar resonance structures, which might be in favor of the postulated nucleophilic addition reaction. It is assumed that for the excited singlet state the carbonyl oxygen at C-4 becomes more negative and the C-6 more positive. This model is not supported by the results of Burr *et al.* (1972). They stated that electron densities at C-5 and C-6 reported for the excited singlet states of cytosine and uracil make C-5 more positive than C-6. Nevertheless, the higher polarity of pyrimidines in their excited singlet states leads to dipole–dipole interactions between the dipole of an excited base and the dipole of the surrounding solvent molecules in the ground state. The reorientation of the solvent molecules, a process that can easily be finished within the lifetime of the first excited singlet state, is the rate-determining step. It brings the excited pyrimidine molecule and at least one solvent molecule in a suitable spatial position in which both species become covalently linked. Since photohydration is independent of the viscosity of the solvent, the reaction proceeds via non-diffusion-controlled processes. It is concluded that uracil and cytosine undergo photohydration by pathways which predominantly involve the excited singlet state.

According to different reports (Becker *et al.*, 1967; Burr and Park, 1968*a,b*; Burr *et al.*, 1968*a*, 1972; Hauswirth *et al.*, 1972; Khattak *et al.*, 1972), the photohydration of uracil and cytosine is a pH-dependent reaction. Based on the observation that the rate of photohydration is a sigmoid function of the pH of the solution, Becker *et al.* (1967) and Burr *et al.* (1972) concluded that, for uracil and cytosine derivatives, the neutral excited species react faster with water; cationic and anionic forms react more slowly. The apparent pK_a of the excited species follows from the inflection point of the sigmoid curve. The pK_a strongly differs from the pK_a of the ground-state molecules. Since such strong shifts in base–acid equilibrium are known when molecules undergo excitation from the singlet ground state to the first excited singlet state, the observed strong pK_a shift is a further indication for the proposed singlet mechanism.

2.1.2d. *Molecular Biological Effects of the Photohydrates*

The biological importance of photohydrates has been studied but appears not to be very well understood at present. According to different investigators (Singer and Fraenkel-Conrat, 1966; Carpenter and Kleczkowski, 1969; Remsen *et al.*, 1970), the photohydrates of uracil and cytosine may have a role in the inactivation of viruses. Generally, the quantum yield of the photohydration reaction is lower for viral RNA *in situ* than for isolated RNA. The lower frequency of photohydrates in viral RNA *in situ* suggests two possibilities (1) the conformational state of RNA *in situ* does not favor photohydration, and (2) the protein envelopes of viruses have some protective effect (Remsen *et al.*, 1970). Logan and Whitmore (1966) have presented biochemical proof that the photohydrate of uracil has an influence on the ability of poly(U) to code polyphenylalanine synthesis in cell-free systems. Various results indicate that uridine photohydrates in RNA would be read as cytidine (Grossman, 1963; Wacker *et al.*, 1964*a*, 1968; Ono *et al.*, 1965*a,b*; Pearson and Johns, 1966*a*; Pearson *et al.*, 1966; Ottensmeyer and Whitmore, 1968).

On the other hand, the cytidine photohydrate in RNA would be read as uridine (Ono *et al.*, 1965*a*). Howard and Tessman (1964) have shown that mutational changes in phage S13 after irradiation can be explained by assuming a base change from cytosine to uracil. Experiments by Johns *et al.* (1965) provided the chemical basis for this, since they have shown that uracil is the final result after irradiation of cytosine. In DNA, uracil would presumably be read as thymine. Finally, Witkop (1968) described a photo-reduction mechanism that acts rather selectively on uracil. Thus uracil undergoes, in the presence of $NaBH_4$, a UV-induced reduction process yielding 5,6-dihydrouracil as the final product. This reaction suggests a new possibility for selectively damaging uracil residues in RNA.

2.1.3. Stereochemical Effects Involved in Dimerization and Hydration Reactions of the Nucleic Acids

The photochemical modification of nucleic acids or model polynucleotides appears to occur only under particular conditions of solvent, temperature, pH, etc.—factors that are known for their ability to influence the conformation of a polynucleotide (Lomant and Fresco, 1972*a*). Numerous studies suggest that the formation of photodimers is preferred when the polymer exists in a less ordered state, as is true for thermally denatured DNA (Wacker *et al.*, 1962). On the other hand, base stacking must be guaranteed. This can be concluded from the decrease of the dimerization rate in the presence of organic solvents, which usually produce an unstacking of the bases. Since the photohydration of pyrimidines in DNA is also enhanced in going from the native to the denatured state of DNA, it has been stated that reactive pyrimidines are easily accessible to the surrounding water molecules in the less ordered, denatured state of DNA. Generally, reaction rates of both photodimerization and photohydration are lower when native DNA is used; however, photodimers prevail over photohydrates (Setlow and Carrier, 1963; Setlow *et al.*, 1965; Pearson and Johns, 1966*b*; Pearson *et al.*, 1966; Mund and Venner, 1967). Irradiation of more ordered polynucleotides, e.g., DNA, yields clusters of photoproducts rather than a statistical distribution over the whole macromolecule (Brunk, 1973). It may be interpreted in such a way that dimerization leads to a local unwinding of the double helix, thus forming melted regions in which pyrimidines become preferentially photohydrated. Yields of photoproducts were analyzed by RNase hydrolysis of poly(A·U) followed by chromatographic separation and identification of components (Pearson and Johns, 1966*a*). The clustering of photoproducts was also noted for poly(C) (Lomant and Fresco, 1972*b*). The complex poly(A,U)·poly(U) contains two types of U residues: some that are readily modified upon UV irradiation and are nonessential to the secondary structure (non-base-paired regions) and some that are more resistant to photomodification and are required for helix stability (base-paired regions) (Lomant and Fresco, 1973). The photoreactivities of the presumed double helical and single-stranded forms of poly(rT) have been compared (Tramer *et al.*, 1969). The quantum yield for thymine dimerization at 254 nm is twice as high for the bihelical form as for the single-stranded form.

Results are available that support the notions that U residues in polynucleotides become more resistant to the effects of UV irradiation when they are in an ordered structure. It was shown that, in the ordered structure of poly(A·U), the rate of hydrate formation was reduced by a factor of 10 relative to poly(U) in the random coil (Pearson *et al.*, 1966). De Boer *et al.* (1967) irradiated poly(U) at 1°C in the presence of either spermine or Mg^{2+} ions. The more ordered structure of poly(U) favors dimers and suppresses hydrates. Irradiation of the melted structure at 35°C yields a higher rate of photohydrates, while dimer formation becomes strongly reduced. To obtain more information on the processes occurring in disordered structures, DNA was irradiated at elevated temperature where it is thought to exist in the random coiled state (Glisin and Doty, 1967). Under such conditions, the formation of interstrand dimers has been observed.

In a number of papers it has been shown that pyrimidine dimers formed by two pyrimidine bases located on the same strand induce local conformational changes that are evidenced by a decrease of the melting temperature (Venner and Zimmer, 1964; Bagchi *et al.*, 1969; Hayes *et al.*, 1971; Uliana *et al.*, 1971; Shafranovskaya *et al.*, 1973; Rahn and Stafford, 1974; Vorličkova and Paleček, 1974). Recent reports discuss this effect as being due to a local change of DNA conformation in the vicinity of dimers from a B-type to a C-type structure (Lang and Luck, 1973; Lang, 1974*a,b*). The UV inactivation of several tRNAs, i.e., the loss of amino acid acceptor capacity, is assumed to be caused by conformational changes (Buc and Scott, 1966; Harriman and Zachau, 1966; Sarin and Johns, 1968). This could be concluded from the fact that UV inactivation of purine anticodons proceeds to the same extent as the UV inactivation of pyrimidine anticodons. The photochemistry of poly(dA) supports the idea that the presence of photoproducts can alter its secondary structure (Pörschke, 1973*a,b*).

2.1.4. Dimer Photoproducts of the Noncyclobutane Type

Pyrimidine derivatives form a second type of UV-induced homodimers and heterodimers generally referred to as "pyrimidine adducts" (Varghese,

1972*a*). Thymine, uracil, and cytosine, either as free bases or as nucleosides, yield such adducts. They are easily generated from the nucleosides on irradiation as a thin, solid film or in aqueous frozen solutions with UV radiation at 254 nm. A common property of these adducts is the strong displacement of the characteristic absorbance maximum of the monomers at about 260 nm toward the red ($\lambda_{max} \sim 300$ nm).

This fact strongly indicates that photoproducts other than cyclobutane dimers must be formed. Those photoproducts known as pyrimidine adducts are generated to a lesser extent, however, than are cyclobutane-type dimers. Moreover, pyrimidine adducts are generally not very stable. Probably both facts have caused the delay of their detection.

Pearson *et al.* (1965) first described a thymidyl–thymine dimer that was not identical with a cyclobutane dimer. Since 1967, a number of publications have appeared dealing with the formation of thymine photoadducts on UV irradiation of thymine in frozen aqueous solution and DNA in fluid aqueous solution (Varghese and Wang, 1967*b*, 1968*a,b*). On the basis of data from mass spectra, nuclear magnetic resonance, infrared, and fluorescence spectra, and crystallographic analysis, the structure given in Fig. 3 was proposed (Fenselau and Wang, 1969; Karle *et al.*, 1969; Hauswirth and Wang, 1973). Cohn *et al.* (1974) recommended abbreviations for pyrimidine adducts.

UV irradiation of uracil in frozen aqueous solution yielded a uracil–uracil adduct (Khattak and Wang, 1969). A cytosine–cytosine adduct apparently arises in a similar manner; hence, the cytosine adduct has been detected on UV irradiation of frozen aqueous solutions of cytidine, deoxycytidine, poly(rC), and poly(dG · dC) (Rhoades and Wang, 1971*a,b*). The chemical stability of homophotoadducts decreases in the order thymine > uracil > cytosine. Evidence for the formation of an adduct from two nonidentical bases was obtained by the identification of photoproducts from UV-irradiated DNA and from acid hydrolysates of equimolar mixtures of cytosine and thymine nucleosides or nucleotides irradiated in frozen aqueous solution. A uracil–thymine adduct was identified from the frozen solution irradiation of a mixture of uracil and thymine. Thymidine also

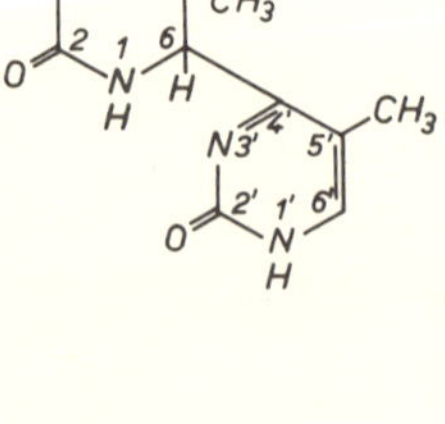

Fig. 3. Formula of the thymine–thymine adduct. According to Varghese and Wang (1968*a*).

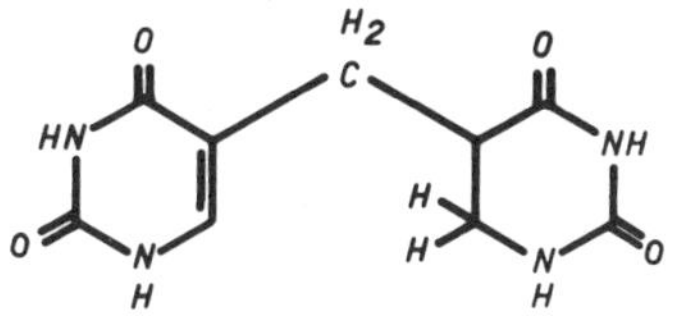

Fig. 4. Formula of the "spore" photoproduct.

forms adducts with uridine and cytidine (Wang and Varghese, 1967; Rhoades and Wang, 1970; Varghese and Day, 1970). In addition to the cyclobutane-type dimer and dimer adduct, frozen solution irradiation of thymine forms a trimer containing a cyclobutane and an oxetane ring system (Smith, 1963; Flippen *et al.*, 1971; Wang, 1971; Wang and Rhoades, 1971). By means of kinetic measurements, it was demonstrated that thymine dimer adducts and higher aggregates both possess the same photochemical precursor, i.e., an oxetane derivative (Rahn and Hosszu, 1969*b*). The adduct is not formed when thymine is irradiated in solution in the presence of triplet-state sensitizers (Lamola, 1968). This may be seen as an indication that thymine undergoes photoadduct formation by pathways that involve the excited singlet state.

UV irradiation of thin, solid films of DNA at $-78°C$ yields another photoproduct that is not identical to any of those mentioned above (Varghese, 1970*b*, 1971*a*). This product is detectable only at low temperature and under complete exclusion of water. EPR investigations indicate the participation of radical intermediates (Varghese, 1970*c*). This photoproduct has been identified as the "spore photoproduct" (Fig. 4) (Varghese, 1970*c*, 1971*a*; Jagger *et al.*, 1970).

Photoproducts of the noncyclobutane type have also been indicated in tRNA (Section 2.1.6). From a theoretical view, there is no reason to doubt that photoadducts may be formed between pyrimidines of DNA and mRNA during the transcription process; however, until now, only DNA–protein cross-linking has been detected.

2.1.5. Photoalkylation

Evidence derived from studies of the UV irradiation of purines in the presence of various proton-donating compounds, i.e., alcohols, amines, or amino acids, suggests that photoalkylation occurs in dilute aqueous solution (Conolly and Linschitz, 1968; Linschitz and Conolly, 1968; Elad and Rosenthal, 1969; Evans and Wolfenden, 1970; Elad and Salomon, 1971; Stankunas *et al.*, 1971; Salomon and Elad, 1974). The purine bases are much more sensitive to attack by photoalkylating agents than are pyrimidine bases. Steinmaus *et al.* (1969, 1971) have presented unambiguous chemical proof for the structures of the major photoproducts of adenine, adenosine, guanine, and hypoxanthine using isopropanol as the photoalkylating agent (Fig. 5). According to their results, position C-8 or C-6 of purines, if not substituted prior to irradiation, is attacked preferentially. The reaction pathway proceeds via radical intermediates (Nicolau, 1972; Taylor *et al.*, 1969; Stermitz *et al.*, 1970). Photoalkylation is more efficient in denatured than in native DNA; i.e., the photoaddition of alkyl radicals to purines is enhanced when the bases are more accessible to the alkylating agent, as in the denatured form (Ben-Ishai *et al.*, 1973). These results suggest that alkylation of the purines by alkyl radicals is sterically hindered in native DNA. Secondary structural changes in DNA, such as those produced by denaturation, single-strand breaks, and dimerization of adjacent pyrimidines, may expose purine residues in DNA to attack by alkyl radicals. On the other hand, the chemical dark alkylation process is known to involve the N-7 position of guanine. This result supports the idea that dark alkylation proceeds by another pathway, e.g., via polar intermediates (Lawley, 1966). A study of the stability to UV light of uracil and some of its derivatives in the presence of amines and amino acids shows that uracil components also become photoalkylated during irradiation (Smith, 1969; Gorelic *et al.*, 1972; Leonov *et al.*, 1973). Various sensitizers, e.g., acetone, acetylated polystyrene, and di-*t*-butylperoxide, are able to generate photoalkylated products by a sensitized photoreaction (Leonov *et al.*, 1973). Salomon and Elad (1974) reported on a sensitized photoalkylation of uracil without any production of photodimers. The photoreaction of bases with amines and amino acids can be considered as a simple model for the photochemical production of nucleic acid–protein cross-links (Yang *et al.*, 1971).

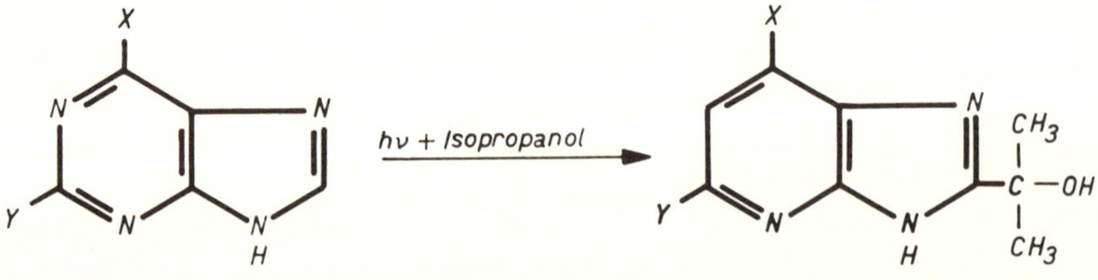

Fig. 5. Photoreaction of purine derivatives with isopropanol, adenine (X = NH₂, Y = H), guanine (X = NH₂, Y = OH), hypoxanthine (X = OH, Y = H). According to Steinmaus *et al.* (1969).

2.1.6. Minor Bases

The biochemical and genetic functions of minor bases (rare bases) to be found in various DNAs and RNAs are still largely unknown. Photochemical investigations could contribute to our understanding of the biological importance of minor bases if two prerequisites could be met: (1) an improvement in the techniques for their qualitative and quantitative analyses and (2) a way to selectively damage the minor bases in nucleic acids. Both attempts have been reported in the literature.

First of all, efforts need to be directed toward an unambigous detection of minor bases or photoproducts of them among a large amount of usual pyrimidines and purines. One possible assay might be based on the fact that the minor bases 5-methylcytosine and 5-methyldeoxycytidine are fluorescent in aqueous solution at room temperature and at neutral pH (Gill, 1970, 1971; Gill *et al.,* 1974). 5-Methyldeoxycytidine, 10^{-4} M, pH 7.5, 20°C, has a fluorescence quantum yield of 5×10^{-4}, which was shown to be greater than those of the major bases (Gill, 1970). Moreover, 5-methyldeoxycytidine fluorescence can be observed when the nucleoside is incorporated into synthetic DNAs. Fluorescence, however, is sensitive to the DNA structure and composition.

Tomasz and Chambers (1966) published the first comprehensive paper that dealt with the photochemical properties of pseudouridine (ψ). UV irradiation at 254 nm of the oligonucleotide Tp ψ pCpGp, which contains ψ as a constituent, produces a chain break by the splitting of a sugar–phosphate bond (TpψpCpGp → TpY + 5-formyluracil + pCpGp; Y is unknown). Since most of the possible photochemical reactions of nucleic acid bases can be produced at that wavelength, the probability for a selective damage of ψ appears to be rather low. Another naturally occurring fluorescent minor base, 4-thiouracil, has been indicated in tRNA [Val] (Secrist *et al.,* 1971). Its fluorescence can be used as a probe for secondary and tertiary structural changes taking place in tRNA [Val] (Pochon *et al.,* 1971). The long wavelength absorption band of 4-thiouracil exhibits maximal absorbance at 335 nm. From irradiation at about 335 nm of a mixture of 4-thiouracil and cytosine and their nucleosides, a noncyclobutane dimer has been identified (Favre and Yaniv, 1971; Bergstrom and Leonard, 1972; Favre *et al.,* 1972; Petrissant and Favre, 1972). It is formed from the single bases or nucleosides (Favre *et al.,* 1969; Leonard *et al.,* 1971; Bergstrom and Leonard, 1972; Favre, 1974), the synthetic copolymers (Favre and Fourrey, 1974), and tRNA (Favre *et al.,* 1969; Yaniv *et al.,* 1971; Favre *et al.,* 1972) on UV irradiation in aqueous solution. Under certain conditions of irradiation, no other constituents of tRNA, except 4-thiouracil, will be excited. UV irradiation of several *Escherichia coli* tRNAs at 335 nm has been shown to bring about a specific modification of 4-thiouridine. The photoreaction involves the sulfur base residue located in position 8 and the cytidine residue located in position 13 of *E. coli* tRNA^{Val}, tRNA^{Met}, and tRNA^{Phe} (Favre *et al.*, 1969). The formation of a photoadduct whose components are nonadjacent in the primary structure of tRNA has been described (Nino *et al.*, 1969; Yaniv *et al.*, 1971). Moreover, the formation of the photoadduct depends on the integrity of the tRNA structure. No product was obtained in an RNase digest of tRNA^{Val}, or when tRNA^{Val} was irradiated either in distilled water or at pH values higher than 10 or lower than 3.5 (Favre *et al.*, 1971). Photochemically modified tRNAs show no change in the rate and extent of repairing capacity of the pCpCpA 3′ terminus of tRNA by purified *E. coli* tRNA nucleotidyl transferase (Carré *et al.*, 1974). Furthermore, no difference was observed between normal and irradiated tRNA for ribosome binding in the presence of poly(U·G) (Yaniv *et al.*, 1971).

Sawada (1974) has shown a covalent binding between 4-thiouridine and RNase caused by UV irradiation at 335 nm. The cystinyl residues in the protein participate in the disulfide bonds formed between the cysteinyl residues and 4-thiouridylic acid or its photoproducts (Sawada and Kanbayashi, 1973). This reaction causes a variation in conformation of the RNase–4-thionucleotide complex (Samejima *et al.*, 1969). The selective excitation of 4-thiouracil in tRNA is an exception up to now. For the biochemical study of the minor bases in DNA, it would be advantageous if one could modify them selectively. In attempting this, photochemical methods should not be excluded.

2.2. Nucleic Acid Base Analogues

There are numerous pyrimidine and purine base analogues that do not occur in nucleic acids in their natural state. Nevertheless, their incorporation into DNA and RNA of mammalian cells, bacteria, and viruses has been deserved. Biochemical effects resulting from the incorporation of a base analogue instead of a normal one have become the objects of widespread investigations. We would like to restrict ourselves to photochemical and photobiological aspects arising from the fact that, as a rule, the photochemical properties of a base analogue deviate from those of its parent compound.

2.2.1. Halogeno Derivatives of Uracil

The incorporation of 5-bromouracil into DNA of cells (Köhnlein and Hutchinson, 1969; Mönkehaus, 1974) causes an increased sensitivity to the lethal effect of UV irradiation in the case of bacteria (Dellweg *et al.,* 1964; Smith, 1964*c*; Leutzen and Walker, 1970; Hutchison, 1973), phages (Yan, 1969; Stephan *et al.,* 1970; Mönkehaus and Köhnlein, 1972), and mammalian cells (Scaife, 1970; Regan *et al.,* 1971). The 254-nm irradiation of 5-bromouracil in aqueous solution yielded various photoproducts, e.g., 5,5′-diuracil, uracil, glyoxaldiurene, barbituric acid, oxalic acid, iso-orotic acid, parabanic acid, urea, ammonia, and glyoxal, which were all quantitatively isolated and identified (Ishihara and Wang, 1966*a,b*). 5-Bromouracil forms at least two further types of photoproducts in DNA, i.e., a uracil dimer and a uracil–thymine heterodimer (Dellweg *et al.,* 1964; Dellweg and Wacker, 1964; Smith, 1964*c*). Haug (1964) proposed the formation of a 5-bromouracil–thymine dimer; however, its existence could not be confirmed in later work (Peter and Drewer, 1970, 1971). The process of UV-induced bromine dissociation proceeds more easily in native DNA, while the reaction rate is considerably lower when thermally denatured DNA or apurinic acid is irradiated (Dellweg and Wacker, 1964).

Conformational aspects have been discussed by Lion and Köhnlein (1974). 5-Iodouracil behaves quite similarly to 5-bromouracil. It has been observed that aqueous solutions of 5-iodouracil are very sensitive to UV radiation under conditions where 5-bromouracil is unaltered. Alloxanic acid, isodialuric acid, and formaldehyde have been identified as photoproducts (Rupp and Prusoff, 1964, 1965*a,b*; Gilbert *et al.,* 1971). Probably the photochemically reactive precursor for 5-bromouracil and 5-iodouracil is the triplet state (Rothman and Kearns, 1967; Danziger *et al.,* 1968; Langmuir and Hayon, 1969). The primary reaction is shown to be a homolytic carbon-halogeno bond cleavage (Fig. 6), a process leading to the formation of uracil and halogeno radicals (Rupp and Prusoff, 1965*b*; Ishihara and Wang, 1966*b*; Gilbert *et al.,* 1971). This reaction is followed by hydrogen abstraction from appropriate hydrogen donors to yield uracil and hydrobromic acid (Gilbert *et al.,* 1971; Campbell *et al.,* 1974). A comparison has been made between 5-bromouracil photolysis in organic, hydrogen-donating solvents and 5-bromouracil within the DNA of bacteria or phage. It has been concluded that the much higher quantum yields observed for chain breaks in the photolysis of DNA containing 5-bromouracil compared to photolysis of 5-bromouracil in aqueous solution

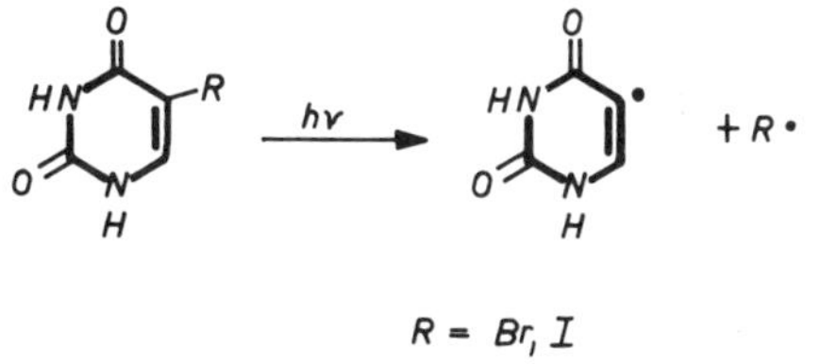

Fig. 6. Initial step of photochemically induced splitting of the halogeno-carbon bond of 5-halogen derivatives of uracil. According to Gilbert *et al.* (1971).

are due to the high local concentration of hydrogen donors within the DNA molecule, i.e., sugar residues (Gilbert and Christallini, 1973; Campbell *et al.,* 1974). This reaction may produce single-strand breaks and even double-strand breaks in UV-irradiated DNA (Stephan *et al.,* 1970; Mönkehaus and Köhnlein, 1973).

The incorporation of the pyrimidine analogue, 5-fluorouracil, into RNA of tobacco mosaic virus sensitizes the intact virus (Becarevic *et al.,* 1963; Lozeron and Gordon, 1964) and the infectious RNA (Lozeron and Gordon, 1964) to UV light. The principal photoproduct formed when 5-fluorouracil is irradiated with UV mainly at 254 nm has been identified as 5-fluoro-6-hydroxyhydrouracil (Lozeron *et al.,* 1964; Fikus *et al.,* 1965). The photoproduct is stable at acid or neutral pH, but decomposes in alkali. At present, no conclusive data on the photolysis of 5-chlorouracil are available.

2.2.2. Alkyl Derivatives of Uracil

It has been shown that 5-ethyluracil can substitute for thymine in bacterial DNA (Pichowska and Shugar, 1965) and that 5-ethyldeoxyuridine, a nonmutagenic thymidine base analogue with antiviral activity, is readily incorporated into phage DNA (Pietrzykowska and Shugar; 1966). A study of the photochemistry of phage T3 in which 65% of the thymine residues were substituted by 5-ethyluracil revealed that the sensitivity to radiation at 254 nm, as well as subsequent photoreactivation and dark reactivation, was identical with that of the normal phage. Irradiation at 254 nm leads to a cleavage of the 5-ethyl group with the formation of deoxyuridine, which subsequently becomes photohydrated. A photointermediate was isolated by thin-layer chromatography that, on further irradiation at 254 nm, was quantitatively converted to uridine, with the simultaneous elimination of ethylene. The photohydration process occurring on irradiation at 254 nm fulfills criteria typical for a singlet reaction. On UV irradiation at 254 nm, photohydration also occurs for two higher 5-alkyl analogues, 5-propyluridine, and 5-isopropyluridine. Irradiation of 5-ethyluridine at 265 nm yields a cyclobutane-type photodimer, similar to that formed by thymine or thymidine. No photodimer formation was observed for higher alkyl derivatives, possibly because of steric hindrance by the larger 5-alkyl substituents (Pietrzykowska and Shugar, 1968, 1970; Gauri *et al.,* 1971; Krajewska and Shugar, 1971). If the ring nitrogen atoms for 5-alkyluracil become alkylated, the yield of photodimer formation is enhanced (Gauri *et al.,* 1969).

The photodimerization reaction of 5-ethyluracil becomes quenched if molecular oxygen or paramagnetic metal ions are present in solution. The foregoing observations are consistent with the idea that photodimerization involves the triplet state of 5-alkyluracil compounds.

2.2.3. Azapyrimidines and Azapurines

The aza derivatives of pyrimidines and purines, when incorporated into DNA or RNA of bacterial cells or plant cells, have been reported to induce various biochemical effects (see Section 2.2.4) (Table 4). On the other hand, remarkable deviations have been indicated in the photostability of aza bases

TABLE 4. Biochemistry and Genetics of Aza Analogues of Nucleic Acid Bases[a]

Aza analogue[b]	Incorporation into nucleic acids	Biochemical and genetic effects
5-Azacytosine	DNA, RNA (Doskočil and Šorm, 1970, 1971)	Protein synthesis inhibited (Doskočil and Šorm, 1970) Cancerostatic (Kalousek *et al.*, 1966) Antimetabolic (Šorm *et al.*, 1966)
6-Azacytosine	DNA, poly(C) (Škoda, 1969)	Protein synthesis inhibited (Grünberger and Šorm, 1963) Pyrimidine synthesis inhibited (Škoda, 1969) Cancerostatic (Šorm and Škoda, 1964)
8-Aza-adenine	Probably (Jacob and Kittler, 1970)	
8-Azaguanine	RNA (Matthews and Smith, 1956)	Protein synthesis inhibited (Grünberger and Šorm, 1963)
6-Azauracil	DNA, small or not (Horvath *et al.* 1969; Škoda, 1963) RNA of leaves (Matolcsy *et al.*, 1969)	Pyrimidine synthesis inhibited (Škoda, 1963) Cancerostatic (Šorm and Škoda, 1964) Virostatic (Rada and Zavada, 1962)
6-Azathymine	DNA (Günther and Prusoff, 1962) RNA of bean leaves (Matolcsy *et al.*, 1969)	Virostatic (Sekely and Prusoff, 1966)

[a] From Kittler and Löber (1973).
[b] The nucleosides are partially used.

in comparison to their natural analogues when both are exposed to UV radiation at 254 nm (Günther and Prusoff, 1962; Wacker *et al.*, 1964*b*; Kittler and Berg, 1967; Löber and Kittler, 1973).

It has been shown that 6-azauracil is less sensitive to UV irradiation than uracil, the primary photoproduct being a photohydrate, 6-hydro-5-hydroxy-6-azauracil (Kittler, 1968; Kittler and Löber, 1969). The properties of heat and acid reversal were used to confirm the formation of a 6-azauracil hydrate which, however, deviates in its structure from the structure of uracil hydrate by exchanging the hydroxy group from the 6- to the 5-position of the azapyrimidine ring (Fig. 7). This result is in good agreement with data obtained from quantum chemical calculations where the calculated electron density distribution in the excited state coincides with a nucleophilic addition reaction at the 5- and 6-positions (Jakubetz *et al.*, 1973). It even supports the higher reactivity of position 5 to attach hydroxy groups (Kittler and Löber, 1971). In fluid aqueous solution, the photohydrate of 6-azauracil arises by way of the triplet state. This is confirmed by triplet quenching and triplet sensitization experiments (Kittler and Löber, 1968). UV-induced decomposition of 5-halogeno derivatives of 6-azauracil, i.e., 5-bromo-, 5-iodo-, or 5-chloro-6-azauracil, has been described by Kittler (1969). It was observed that a splitting of the carbon-halogeno bond is the primary reaction that is followed by the photohydration of the 6-azauracil intermediate. The photokinetics of both processes have been measured polarographically (Kittler and Berg, 1968).

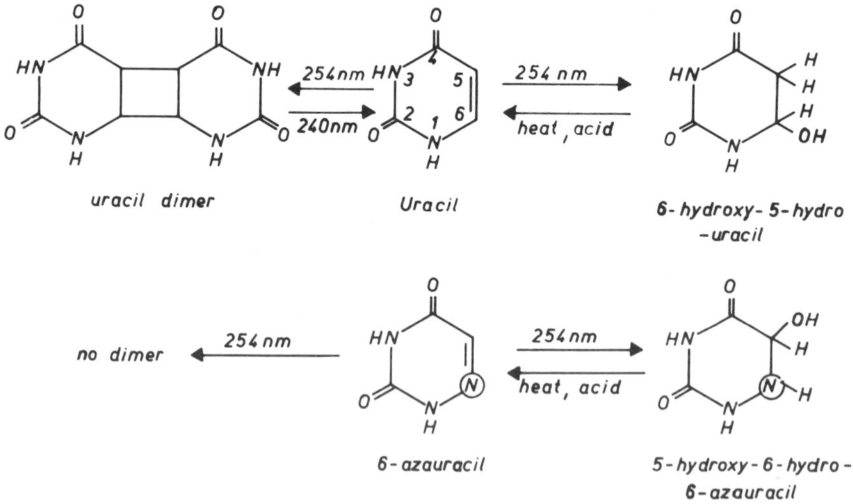

Fig. 7. Photoreactions of uracil and 6-azauracil in aqueous solution. According to Kittler and Löber (1973).

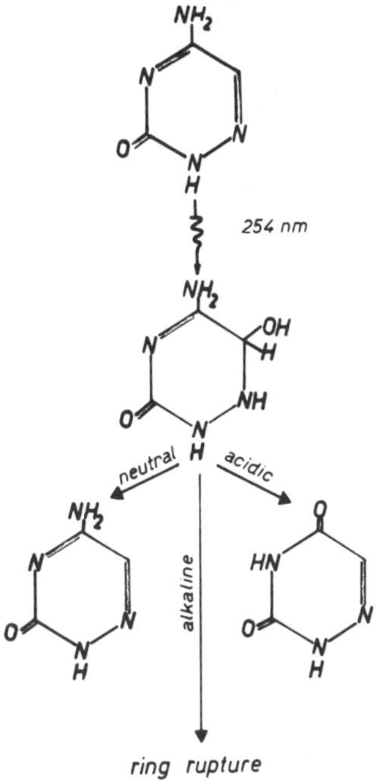

Fig. 8. Photohydration of 6-azacytosine in aqueous solution and subsequent dark reactions, i.e., deamination reaction yielding 6-azauracil (right side, bottom), ring rupture processes (middle), and dark reversal yielding the starting compound (left side, bottom). According to Kittler and Löber (1973).

It has been shown that 6-azacytosine is more sensitive to UV irradiation than cytosine (Kittler, 1970, 1972). There is evidence, such as reversible absorbance changes, to indicate that 6-azacytosine forms a hydrate on irradiation in aqueous solution. From UV-irradiated aqueous solutions of 6-azacytosine, Kittler (1970, 1972) isolated a photoproduct that has been identified as 6-hydro-5-hydroxy-6-azacytosine from infrared spectra and quantitative chemical analysis. Under comparable irradiation conditions, 6-azacytidine forms a similar photohydrate. Polarographic and chromatographic measurements do not favor a photochemical cleavage of the N-glycosidic bond. The photoreaction of 6-azacytosine cannot be sensitized with triplet sensitizers, i.e., acetone, suggesting that 6-azacytosine has a high-energy triplet state or that the photoreaction does not involve the triplet state at all. The observation that paramagnetic ions, i.e., Mn^{2+}, only partially quench the photohydration reaction at room temperature suggests that both triplet and singlet forms are the photoreactive precursors. The properties of alkali and acid reversal are generally used to follow photoproduct formation. 6-Hydro-5-hydroxy-6-azacytosine is rather unstable. The kinetics of the reversal have been studied (Kittler, 1972). The main conclusions derived from these investigations are as follows: (1) the parent compound is reformed on heating at neutral pH; (2) reversal occurs at acidic pH, but 6-azacytosine becomes deaminated and 6-azauracil appears as final product; and (3) a ring opening takes place under alkaline conditions yielding, *inter alia,* a urea derivative (Fig. 8). Although theoretical considerations based on quantum chemical data do not exclude dimer formation for 6-azauracil and 6-azathymine (Pullman, 1968), neither has been identified as yet. This may be due to the fact that four-membered rings containing nitrogen atoms are generally not favored for energetic reasons. There is evidence that 8-azapurines are more sensitive to UV irradiation at 254 nm than other natural bases, while natural purine bases are about 10 times less sensitive than pyrimidine bases (Kittler and Berg, 1967; Kittler, 1968). The proposed photodecomposition of 5-azacytosine and 8-azapurine involves a ring-opening process.

2.2.4. Molecular Biological Effects of Photoproducts of Base Analogues

5-Bromouracil can replace thymine in the DNA of phage, bacteria, and other cells without substantially altering their biological activity. Thus it has been found that 5-bromouracil is as effective as thymine in protecting *E. coli* 15 TAU against thymineless death (B. J. Smith, 1966). 5-Bromouracil was incorporated into the DNA of the *Serratia* phage K and replaced about 28% of the thymine. The frequency of two types of plaque mutants in the 5-bromouracil-substituted phage K population was increased significantly above the frequencies in 5-bromouracil-free phage. The UV inactivation rate and the frequency of UV-induced mutations were increased if 5-bromouracil was incorporated prior to irradiation (Pohl and Kaplan, 1968). 5-Bromouracil-induced dark mutagenesis of phage C17 *am*08, in relation to the recombination systems of phage or bacteria, was studied. The mutations investigated were *am* → *am*+ (Pietrzykowska, 1973). A hypothesis on a common mechanism of 5-bromouracil-induced and UV-induced mutagenesis was proposed, which involves a recombination repair process.

An inhibitory effect of 5-bromouracil on DNA synthesis was observed to a variable extent in various strains of *E. coli.* Postirradiation inhibition of the rate and extent of total DNA replication by 5-bromouracil also varied according to strain (Hewitt *et al.,* 1967). The results of Jones and Dove (1972) indicated (1) that 5-bromodeoxyuridine substitution in DNA reduces its efficiency as a template for RNA synthesis *in vivo* and (2) that 5-bromodeoxyuridine substitution in DNA sensitizes RNA synthesis *in vivo* to UV light. Miscoding caused by 5-fluorouracil has been observed for an amber alkaline phosphatase mutant in the dark (Rosen *et al.,* 1969). Guschelbauer *et al.* (1965) have shown that the increase in sensitivity of biological systems where 5-halogenouracils have been incorporated before irradiation is not simply correlated to an increase of the photochemical quantum yields. According to Dellweg *et al.* (1964), DNA damage resulting from the UV sensitization of incorporated 5-halogenouracil derivatives is not repaired by photoreactivation. The idea of irreparable lesions produced by the sensitizing effect has also been offered in other papers (Howard-Flanders and Boyce, 1966; Dellweg and Wacker, 1964; Szybalski, 1967). Other writers have reached conclusions that are contradictory to this picture. Photochemical damage produced by UV irradiation was studied in normal and 5-bromouracil-substituted DNA of mammalian cells in relation to repair processes known to occur in these cells. The main photoproduct was uracil, accompanied by a single-strand break in the DNA chain. Uracil was excised from the DNA and the break was rejoined at a rate much faster than the excision of pyrimidine dimers from unsubstituted DNA (Smets and Cornelis, 1971). At the same time, irreparable damage was induced in 5-bromouracil DNA as double-strand breaks and DNA–protein cross-links. These types of damage were not observed in normal cells at doses of biological interest and may therefore account for the sensitizing effect of the halogenated analogue. The occurrence of double-strand breaks has been observed by DNA analysis on sucrose density gradients (Stephan *et al.*, 1970; Regan *et al.*, 1971; Mönkehaus and Köhnlein, 1974). UV irradiation of phage T1 containing 5-bromouracil yielded double- and single-strand breaks at an increased rate. UV irradiation in the presence of a radical scavenger (cysteamine) prevented the occurrence of double- but not of single-strand breaks (Stephan *et al.*, 1970). Cysteamine, on the other hand, was able to prevent the production of single-strand breaks in 5-bromouracil DNA of phage λ (Mönkehaus, 1974). 5-Bromouracil-containing human kidney T cells showed more DNA strand breaks and the DNA was subsequently degraded to a greater extent. The incorporation of new bases was consequently higher in such degraded strands (Scaife, 1970). On the other hand, it was found that the presence of UV-irradiated 5-bromouracil-DNA reduces the host cell reactivation of UV-induced damage in phage λ (Yan, 1969).

The alkyl derivatives of uracil are known to exhibit antiviral (Swierkowski and Shugar, 1969) and immunosuppressive (Gauri *et al.*, 1969) activity. Their incorporation into DNA of phage, bacteria, and mammalian cells caused no mutagenic effects (Pietrzykowska and Shugar, 1968; Gauri *et al.*, 1969). No specific photobiological effects due to the presence of the alkyl derivatives of uracil in DNA have yet been observed.

Azapyrimidines and azapurines act as inhibitors of different biochemical processes mainly by inhibiting pyrimidine and protein biosyntheses. Furthermore, they show virostatic, cancerostatic, and antimetabolic effects (Table 5). The most important characteristic of aza analogues,

TABLE 5. Relationship between Photochemical and Photobiological Behavior of Aza Derivatives and Their Biochemical Incorporation into Nucleic Acids *in Vivo*[a]

Aza derivative	Incorporation	Photobiological UV sensitivity compared with untreated control	Photochemical UV sensitivity compared with normal constituents
5-Azacytidine	30% into DNA and RNA of *Escherichia coli* K12 C600 and *Escherichia coli* B (Doskočil, 1968; Pačes *et al.* 1968)	*Escherichia coli* K12 C600, increased	Increased (Kittler, unpublished)
	2–3% into DNA of phage λcb$_2$ (Doskočil, 1968)	Phage λcb$_2$, unaffected	
6-Azacytosine	Not described	*Escherichia coli* K12 C600, slightly increased *Escherichia coli* B, increased	Increased (Kittler, 1972)
6-Azacytidine	Into synthetic polyribonucleotides (Škoda and Šorm, 1964) Into code triplets of tRNA (Škoda, 1968)	*Escherichia coli* K12 C600, increased	Increased (Kittler, 1972)
6-Azathymine	15% into DNA of *Streptococcus faecalis* (Günther and Prusoff, 1962) 7% into DNA of *Enterococcus stei* (Wacker and Jacherts, 1962)	*Escherichia coli* K12 C600, not affected *Streptococcus faecalis,* decreased *Enterococcus stei,* decreased	Decreased (Günther and Prusoff, 1962)
6-Azauracil	No incorporation into bacteria (Škoda, 1963) In small amounts into DNA of bean leaf tissue (Horvath *et al.,* 1969)	*Escherichia coli* K12 C600, not affected *Streptococcus faecalis* (Günther and Prusoff, 1967), decreased	Decreased (Kittler and Löber, 1969)
8-Aza-adenine	Probable (Jacob and Kittler, 1970)	*Bacillus cereus,* increased *Escherichia coli* B, increased	Increased (Kittler and Berg, 1967)

[a] From Kittler *et al.* (1975).

however, is their ability to substitute for normal nucleic acid bases in cellular DNAs and RNAs. The question has been discussed of how differences in the photostability between natural nucleic acid bases and aza bases are reflected under *in vivo* conditions; e.g., aza analogues are incorporated into bacterial nucleic acids prior to UV irradiation (Jacob and Kittler, 1970; Kittler and Löber, 1973; Kittler *et al.*, 1975). As reported in Section 2.2.3, photochemical experiments showed that 5-azacytosine (I), 5-azacytidine (II), 6-azacytosine (III), 6-azacytidine (IV), and 8-aza-adenine

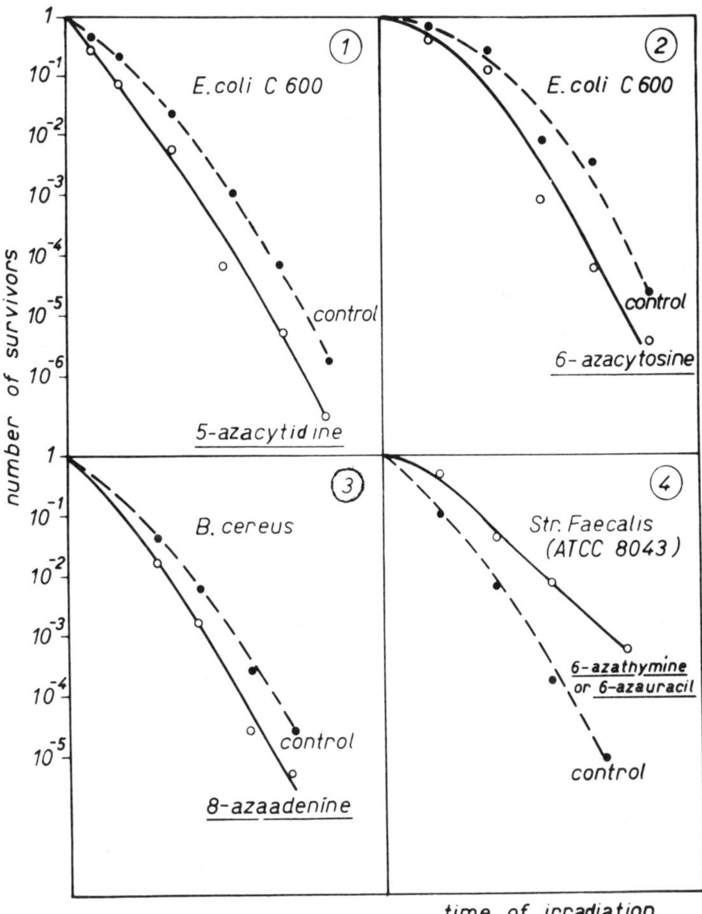

Fig. 9. Survival curves of UV-irradiated bacteria grown in media supplemented with aza analogues prior irradiation. Data for (1) and (2) according to Kittler *et al.* (1975); data for (3) and (4) according to Jacob and Kittler (1970) and Günther and Prusoff (1967), respectively.

(V) are more sensitive and 6-azathymine (VI) and 6-azauracil (VII) less sensitive to UV compared to corresponding natural nucleic acid constituents. The changed UV sensitivities are reflected in the survival curves after UV irradiation at 254 nm inasmuch as aza compounds become incorporated into the nucleic acids *in vivo* (Table 5). This explains the increase of UV sensitivity of *E. coli* K12 C600, *E. coli* B, and *Bacillus cereus* supplemented with I, II, III, IV, and V (Kittler and Löber, 1973; Kittler *et al.*, 1975) and the decrease of UV sensitivity of *Streptococcus faecalis* supplemented with VI and VII (Günther and Prusoff, 1967) (Fig. 9). Azapyrimidines that decrease UV sensitivity may do so by reducing the ability to form dimers in irradiated DNA. Since corresponding azacytosines and azacytidines show no differences in their photochemical stabilities, the higher UV sensitivity found in the survival rates of azacytidine-treated bacteria probably has its origin in a facilitated incorporation of the ribosylated compounds.

3. PHOTOADDITION REACTIONS OF NUCLEIC ACIDS AND THEIR MOLECULAR BIOLOGICAL EFFECTS

3.1. Nucleic Acid–Protein Cross-Links

Ultraviolet light produces lethal and mutagenic effects on living systems and, without doubt, the primary target for damage is DNA. In the past two decades, research in this field has led to a large amount of literature, and the general feeling today is that many types of photoproducts are involved, with their relative importance depending on many cellular factors. There is, however, another class of photoproducts not discussed in previous sections that produce the covalent binding of proteins to DNA (Alexander and Moroson, 1962; Smith, 1962, 1964a). Since nucleic acids and proteins are bound to each other *in vivo,* photochemical reactions involving both components can be expected. Of all the amino acids, cysteine reacts most strongly with uracil, to form 5-*S*-cysteinyl-6-hydrouracil and 5,6-dihydrouracil as major photoproducts (Smith and Aplin, 1966; Smith, 1969) (Fig. 10). Presumably the cysteine competes with other amino acid residues in the protein for excited heterocyclic bases in DNA. Thus Smith (1969) found that gelatin, which contains no SH groups, cross-links to some extent to DNA. Ten of 22 sulfur-containing and aromatic amino acids have been checked and found to cross-link preferentially to uracil and thymine (Smith, 1969, 1970; Varghese, 1973, 1974a). Sulfhydryl compounds such as cysteine and glutathione enhance the photochemical reactivity of 5-bromouracil. Photochemical addition of the sulfhydryl compound to the 5-bromouracil

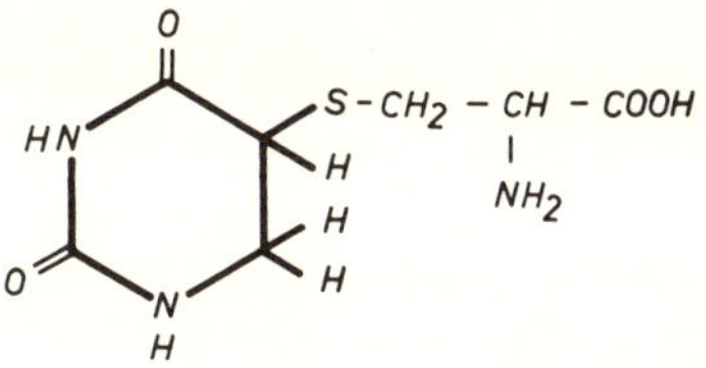

Fig. 10. Formula of the uracil–cysteine photoproduct. According to Smith and Aplin (1966).

residue takes place (Varghese, 1974*b,c*). Lysine, arginine, cysteine, and cystine have been found to be reactive in forming thymine–amino acid heteroadducts (Schott and Shetlar, 1974). The results suggest that lysine and arginine may be determinants in the UV-induced cross-linking of histones to DNA in DNA–histone complexes. Cysteine adds to single-stranded and double-stranded polynucleotides (Smith and Meun, 1968). Dihydrothymine and 5-*S*-cysteinyl-6-hydrothymine have been reported after the photolysis of a solution of thymine and cysteine (Smith, 1970). The addition reactions arise from the triplet state of the pyrimidine, i.e., thymine and uracil. Evidence for the triplet reaction is based on the observed quenching of the triplet state by cysteine (Fisher *et al.*, 1974). Jellinek and Johns (1970) concluded that the initial step, after absorption of a photon by the pyrimidine, was intersystem crossing to the triplet state. Dimers are formed after collision of a triplet molecule with a ground-state base, but cysteine can compete for triplets. Probably a hydropyrimidine radical intermediate is formed, which then reacts with cysteine to give the addition product (Fisher *et al.*, 1974). The photosensitizers acetone and acetophenone, when singlet excited with light at 313 nm, efficiently undergo intersystem crossing to their triplet states and are able to transfer their triplet energy to thymine to form thymine triplets, which then form the addition product. Riboflavin, with a lower triplet energy than thymine, is a poor sensitizer. These results implicate the pyrimidine triplet as a precursor of the products with cysteine.

The biological importance of nucleic acid–protein cross-linking has been well documented. Thus it was observed that with increasing doses of UV irradiation, decreasing amounts of DNA free of protein could be extracted from irradiated cells (Smith, 1962). This phenomenon of decreased extractability of DNA has been observed after the UV irradiation of bacteria, mammalian cells, salmon sperm heads, and mixtures of protein and DNA (Smith, 1964*a,b*, 1968, 1974, 1976; Smith *et al.*, 1966). One way to demonstrate the biological role of a photoproduct in DNA is to determine if its production varies with radiation exposure in the same manner as the efficiency for killing, when cells are pretreated under various conditions which influence their sensitivity to UV irradiation. That the photochemical formation of DNA–protein adducts plays a significant role in the killing of UV-irradiated cells has been shown for several experimental conditions: (1) the amount of DNA cross-linked to protein in *E. coli* 15 TAU cells changed with the time of thymine starvation in the same manner as UV-induced cell killing; (2) the amount of DNA cross-linked to protein in *E. coli* B/r cells irradiated in frozen cell suspensions varied in yield as a function of temperature in the same manner as did UV-induced cell killing, and, in particular, thymine dimers did not satisfactorily explain those findings; (3) a decrease in the yield of thymine dimers, but an increase in sensitivity to killing by UV irradiation, and an increased yield of DNA–protein cross-links were observed when cells of *E. coli* B/r were UV irradiated in the dry state as compared to irradiation in solution; (4) when DNA thymine was replaced by 5-bromouracil, cells of *E. coli* B/r became markedly sensitive to the formation of UV-induced DNA-protein cross-links.

3.2. Furocoumarins

Furocoumarins (Fig. 11) as well as coumarins can produce various biological effects (Table 6). But furocoumarins are especially known for the photobiological effects that they exert on irradiation with long-wavelength UV radiation at 365 nm, where, as a rule, other cell constituents do not strongly absorb. To examine the possibility of photoreactions between furocoumarins and cell constituents, the photoreactions between furocoumarins and flavin mononucleotide were studied (see review of Musajo and Rodighiero, 1972). Some photoproducts were isolated that contained a furocoumarin and a flavin moiety. Such photoreactions between furocou-

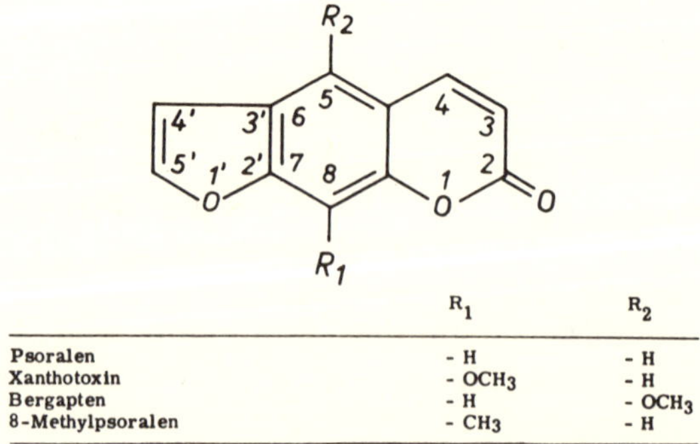

	R_1	R_2
Psoralen	- H	- H
Xanthotoxin	- OCH_3	- H
Bergapten	- H	- OCH_3
8-Methylpsoralen	- CH_3	- H

Fig. 11. Formulas of some photochemically active furocoumarins.

TABLE 6. Photobiological Effects of Furocoumarins

In vivo effects	References
Skin erythem	Pathak *et al.* (1960)
	Caporale *et al.* (1967)
Skin sensitization and photoadduct formation	Musajo *et al.* (1966)
	Pathak *et al.* (1960)
	Caporale *et al.* (1967)
	Pathak and Krämer (1969)
	Rodighiero *et al.* (1970*a,b*)
	Musajo and Rodighiero (1972)
Skin sensitization and cross-linking	Pathak *et al.*, (1967)
	Rodighiero *et al.* (1970*a*)
	Dall'Acqua *et al.* (1971, 1972, 1974*a*)
	Marciani *et al.* (1971)
	Trosko and Isoun (1971)
	Bordin *et al.* (1973)
Skin pigmentation	Sehgal (1971)
Inhibition of infective capacity of Ehrlich ascites tumor cells	Musajo *et al.* (1966, 1974)
Inhibition of protein synthesis in Ehrlich ascites tumor cells	Bordin *et al.* (1973)
Inhibition of DNA synthesis and regression of psoriasis lesions	Walter and Voorhees (1973)
	Parrish *et al.* (1974)
Inactivation of Chinese hamster cells and cross-linking	Ben-Hur and Elkind (1973)
Formation of giant cells by mammalian cells adapted to *in vitro* growth	Colombo *et al.* (1965)
Formation of polynuclear cells in sea urchin eggs fertilized with sperm irradiated in the presence of psoralen	Colombo (1967)
Effect on virus-producing Graffi leukemia cells	Bordin and Baccichetti (1974)
Inactivation of DNA viruses	Musajo *et al.* (1965)
Mutagenic action on *Sarcina lutea*	Mathews (1963)
Production of *r* mutants of phage T4	Drake and McGuire (1967)
Mutagenic action on *Drosophila melanogaster*	Nicoletti and Trippa (1967)
Suppressor mutations and reversions of *Escherichia coli*	Igali *et al.* (1970)
Inhibition of the template activity of DNA and RNA polymerases	Chandra and Wacker (1968)
	Cole (1970)
Inactivation of bacteria	Fowlks *et al.* (1958)
	Oginsky *et al.* (1959)
	Dall'Acqua *et al.* (1974*b*)
Lack of photoreactivation in bacteria	Chandra *et al.* (1971*a*, 1971*b*)
Indications for dark repair in:	
1. Guinea pig skin	Baden *et al.* (1972)
2. *Saccharomyces cerevisiae*	Chandra *et al.* (1974)
3. *Escherichia coli*	Chandra *et al.* (1973), Cole (1973)
Inhibition of pyrimidine dimerization	Bridges (1971)

marins and coenzymes, however, do not explain all biological effects induced by furocoumarins.

Perhaps an important discovery on the mode of action of furocoumarins was the finding that those substances photoreact with DNA upon UV irradiation at 365 nm. The photoreaction with DNA and RNA was occasionally described as a photodynamic process (Wacker *et al.*, 1964*b*; Chandra, 1972), and, moreover, Drake and McGuire (1967) have shown that, despite their considerably different molar efficiencies, the photoinduced mutagenicities in phage T4 in the presence of thionine, which may act as a photodynamic sensitizer, and psoralen are essentially identical. However, two facts concerning the action of furocoumarins are clearly opposed to criteria very specific for photodynamic action: (1) Furocoumarins do not act as photocatalysts like acridine orange or methylene blue. On irradiation at 365 nm, they become covalently linked to nucleic acids. (2) The photoaddition of furocoumarins to nucleic acids proceeds without consumption of molecular oxygen (Krauch *et al.*, 1967*a*). Furocoumarins photoreact with nucleic acids under irradiation with long wavelength UV radiation, forming cyclobutane-type adducts with the pyrimidine bases. The reactive site of thymine or uracil is the 5,6 double bond, while furocoumarins can react with either their 3,4 and/or their 4′,5′ double bond (Dall'Acqua *et al.*, 1970; Krämer and Pathak, 1970; Musajo and Rodighiero, 1970, 1972). Two types of monofunctional photoaddition reactions are considered. The major types of photochemical changes furocoumarins undergo are (1) cycloaddition of the furane ring moiety to pyrimidines (photoproduct A, fluorescent) and (2) cycloaddition of the pyrone ring moiety to pyrimidines (photoproduct B, nonfluorescent) (Fig. 12). Experimental findings indicated that the formation of photoproduct A prevails over the formation of photoproduct B. This fact is supported by data on bond orders and π-electron density distributions in psoralen as obtained from quantum chemical calculations, which indicate that the 3,4 position of the pyrone ring is favored to undergo photocycloaddition more than the 4′,5′ position of the furane ring (Song and Gordon, 1970; Song *et al.*, 1971; Henry and Hunt, 1971; Moore *et al.*, 1973). The photocycloaddition of coumarins at the 3,4 double bond with the thymine 5,6 double bond has been correlated with partially localized π-electron densities of the triplet state, high spin densities, and reactivity indices. The model energy localization in the triplet state is reinforced by the fact that several psoralen derivatives and model coumarins exhibit weak fluorescence but relatively strong phosphorescence at low temperature (Mantulin and Song, 1973). Photo-3,4-cycloaddition of psoralen and pyrimidine shows slower reaction rates in the presence of molecular oxygen and various paramagnetic ions, e.g., Co^{2+}, Mn^{2+}, and Ni^{2+} (Bevilacqua and Bordin, 1973). On the basis of molecular orbital calculations and experimental results, the triplet state of psoralen and probably also other furocoumarins can normally be assumed as the reactive precursor. It is known that the psoralen–thymine photoadducts, which are formed by irradiation at 365 nm, are decomposed by postirradiation at 254 nm, restoring the parent compounds (Dall'Acqua *et al.*, 1969). Therefore, attempts were made to reirradiate bacterial cells damaged by irradiation at 365 nm in the presence of psoralen, using 254-nm UV radiation. No reactivation, however, has been obtained in this way. Experiments to study the photosplitting of furocoumarin moieties from DNA showed that splitting occurs to a very limited extent (Dall'Acqua *et al.*, 1969; Chandra *et al.*, 1971*a*; Rodighiero *et al.*, 1971).

Psoralen was found to react with native and denatured DNA and RNA to a comparable extent (Marciani *et al.*, 1973*a*). On the other hand, the photoreaction between DNA and either bergapten, xanthotoxin, or 8-methylxanthotoxin predominantly takes place with DNA in its double helical state. Like many other organic ligands, e.g., acridines and aromatic hydrocarbons, furocoumarins are able to form intermolecular complexes when added to an aqueous solution of nucleic acids. From various properties of the complexes, it seems probable that the furocoumarin molecules become intercalated between the stacked bases. Complex formation is discussed as an important preliminary condition for the subsequent photoreaction when the solution is irradiated (Cole, 1970; Musajo and Rodighiero, 1970). The bifunctional character of a furocoumarin enables it to photoreact first with its 4′,5′ double bond and second with its 3,4 double bond; the two reacting pyrimidine bases must belong to two different strands of DNA. In this way, a cross-link is formed between the two strands. The formation of interstrand cross-links is accompanied by a T_m increase after a long period of irradiation in the presence of furocoumarin. Bergapten was found as most effective in producing cross-links, followed by psoralen, xanthotoxin, and mono-, di-, and trimethylpsoralen derivatives.

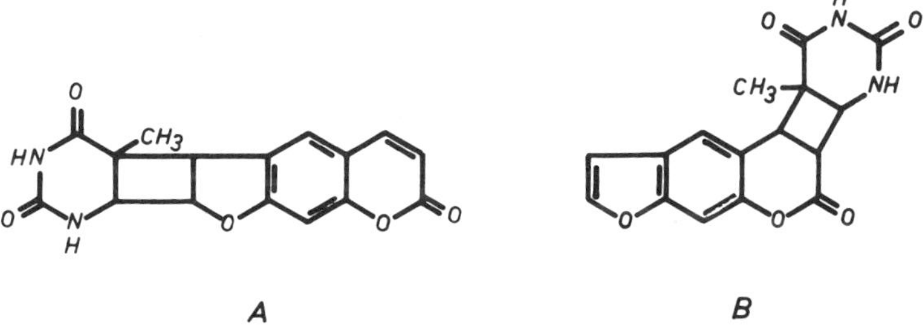

Fig. 12. Monoadducts of psoralen with thymine: (A) fluorescent, (B) nonfluorescent. According to Mussajo and Rodighiero (1970).

Angelicin and 7-methyl-allo-psoralen, both angular derivatives, although photobinding to DNA, appear unable to form cross-links (Dall'Acqua *et al.*, 1971). Despite the fact that the structural arrangement of photoreactive sites in angular furocoumarins is not favored for cross-linking, the formation of cross-links, although to a lesser extent, has been observed in single-stranded form of DNA and RNA (Marciani *et al.*, 1973*b*).

Dall'Acqua *et al.* (1968) reported on diffusion and viscosity measurements which yielded no indication for chain breaks or conformational changes after photoaddition. On the contrary, CD measurements clearly indicate that conformational changes of DNA must occur after the photobinding of furocoumarins (Kittler and Zimmer, 1976). Thus from the increase in the positive CD maximum, it is concluded that monofunctional as well as bifunctional photobinding to double-stranded native DNA caused a decrease in DNA helicity. On the other hand, from the decrease in the negative CD maximum, it is concluded that cross-link formation alone produces an additional distortion of base–base interaction in DNA. This effect increases with increasing AT content of the polymers. Cross-link formation induced by the photoaddition of xanthotoxin to native double-stranded DNA is favored for DNA in its B conformation. Nucleic acids possessing A-like structures, i.e., double-stranded RNA, are quite unsuited to photobinding of xanthotoxin. Moreover, if xanthotoxin had already bound to B-state DNA, no conformational transition to the A state took place, as usually occurs upon the addition of ethanol to a DNA solution.

An important question put forward in various investigations is whether the photoreactions of furocoumarins and nucleic acids can explain the photobiological effects of furocoumarins. There is a clear correlation between such photoinduced damage to DNA and killing of bacteria and mutation. A large amount of data exists for the skin sensitization effect. The furocoumarin derivatives that photosensitize the skin are able to photoreact with pyrimidine bases, and with nucleic acids in the skin. The action spectrum of xanthotoxin for the photoreaction with DNA is in accordance with its action spectrum for the production of erythema on human skin. Moreover, there exists a correlation between skin sensitization by furocoumarins and cross-link formation to DNA. The formation of cross-links is discussed as the main reason for the inactivation of Chinese hamster cells by furocoumarins under UV irradiation. Photoaddition of psoralen to DNA conditioned an inhibition of the infective capacity of Ehrlich ascites tumor cells. Photoreactions between furocoumarins and nucleic acids can take place *in vivo* for various other biological systems. A promising attempt has been undertaken in the clearing of generalized psoriasis lesions by means of 8-methoxypsoralen, taken orally, and long wavelength UV light. Its mode of action is discussed in terms of an inhibition of epidermal DNA synthesis (Table 6).

5- or 8-Hydroxypsoralen does not photobind to DNA. According to Song *et al.* (1975), the photodissociation process of the hydroxy group in the excited singlet state effectively competes with the singlet–triplet intersystem crossing, thus reducing the photoreactive triplet population.

4. SENSITIZED PHOTOREACTIONS OF NUCLEIC ACIDS AND NUCLEIC ACID COMPONENTS

One of the major features derived from studies of sensitized photoreactions is that sensitizer molecules absorb light and transmit the excitation energy to substrate molecules. Although these studies have shown a variety of possible reaction pathways, we will restrict ourselves to those in which no consumption of sensitizer molecules takes place during the reaction cycle, and in which the sensitizer molecules act as true photocatalysts. On the other hand, there is experimental evidence that excited nucleic acid bases are able to transfer their excitation energy to low-lying electronic states of other species, e.g., dye molecules, and it has been suggested that this could be used as a method for protecting nucleic acids from photochemical change.

4.1. Ketone Sensitization

The basis of ketone-sensitized photoreactions is as follows: Ketone molecules acting as sensitizers (energy donors) are excited to the first singlet state by absorption of a photon of light. Subsequent intersystem crossing populates the corresponding triplet state. Singlet and triplet excitation of the nucleic acids (energy acceptors) must be prevented by the use of appropriate light filters. Various ketones are known for their high probability of populating the triplet state by an intermolecular singlet (S_1)–triplet (T_1) energy transfer process. A special arrangement of the singlet and triplet levels of donor and acceptor molecules is a prerequisite for the functioning of the energy transfer mechanism. If the triplet state of the donor lies above that of the acceptor, a triplet–triplet transfer will be possible if both species are sufficiently close together (Lamola, 1970) (Fig. 13). Sensitizers, i.e., acetone, acetophenone, propiophenone, and benzophenone, illuminated with UV radiation at 313 nm induce a sensitized triplet excitation of nucleic acid bases (Greenstock and Johns, 1968; Lamola, 1969). The ketone sensitiza-

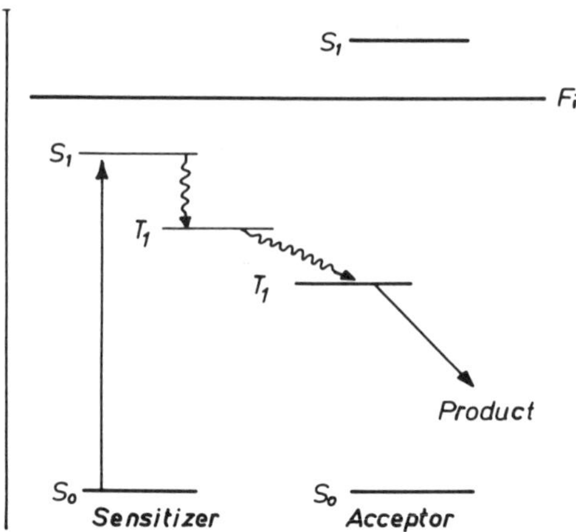

Fig. 13. Energy level scheme showing the triplet–triplet sensitization. S_0, singlet ground state; S_1, first excited singlet state; T_1, first excited triplet state; Fi, filter.

tion mechanism thus appears to produce predominantly photoreactions in which triplet molecules are the reactive precursors, e.g. dimerization reactions of thymine (Wierzchowski and Shugar, 1961*b*; Greenstock and Johns, 1968), uracil (Krauch *et al.,* 1967*b*; Greenstock and Johns, 1968), cytosine (Varghese, 1972*b*), and orotic acid (Charlier and Helénè, 1967). The quantum yield of a sensitized photoreaction increases with the difference between the triplet levels of donor and acceptor molecules. The triplet level of acetone is higher than the triplet levels of all natural nucleic acid bases (Lamola, 1970). This fact should explain the variety of photoproducts described in the case of acetone sensitization, i.e., homo- and heterodimers of thymine, cytosine, and orotic acid (Ben-Ishai *et al.,* 1968). This would be consistent with the finding that the *trans-anti* isomer of thymine may be formed by acetophenone sensitization (Ben-Hur and Rosenthal, 1970). A study of the ketone sensitization of DNA indicated the formation of DNA single-strand breaks (Zierenberg *et al.,* 1971). The fact that ketone sensitization mainly produces dimerization photoproducts can be used for refined biochemical studies. The fact that acetone sensitization preferentially inhibits poly(U)-dependent incorporation of phenyalanine during protein synthesis in cell-free systems strongly suggests photochemical changes of the uridylyl part. Moreover, acetone sensitization decreases the binding of GUU codons to ribosomes (Chandra *et al.,* 1971*c*) and inhibits the activity of amino acid acceptors in several tRNAs (Chambers *et al.,* 1969). Inactivation of T4 phages as a consequence of acetophenone sensitization is believed to be caused only by the formation of thymine dimers (Meistrich *et al.,* 1970). Biochemical repair mechanisms, i.e., photorepair and excision repair processes, are able to eliminate DNA damage that appeared after acetone photosensitization in *E. coli* (Menningmann and Wacker, 1970).

4.2. Dye Sensitization

4.2.1. Fluorescence of Dye–Nucleic Acid Complexes

A number of drugs are capable of forming intermolecular noncovalent complexes with nucleic acids, oligonucleotides, and polynucleotides. The ability of most of the drugs to aggregate gives rise to the possibility of binding of the drugs in the monomeric as well as in the polymeric form, according to type I and type II complexes, respectively (Peacocke and Skerrett, 1956). In this chapter, only type I complexes will be taken into consideration. Many authors who have discussed the problem of monomer binding consider that the dyes interact not only with the charged phosphates but also with the heterocyclic bases. This includes the possibility that drug molecules are inserted between the bases to some extent (Lerman, 1961; Pritchard *et al.*, 1966). Photochemical processes, which include nucleic acid bases and drugs, are favored for intercalated structures. The ability of a number of intercalative drugs to fluoresce allows the study of the initial steps of the dye sensitization processes by fluorescence spectroscopy.

Fluorescent ligands can be classified into two groups:

1. In the first group are ligands enhancing their fluorescence regardless of nucleic acid base composition, e.g., the dye acridine orange. In the past, various interpretations were given to explain the fluorescence increase observed as a function of the level of drug binding (Table 7). The intrinsic simplicity of the postulate that the nonfluorescent dye aggregates originally present in the solution are dissociated in favor of a sandwich-like stacking between the base pairs is not very attractive to explain the fluorescence effect at very low dye concentration (2×10^{-7} M) where aggregation phenomena are absent (Löber and Achtert, 1969).

Another question that arises is whether light energy primarily absorbed by nucleic acid bases can be transferred to bound dye molecules, giving rise to an enhanced fluorescence. Indeed, it was pointed out that excitation energy is transferred from a DNA triplet to a dye singlet (Basu and Greist, 1963; Isenberg *et al.*, 1964), yielding a delayed fluorescence emission of dye molecules (Isenberg *et al.*, 1964). However, it has been shown that fluorescence of bound acridine orange is even enhanced if excitation is performed with wavelengths longer than 450 nm, where singlet and triplet

TABLE 7. Fluorescence Enhancement of Organic Ligands Bound to Nucleic Acids and Related Macromolecules

Acceptor for organic ligands	References	Acceptor for organic ligands	References
Proflavine		Quinacrine	
Poly[d(A · T)]	Weisblum and De Haseth (1972)	AT DNA	Michelson *et al.* (1972), Weisblum and De Haseth (1973), Weisblum (1973), Borisova *et al.* (1974), Selander (1974*a,b*)
2,7-Di-*t*-butylproflavine			
DNA	Müller *et al.* (1973), Distéche and Bontemps (1974)	Ethidium bromide	
		Poly[d(A · T)]	Schreiber and Daune (1974)
Acridine orange		DNA	Le Pecq *et al.* (1964), Le Pecq and Paoletti (1967), Bittmann (1969), Burns (1969), Sela (1969), Löber *et al.* (1974)
Mononucleosides	Basu and Greist (1963);		
Mononucleotides	Tomita (1968); Yamabe (1969);		
Poly[d(A · T)], poly(dG) · poly(dC)	Schreiber and Daune (1974)		
		RNA	Bittmann (1969), Sela (1969), Maelicke (1970), Rigler *et al.* (1970)
DNA	Loeser *et al.* (1960), Boyle *et al.* (1962), Weill and Calvin (1963), Borisova and Tumerman (1964), Löber (1965), Rigler (1966), Löber (1968), Faddejeva (1969), Löber and Achtert (1969), Ichimura *et al.* (1971), Fredericq and Houssier (1972), Romanovskaja *et al.* (1972), Kubota (1973)	Chloroquine	
		DNA	Sutherland and Sutherland (1969*b*)
		Berberine	
		DNA	Hahn and Krey (1971)
		Hydroxystilbamidine	
Chromatin	Borisova and Minyat (1969), Killander and Rigler (1969), Kubota (1970*a,b*)		Festy and Daune (1973)
		Bibenzimidazole	
		DNA	Weisblum and Haenssler (1974)

excitation of nucleic bases is not very probable. Experiments have been undertaken to characterize the absorption and fluorescence spectra of dyes capable of binding to DNA under different environmental influences (Löber *et al.*, 1972, 1974*b*). The method applied is based on the idea that a dye molecule passing from its free water-solvated state to the DNA-bound state changes its environment so that the surroundings become less polar. The environmental change, which induces distinct alterations of the spectral properties of the dye, can be simulated by investigating the dye spectra in various organic solvent–water mixtures. Striking similarities in spectroscopic properties of dyes bound to nucleic acids by intercalation and the same dyes dissolved in organic solvents have been shown. Alterations in fluorescence intensity with increasing amounts of organic solvents in an organic solvent–water mixture are of particular interest. Fluorescence of acridine orange, proflavine, quinacrine, and ethidium increases with an increasing amount of organic solvent and upon excitation with radiation at appropriate wavelengths. A plausible hypothesis to explain this phenomenon has recently been developed. It is based on changes in separation of the corresponding singlet and triplet states of the dyes: a decrease of the singlet–triplet energy splitting in aqueous medium facilitates intersystem crossing, and thereby a significant lowering of the fluorescence quantum yield (Seliskar and Brand, 1971).

2.   In the second group of dyes are those whose fluorescence becomes quenched upon complexing with nucleic acids (Table 8). The stereochemical properties of the dye–nucleic acid complexes are such that any process of intercalation demands a direct interaction of the nucleic acid bases with the dye molecules. Although the spectroscopic processes of fluorescence quenching of dyes belonging to that category have not been completely explained, there is good evidence that guanine moieties are involved in the quenching of fluorescence of acriflavine (Tubbs *et al.*, 1964), proflavine (Weisblum and De Haseth, 1972), and quinacrine (Michelson *et al.*, 1972; Weisblum and De Haseth, 1972). A charge transfer process taking place between the excited dye molecule and guanine, which is known to have good electron-donating properties, is discussed as being responsible for the quenching effect. Guanine possesses a unique quenching ability when compared with other bases. In contrast, polynucleotides containing no guanine, i.e., poly(dA·dT), poly(dA)·poly(dT), and poly(A)·poly(U), enhance the fluorescence (Michelson *et al.*, 1972; Weisblum and De Haseth, 1972). In the specific case of natural DNAs, enhancement is restricted to regions rich in AT base pairs. The ability of GC base pairs to quench quinacrine fluorescence, for example, predominates over the apparent tendency of AT base pairs to enhance quinacrine fluorescence. Thus the overall effect consists of a fluorescence quenched by most of the natural DNAs. This

TABLE 8. Fluorescence Quenching of Organic Ligands Bound to Nucleic Acids and Related Macromolecules

Acceptor for organic ligands	References	Acceptor for organic ligands	References
Proflavine		**Riboflavine**	
Poly[d(G·C)]	Weisblum and De Haseth (1972)	DNA	Roth and McGormick (1967)
DNA	Weill and Calvin (1963), Löber (1968), Tomita (1968), Ellerton and Isenberg (1969), Löber (1969), Löber and Achtert (1969), Thomas *et al.* (1969), Bidet *et al.* (1970) Löber (1971)	**Thionine, methylene blue**	
		DNA	Tomita (1968)
		LSD	
		DNA	Yielding and Sternglanz (1971)
RNA	Tomita (1968), Schoentjes and Fredericq (1972)	**Daunomycin**	
		DNA	Calendi *et al.* (1965)
Chromatin	Lawrence and Louis (1972)	**Aromatic hydrocarbons**	
Acriflavine		DNA	Boyland and Green (1962), Liquori *et al.* (1962), Boyland and Green (1964)
DNA	Tubbs *et al.* (1964), Löber (1965), Chan and Van Winkle (1969)		
RNA	Chan and Van Winkle (1969), Surovaya and Trubitsyn (1972), Borisova *et al.* (1973)	**Aminoacyl tRNA synthetase**	
		tRNA	Helénè *et al.* (1969), Bruton and Hartley (1970), Rigler *et al.* (1970)
Chromatin	Minyat *et al.* (1970)		
Rivanol. acridine yellow		**Tyrosine, tyramine, tryptophan**	
DNA	Löber (1965), Löber and Achtert (1969)	DNA	Helénè *et al.* (1971)
Quinacrine			
DNA	Weisblum and De Haseth (1972),		
(GC DNA)	Michelson *et al.* (1972)		

observation of a mixed effect, in which AT enhances and GC quenches, suggests a possible explanation for the occurrence of fluorescence banding patterns in chromosomes.

4.2.2. Fluorescence Labeling of Chromosomes

Fluorescent basic dyes have found extensive application in studies of biological systems (Fig. 14). They are routinely used to mark cellular regions that are rich in nucleic acids, for light microscopic observations. Moreover, various fluorescent dyes are able to produce chromosomal fluorescence banding patterns. Differential fluorescence of chromosomes has been used as a valuable tool in cytogenetic research to identify chromosomes and to diagnose hereditary diseases (Caspersson *et al.*, 1972). The stains quinacrine and quinacrine mustard (Caspersson *et al.*, 1968; Barr and Ellison, 1971; Hollander and Borgaonkar, 1971; Vosa, 1971; Yamasaki, 1973), proflavine and its derivatives (Caspersson *et al.*, 1972; Distèche and Bontemps, 1974), and the bibenzimidazole derivative Hoechst 33258 (Hilwig and Gropp, 1972; Weisblum and Haenssler, 1974) are used most frequently. In most cases, the dyes producing banding show a pronounced binding affinity to DNA and deoxyribonculeoprotein. Therefore, it seems reasonable to interpret the specificity in the banding properties on the basis of complex formation of those dyes with chromosomal nucleic acids and chromosomal nucleoproteins, taking into consideration the modifications of the chromosome material arising as a consequence of fixation and pretreatment procedures preceding the staining step. Apparently, several factors

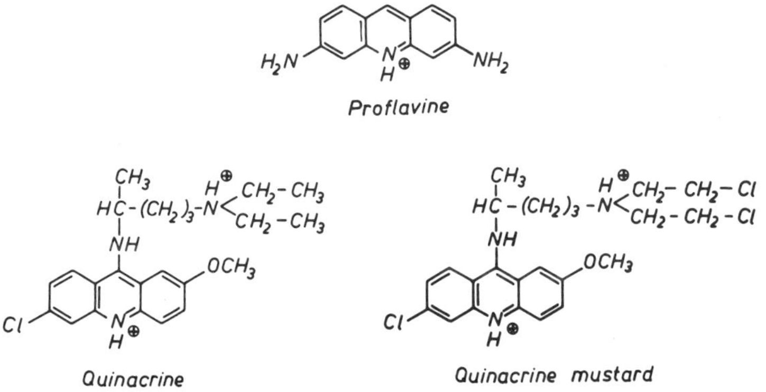

Fig. 14. Acridine dyes capable of producing chromosomal fluorescence banding patterns.

may regulate the appearance of a bright fluorescent band that indicates areas of AT-rich DNA in chromosomes:

1. Nonhomogeneous chromatin condensation, when AT-rich DNA regions become only moderately compact. These regions are preferred for dye binding, e.g., by an intercalation mechanism (Löber *et al.*, 1976).
2. Preferred removal of proteins induced by different chemical treatments from less compact AT-rich chromosome regions. It increases the number of binding sites accessible for DNA–dye binding.
3. AT specificity of intercalative binding of acridine dyes (Kleinwächter and Koudelka, 1964; Kleinwächter *et al.*, 1969; Ramstein and Leng, 1975).
4. Fluorescence enhancement, either general or AT induced, by the environmental influence of organic residues, where organic solvent molecules and nucleic acid constituents act in a similar manner (Fig. 15).

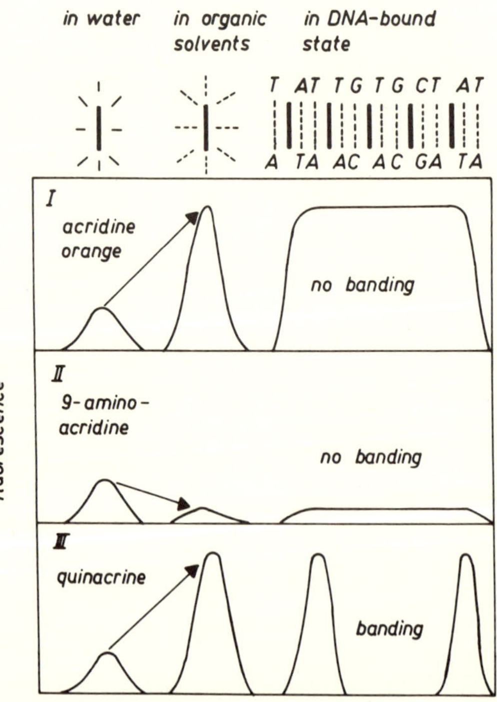

Fig. 15. Schematic representation of the fluorescence changes of a dye molecule in passing from an aqueous environment (dye dissolved in water) to an organic environment (dye dissolved in an organic solvent or attached to DNA). Guanine-specific quenching was taken into consideration in part III. Arrows indicate the fluorescence changes when passing from an aqueous to an organic environment. According to Löber *et al.* (1976).

For some dyes (proflavine, quinacrine), bright fluorescent bands occur only in the absence of quenching by guanine. The enhancement of fluorescence of some dyes by AT-rich regions of chromosomes offers a plausible explanation for this (Weisblum, 1973; Löber, 1975; Löber *et al.*, 1976). In full accord with this assumption is the fact that fluorescence enhancement is more pronounced if the dye molecule intercalates into double-stranded DNA (Weisblum and De Haseth, 1972), which guarantees an enhanced shielding from the polar environment, as compared with less ordered forms of DNA. From this point of view, the importance of possible reannealing of repetitive DNA after denaturing and staining should be mentioned. Bright green fluorescence of acridine orange in chromosome regions where repetitive DNA is expected supports this idea (Stockert and Lisanti, 1972; Comings *et al.*, 1973; De la Chapelle *et al.*, 1973).

4.2.3. Migration of Excitation Energy from a Nucleic Acid Triplet to a Dye Singlet

Tubbs *et al.* (1964) reported the possibility of a transfer of excitation energy from an excited dye molecule to a guanine residue of DNA, which results in a quenching of the fluorescence. On the other hand, light energy, which was primarily absorbed by nucleic acid bases via a singlet–singlet absorption process, can be internally converted between DNA singlets and triplets; this is called "intersystem crossing." The switch from a singlet manifold to triplet manifold populates the triplets of nucleic acid bases so that a transfer of excitation energy to dye singlets may occur. The near proximity of dye molecules and nucleic acid bases in the intercalated complex is in favor of energy transfer processes. First, this was indicated by the greater fluorescence quantum yield of the intercalated acridine orange–DNA complex obtained upon excitation with UV radiation as compared with the fluorescence quantum yield observed upon excitation with visible light. The distinction between these two cases is that, in the former case, radiation absorbed by the heteroaromatic nucleic acid bases contributes to the fluorescence emission of bound acridine orange molecules, whereas in the latter case fluorescence emission is caused only by light absorption by the bound acridine orange molecules (Weill and Calvin, 1963). Isenberg *et al.* (1964) observed a delayed fluorescence emission of complexes formed between various dyes (proflavine, acridine orange, quinacrine, acriflavine, acridine yellow) and DNA. The mechanism on which the appearance of delayed fluorescence is based is an internal conversion of excitation energy from DNA singlets to DNA triplets. These triplets wander among the bases and will be transferred to an attached dye molecule by means of a triplet–

singlet resonance transfer. The existence of a delayed fluorescence emission has been confirmed for complexes of DNA with ethidium bromide (Le Pecq and Paoletti, 1967), chloroquine (Sutherland and Sutherland, 1969*a*), and acridines (Kubota *et al.,* 1969; Kubota, 1970 *a,b*; Parker and Joyce, 1973).

4.2.4. Migration of Excitation Energy from a Nucleic Acid Triplet to a Dye Triplet

Besides the mechanism reported in Section 4.2.3, delayed fluorescence of dyes bound to nucleic acids could be induced by a transfer of excitation energy among the bound dye molecules. Finally, two dye triplets annihilate in favor of a single dye singlet. In this case, delayed fluorescence of dye molecules arises from two dye triplet molecules placed very close together (Kubota, 1970*a,b*). Otherwise, a transfer of excitation energy from a DNA triplet to a dye triplet, as reported for complexes of DNA with 9-amino-acridine and chloroquine (Galley, 1968; Sutherland and Sutherland, 1969*a,b*), induces a delayed phosphorescence. Its occurrence implies, in accordance with the intercalation hypothesis, a close spatial proximity of donors (DNA base triplets) and acceptors (dye triplets).

4.2.5. Protective Effect of Bound Dyes

The depopulation of DNA triplets by a transfer of the excitation energy to an attached dye molecule is supported by the finding that the rates of photochemical reactions are lowered when DNA triplets are involved. Of particular interest was the observation of an inhibition of pyrimidine dimerization reactions by various intercalating dyes (Beukers, 1965; Sutherland and Sutherland, 1969*a*,b). Perhaps the demonstration of protective effects of acridine dyes against UV inactivation of phage 1ϕ7 (Zavilgelskij *et al.,* 1964), phage ϕX174 (Beukers, 1965), and *E. coli* (Webb and Petrusek, 1966; Alper and Hodgins, 1969) is a convincing indication of the prevention of pyrimidine dimer formation. However, bound acridine molecules apparently have no effect on UV-induced dimer splitting (Bersohn and Isenberg, 1964).

4.2.6. Sensitized Splitting of Pyrimidine Dimers

Contrary to the results obtained with dyes, there is evidence that tryptophan and other indole derivatives, capable of forming intermolecular complexes with bases of nucleic acids (Montenay-Garestier and Hélène,

1968), lead to a photosensitized splitting of pyrimidine dimers upon UV excitation of the indole components. The quenching of the indole fluorescence by pyrimidine dimers suggests that splitting occurs as a result of electron transfer from the excited indole derivative to the pyrimidine dimer (Helénè and Charlier, 1971*a,b*; Santus *et al.,* 1972). The indole system-containing compounds such as serotonin and the oligopeptide Lys-Trp-Lys are also able to photosensitize the splitting effect (Charlier and Helénè, 1975). The specific recognition of single-stranded regions in UV-irradiated DNA by tryptophan-containing peptides, as found by means of fluorescence measurements (Toulme *et al.,* 1974), and their ability to split pyrimidine dimers in DNA on the basis of a sensitized photoreaction provide a plausible model for DNA photoreactivation. However, a general model for photosensitized dimer splitting does not exist at present; moreover, the mechanism seems to depend on the kind of sensitizer used (Lamola, 1972).

4.2.7. The Photodynamic Effect

Although the term "photodynamic action" has been used with somewhat different meanings by several scientists, one should recall the definition given by Blum (1941); for clarity, it should be used only for those dye-sensitized photoreactions in which molecular oxygen is consumed, i.e., dye-sensitized photooxidations. Moreover, the sensitizer dye has the role of a photocatalyst; i.e., it is not consumed in the course of the photoreaction. A number of review articles deal with problems of photodynamic action (Bellin, 1965; Grossweiner, 1969; Spikes, 1967; Spikes and MacKnight, 1970; Lochmann and Michler, 1973). Because the lowest electronically excited states of DNA lie, as a rule, energetically higher than those of dyes, the mechanism of energy transfer from the photoexcited dye to the nucleic acid molecule can hardly be considered as a deciding step for photodynamic processes. It seems more probable that radical reactions take place in which both sensitizing dyes and nucleic acids are involved.

4.2.7a. Two Types of Radical Photoreactions

In a number of contributions, the molecular basis of photodynamic action has been considered on the basis of two types of different radical reactions. In type I photoreaction, monoradicals are involved in the propagation and termination steps. This is discussed in terms of a primarily chemical reaction of excited species with hydrogen donors, to give rise to pairs of monoradicals, e.g., of the reduced sensitizer radical and the oxidized substrate radical. Dissolved molecular oxygen is able to reoxidize sensitizer radicals, thus yielding their initial forms, while oxidized radicals undergo a further irreversible oxidation process, which yields final products. Type I sensitization of oxidation is called "Bäckström-type photosensitized oxidation" or "primary dehydrogenation photosensitized reaction with oxygen." Type I photoreactions are important for the photodynamic degradation of nucleic acids and nucleic acid constituents (Simon and Van Vunakis, 1962; Wacker *et al.*, 1964*b*; Delmelle and Duchesne, 1967, 1968; Knowles, 1967, 1971; Gräslund *et al.*, 1969; Grossweiner, 1969; Chandra and Wacker, 1970; Gollmick and Berg, 1972; Kittler and Löber, 1974) and of model substrates like allylthiourea (Koizumi *et al.*, 1964; Kramer and Maute, 1972*a,b*; Zügel *et al.*, 1972) or *p*-toluene-diamine (Gollmick and Berg, 1968). The papers of Kepka and Grossweiner (1971), Grossweiner and Kepka (1972), Gollmick and Berg (1972), and Berg and Gollmick (1974) can be considered as a strong indication for the participation of reduction–oxidation processes in photodynamic mechanisms involving proteins and nucleic acids. The oxidation half-wave potentials as measured polarographically by means of a paste electrode are shown for various sensitizer dyes and various pyrimidine and purine derivatives in Table 9. From this representation, it seems obvious that the dye acts as a photodynamic

TABLE 9. Oxidation Potential ($\pi_{1/2\,Ox}$) of Sensitizer Dyes and Various Substrates and Photodynamic Decomposition of Bases (+, active; − inactive)[a]

Substrates	$\pi_{1/2\,Ox}$	AO[b] +700 mV	MB[c] +900 mV	TP[d] +950 mV	XX[e] +1200 mV
		Sensitizer dyes			
Uric acid	+250 mV	+	+	+	+
Xanthosine	+800 mV	+	+	+	+
Guanosine	+850 mV	+	+	+	+
Adenine	+950 mV	−	−	−	+
Hypoxanthine	+1000 mV	−	−	−	+
Thymine	+1050 mV	−	−	−	+
5-Hydroxymethylcytosine	+1150 mV	−	−	−	+
7,9-Dimethylguanine	+1150 mV	−	−	−	+
Cytosine	+1200 mV	−	−	−	(+)
Uracil	+1250 mV	−	−	−	−
Orotic acid	+1250 mV	−	−	−	−

[a] From Kittler and Löber (1974).
[b] Acridine orange.
[c] Methylene blue.
[d] Thiopyronine.
[e] bis-Methylanilidotrimethinecyanine perchlorate.

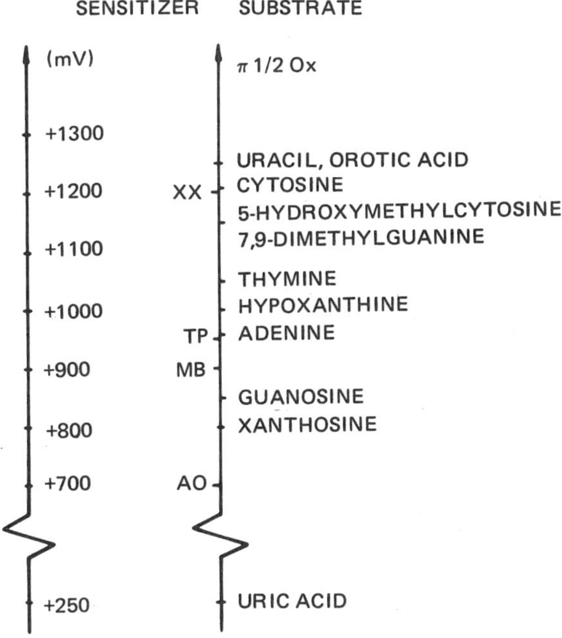

Fig. 16. Oxidation protentials ($\pi_{\frac{1}{2}Ox}$) of photodynamic sensitizers and various purine and pyrimidine derivatives. AO, Acridine orange; MB, methylene blue; TP, thiopyronine; XX, bismethylanilidotrimethinecyanine. According to Kittler and Löber (1974).

sensitizer only if its oxidation potential lies above that for pyrimidines and purines. This fact coincides with the idea that the photodynamic reaction takes place via an oxidation reaction of the substrate (Fig. 16).

This makes it understandable that guanine, which possesses the most negative oxidation potential of all natural nucleic acid bases, is preferentially attacked and that thiopyronine and methylene blue with rather positive oxidation potentials act as good sensitizers. The oxidized dye radical has the role of an oxidizing agent. It was shown by a kinetic study that at least for some sensitizer–substrate combinations, e.g., methylene blue–guanosine and thiopyronine–guanosine, the direct involvement of dye triplets in the reduction–oxidation process need not be considered (Gollmick and Berg, 1972). Nevertheless, a more general scheme needs to consider dye triplets, and three different cases are possible (Gollmick and Berg, 1972) (Fig. 16):

1. If the oxidation potential of the electron donor (substrate) is more negative than the reduction potential of the sensitizer triplet, the

electron acceptor in the course of the photodynamic reaction is the latter. Belonging to that category are photodynamic reactions using as sensitizers and substrates, respectively, thionine and allylthiourea (Kramer and Maute, 1972a), methylene blue and various amino acids (Knowles and Gurnani, 1972), and methylene blue and EDTA (Bonneau *et al.*, 1967; Morita and Kato, 1969).

2. If the oxidation potential of the electron donor lies between the reduction potentials of the sensitizer triplet and the oxidized sensitizer radical, the electron acceptor is the latter. To that category belong photodynamic reactions using as sensitizers and substrates, respectively, thiopyronine (or methylene blue) and allylthiourea (Morita and Kato, 1969) and thiopyronine (or methylene blue) and guanosine (Gollmick and Berg, 1972; Kittler and Löber, 1974).

3. If the oxidation potential of the electron donor is more positive than the reduction potentials of both sensitizer triplet and oxidized sensitizer radical, no photooxidation takes place. This holds true for the system thiopyronine (or methylene blue)–adenine.

Other nucleic acid bases besides guanine might be attacked when sensitizers are used whose oxidation potentials are higher than the oxidation potential of thiopyronine. This condition is fulfilled for the dye XX represented in Table 9. Indeed, this dye was shown to be able to damage most of the natural nucleic acid bases photodynamically. Experiments with isolated bases showed that, besides guanine, adenine and thymine were attacked preferentially. As an exception to case (2), acridine orange acts photodynamically on xanthine and guanosine, although the oxidation potential of acridine orange lies below that of the corresponding substrates.

Currently, the mechanisms for the photodynamic action of proflavine are under discussion, and are based on a two-photon absorption process of the sensitizer. They are thought to produce radicals in thymine and guanine in the presence and absence of oxygen (Van de Vorst and Lion, 1971a,b; Calberg-Bacq and Van de Vorst, 1974; Gräslund *et al.*, 1975). Results were as a rule obtained at 77°K.

In photoreactions that proceed by the type II mechanism, only biradicals take part. Sensitizer molecules in the first excited singlet state or triplet state and molecular oxygen combine to give a short-lived sensitizer–oxygen adduct, which transfers its excited oxygen to unsaturated substrates and the sensitizer in the ground state. Type II photoreaction is called "photosensitized oxygen transfer" or "Schenck mechanism" (Gollnick and Schenck, 1964). A variant related to type II concerns participation of singlet oxygen. The formation of singlet oxygen as an intermediate in photosensitized type II oxygenation with molecular triplet oxygen appears to be well established. In the reactions of olefins with excited oxygen, this reactive oxygen has enhanced dehydrogenating properties, more than the properties of the oxygen sensitizer adduct in the photosensitized oxygen transfer reactions (Gollnick and Schenck, 1964; Schenck, 1970). Singlet oxygen is receiving increasing attention as a reactive species in biological systems. Therefore, it merits discussion in a separate section.

4.2.7b. *Singlet Oxygen and Its Role in Photodynamic Action and Photocarcinogenicity*

The range of reactions and systems in which singlet oxygen is now believed to be directly or indirectly involved is convincing. Singlet oxygen is clearly involved in numerous dye-sensitized photooxygenation reactions of unsaturated compounds, and in the quenching of excited singlet-state and triplet-state molecules (see reviews by Schenck, 1970; Kearns, 1971). This is the reason to suspect that singlet oxygen may also be involved in photobiological processes, i.e., photodynamic action and photocarcinogenicity. A considerable number of natural products are found to be effective sensitizers for photooxygenation reactions involving singlet oxygen, e.g., chlorophyll (Gollnick, 1968), protoporphyrin (Rawls and van Santen, 1970), and pheophytin *a* (Foote, 1968). Recent studies utilizing simple chemical systems have described the interaction of singlet oxygen with natural products such as steroids (Gollnick, 1968), amino acids (Foote, 1968), fatty acids (Rawls and van Santen, 1970), and purines, including DNA itself (Foote, 1968).

Attempts have been made to prove the singlet oxygen mechanism in photodynamic action. Since singlet oxygen can be chemically generated, e.g., in the reaction between hypochlorite and hydrogen peroxide, it should be possible to obtain the equivalent of photodynamic action "without light." The long-lived singlet oxygen was carried from the generator to the sample by a stream of nitrogen. Indeed, it was found that various amino acids were damaged. Singlet oxygen and photodynamic action gave qualitatively similar effects on trypsin and cytochrome *c* when methylene blue was used as a sensitizer in photodynamic experiments (Debey and Douzou, 1970). These results are reinforced by the observation that singlet oxygen induced the damage of tryptophan, histidine, and methionine (Nilsson *et al.*, 1972), and by the reported quenching effect of various amino acids and proteins on singlet oxygen (Matheson *et al.*, 1975).

Perdeuterated water is known to enhance the decay time of singlet oxygen, while azide ions are known as specific quenchers of singlet oxygen. Acridine orange-sensitized photodynamic inactivation of yeast cells was found to be enhanced in going from the normal to the perdeuterated medium, while the inactivation rate was markedly reduced in the presence of azide ions in aqueous suspension. This fact was taken to mean that singlet oxygen molecules participate as the intermediate species in *in vivo* photodynamic action (Ito and Kobayashi, 1974). However, the quenching constants obtained by the methylene blue method of generating singlet oxygen deviate from those obtained by irradiating the oxygenated solution with the 1.06 μm output of a neodymium YAG laser, which generates singlet oxygen directly from triplet oxygen. The pH dependence of thiazine-sensitized singlet oxygen generation has been discussed by Bonneau *et al.* (1975).

With respect to the photodynamic destruction of guanosine monophosphates, the singlet oxygen mechanism is assumed not to play an exclusive role (Nilsson *et al.*, 1972). A comparative study of photodynamic oxidation and radiofrequency-discharge-generated singlet oxygen oxidation of guanosine yielded different photoproducts on chromatographic analysis (Kornhauser *et al.*, 1973). It was shown by Kramer and Maute (1973) that whether a radical mechanism (here termed "type I") or a singlet oxygen mechanism (here termed "type II") dominates depends on the acceptor chosen and its concentration and on solvent conditions. Whereas the photooxygenation of allylthiourea with thionine as a sensitizer takes place via radicals at high allylthiourea concentrations, a change to the singlet oxygen mechanism was observed at low substrate concentrations in pyridine solution.

On the other hand, it is suggested that organisms are protected by carotenoids against the lethal effects of sensitization by their own sensitizers, such as chlorophyll (Foote, 1968). Experiments show that β-carotene will quench singlet oxygen either generated by a sensitizer, e.g., methylene blue, or produced from a chemical reaction (Foote and Denny, 1968; Foote *et al.*, 1970). Thus it seems possible that carotenes play a role in the protection of various substrates, including nucleic acids, from photooxidation.

It is well known that photosensitizing dyes are able to produce skin cancer, and many of the carcinogenic hydrocarbons are known to be good sensitizers. These relationships have been reinforced by Khan and Kasha (1970), who propose a singlet oxygen theory of carcinogenesis. According to their model, the K and L regions on the polycyclic hydrocarbon determine the binding of the hydrocarbon to all the constituents. Then the carcinogenic effect of the molecule is dependent on subsequent generation of singlet oxygen after electronic excitation of the bound aromatic hydrocarbon. Thus singlet oxygen is the tumor-inducing agent in the process of photocarcinogenesis. An alternate theory of carcinogenesis that implicates singlet oxygen and aromatic hydrocarbons has been suggested by Cusachs and Steele (1967). The novel step in their theory is the reaction of singlet oxygen with either the K or L charge density regions of the aromatic hydrocarbon. If singlet oxygen reacts with the L region, the formation of noncarcinogenic transannular peroxides or epiperoxides is expected. If singlet oxygen reacts at a K region, carcinogenic arylhydroperoxides are produced. This theory is consistent with the active K region and inactive L region criterion for carcinogenic activity of aromatic hydrocarbons proposed by Pullman and Pullman (1955).

4.2.7c. *Photodynamic Damage in Nucleic Acids*

Indications have accumulated over the years that guanine is the base in nucleic acids that is preferentially damaged by irreversible oxidation processes in photodynamic experiments (Sussenbach and Berends, 1963, 1964, 1965; Wacker *et al.*, 1963; Zenda *et al.*, 1965; Bellin and Grossman, 1965; Sastry and Gordon, 1966; Fujita and Yamazaki, 1970; Nirmala and Sastry, 1971). There were a number of reports of the ability of photosensitization to change hydrodynamic parameters of DNAs, and this has been interpreted as evidence for the occurrence of strand breaks. Thus salmon sperm DNA is depolymerized on extensive photodynamic treatment with acridine orange, as determined by decreases in viscosity and sedimentation coefficient (Freifelder *et al.*, 1961). The sedimentation coefficient of native bacteriophage T4 DNA decreases only slightly upon photodynamic treatment with methylene blue; subsequent heat treatment causes a marked decrease (Simon and van Vunakis, 1962). Photodynamic treatment with methylene blue has been found to have little effect on the chain length of the polynucleotide poly(U·G) (Simon *et al.*, 1965). DNAs taken from *Proteus mirabilis* and calf thymus show decreased sedimentation coefficients after photodynamic treatment with thiopyronine (Berg and Bär, 1967; Berg *et al.*, 1972). Photodynamic action with methylene blue causes degradation in isolated sections of each DNA chain; on prolonged irradiation, double chain scission occurs, as indicated by decrease in viscosity (Bellin and Yankus, 1966). Photodynamic treatment of guanosine, using methylene blue and acridine orange as sensitizers, showed that a breakage of the *N*-glycoside bond may occur besides the chemical alteration of the guanine part (Sastry and Gordon, 1966; Waskell *et al.*, 1966). In DNA, apurinic sites are unstable in the presence of alkali and give rise to the appearance of single-strand breaks (Tamm *et al.*, 1953). Freifelder and Uretz (1966) found a dose-dependent formation of alkali-labile bonds and correlated this with the photodynamic inactivation of phage T7. Thus is seems possible that, under proper conditions, strand scissions occur in photodynamically treated DNA samples. On the other hand, the formation of strand breaks under *in vivo* conditions cannot be considered as fully proven, since during lysis of cells on alkaline sucrose gradients, alkali-labile apurinic sites might be converted into single-strand breaks (Jacob, 1971). Photodynamic action also produces DNA–protein cross-links in irradiated cells (Smith, 1962; Smith and Hanawalt, 1969).

4.2.7d. *Molecular Biological Effects of Photodynamic Action*

The photodynamic treatment of *in vivo* systems causes deactivation of, for example, plant viruses, bacterial viruses, animal viruses, bacteria, and fungi (see Table 10) (for detailed studies, see Blum, 1941; Spikes, 1967; Spikes and Livingston, 1969; Spikes and MacKnight, 1970; Lochmann and Michler, 1973; Löber and Kittler, 1973). In most cases, little is known about the processes causing inactivation or the kind of defects that might be seen as lethal hits. Proteins (enzymes), polysaccharides, and other constituents of cells are considered to be photodynamically changed, in addition to nucleic acids. Nevertheless, indications are that defects in the structure of DNA are involved in inactivation processes of viruses and bacteria (Ritchie, 1965; Cramer and Uretz, 1966*a,b*; Brendel and Winkler, 1966; Hiatt, 1967*a,b*; Kalab, 1967; Yamamoto, 1967). In addition, as pointed out by various authors (Yamamoto, 1958; Howard-Flanders and Boyce, 1966; Brendel and Kaplan, 1967; Böhme, 1968; Harm, 1970; Geissler, 1968, 1970; Brendel, 1970), photodynamic damage can be eliminated by dark repair mechanisms. It is possible that the sensitizer molecules, which are still present in the biological system after the photodynamic treatment, can act as inhibitors of dark repair processes (Lochmann and Stein, 1968). But there are also systems where dark repair processes are lacking (Uretz, 1964; Patrick *et al.,* 1964; Freifelder and Uretz, 1966; Geissler, 1967). Besides the fact that not all systems are able to repair DNA defects to the same extent, repairing capacity for photodynamically induced defects depends on the kind of sensitizer used and the dose of irradiation. Methylene blue was found to possess a higher photodynamic activity than proflavine (Yamamoto, 1958; Fraser and Mahler, 1961; Calberg-Bacq *et al.,* 1968). An important aspect was the observation that an increased probability for the formation of double-strand breaks in DNA was established when photodynamic sensitization was performed with methylene blue (Berg *et al.,* 1972). Both probes, methylene blue and proflavine, are known for their high binding capacity with nucleic acids. While proflavine

TABLE 10.   Photodynamically Induced Inactivations in Various
Biological Objects

Sensitizers	Biological objects	References
Acridine orange	Tobacco mosaic virus	Chessin (1960)
	Tobacco mosaic virus RNA[a]	Sastry and Gordon (1966)
	Escherichia coli B[b]	Cramer and Uretz (1966a,b)
	Phage T7M[c]	Freifelder and Uretz (1966)
	Escherichia coli	Wallnöfer and Bukatsch (1960)
	Bacillus subtilis	
	Phage of *Staphylococcus aureus*	Janovská and Pillich (1968)
	Saccharomyces cerevisiae	Ito *et al.* (1967)
Acridine orange, methylene blue	*Escherichia coli* (UV^R and UV^S)[d] lysogenic for phage λi	Freifelder (1966)
	Vesicular exanthem	Hackett (1962)
Acridine orange, methylene blue, neutral red	Mouse embryo and mouse L cells[e]	Litwin and Riesterer (1973)
Proflavine	Phage T2	Hessler (1965)
	Phage T3	Witmer and Fraser (1970, 1971a,b)
Proflavine, crystal violet	Actinophages	Noble and Bradley (1972)
Various acridine derivatives	*Proteus vulgaris*	Löber *et al.* (1970)
Proflavine, thiopyronine, methylene blue	Tobacco mosaic virus RNA	Singer and Fraenkel-Conrat (1966)
Neutral red	Tobacco mosaic virus	Orbob (1963)
	Cucumber mosaic virus	
	Alfalfa mosaic virus	
Methylene blue	Phage T2, φX174	Fraser and Mahler (1961)
Methylene blue, riboflavin	Tobacco mosaic virus	Tsugita *et al.* (1965)
Toluidine blue	Phage T2, phage T3	Hiatt (1967a,b)
	Clamidia psittaci (pararickettsia)	Portocala *et al.* (1972)
Thiopyronine	Lysogenic strain of bacteria	Geissler and Wacker (1963)
	Escherichia coli K12 (λ)[d]	Geissler (1968)
	Yeast cells	Lochmann and Stein (1964, 1967)
		Micheler *et al.* (1973)

[a] Protective effect of paramagnetic ions.

[b] Loss of colony-forming ability and capacity for phage T40.

[c] Alkali-labile bond.

[d] Lysogeny.

[e] Decrease of [³H]thymidine incorporation.

is preferentially bound by an intercalation mechanism (Lerman, 1961), methylene blue molecules become bound at the surface of DNA (Michaelis, 1947). It remains unknown at this point if different binding types influence the photodynamic activity of dyes. Nevertheless, a correlation between the binding constants of intercalative bound acridine derivatives and their photodynamic inactivation effectiveness on phage T3 has been found (Balcarová *et al.,* 1971).

As another conclusive indication for the participation of DNA in photodynamic action under *in vivo* conditions, photodynamically induced mutations can be seen (Kaplan, 1950; Matthews, 1963; Böhme and Wacker, 1963; Webb and Kubitschek, 1963; Nakai and Saeki, 1964; Ritchie, 1964, 1965; Zampieri and Greenberg, 1965; Singer and Fraenkel-Conrat, 1966; Calberg-Bacq *et al.,* 1968). This is expected, because various dyes that are used as sensitizers are known as mutagens. However, the mechanism has not yet been clarified. Since many of the studies investigating photodynamic mutations involve sensitizers that are able to bind with DNA (i.e., acridine) or are not able to bind with DNA (i.e., erythrosine), the role of sensitizer–DNA complexes is not very clear. It may be speculated that degradation of guanine makes the production of substitution mutations possible. Otherwise, photodynamically induced single-strand breaks would enable mutations of the deletion type. Finally, it needs to be considered that radical intermediates which appear in the course of a photodynamic reaction might be able to act as chemical mutagens (Delmelle et al., 1966; Delmelle and Duchesne, 1967).

Photodynamic efficiency on yeast and bacteria was found to be enhanced if proper dye combinations were used (Jacob *et al.,* 1967). This result seems to be mainly caused by a better use of available light energy provided that different sensitizers absorb at different spectral regions. Photodynamic treatment of tobacco mosaic viruses with acridine orange yielded a higher efficiency after prior sonication (Elpiner and Shebaldina, 1967). The authors concluded that such a procedure considerably disturbs the protein envelopes of the virus particles.

Sensitization of Ehrlich ascites cancer cells with methylene blue or acridine orange substantially increases the NAD requirement (Hunter *et al.,* 1967). This finding is consistent with the idea that the combination of light and various photosensitizers can cause profound metabolic disturbances that result in severe disruptions of the cellular economy and may result in cell death. Indications are available that Yoshida hepatoma cells contain natural pigments which act as photosensitizers in the presence of oxygen (Santamaria, 1967). Glycolytic and respiratory enzymes become rapidly inactivated. It is doubtful that nucleic acid alterations are involved in this process; nevertheless, photodynamic treatment of intradermal tumors of

white mice with methylene blue or thiopyronine as sensitizer yielded significantly high cure rates (Berg and Jungstand, 1966; Jungstand and Berg, 1967).
Recent experiments indicate that argon ion lasers used in combination with
oral uptake of the photosensitizing dye acridine orange may prove successful
in the treatment of skin cancer induced in mice. The dye has been found to
exhibit greater preference for the cancer cells, and as a consequence is
localized in the tumors, where it is combined with the nucleic acids in the
tumor cells. Irradiation of the tumors with laser light, at energy levels that do
not cause heat damage to healthy tissue, induces the acridine orange in the
tumor cells to break down, forming free radicals that destroy them (see
Spectra-Physics Laser Review, 1975).

5. CONCLUSIONS AND PERSPECTIVES

In the past two decades, the photochemistry of nucleic acids (DNA
and RNA), nucleic acid bases, and model polynucleotides has been
examined under a variety of experimental conditions. In particular, the
formation of photoproducts and strand scissions after either UV or
photosensitizing irradiation was measured and found to be strongly
dependent on the solvent, temperature, pH, relative humidity, and presence
of bound molecules. There are indications of the biological relevance of
photochemical processes on nucleic acids *in vivo.* Various promising investigations are under way that may suggest ways to produce selective effects
in the photodamage of nucleic acids by the choice of proper reaction conditions. For example, it has been shown that the survival curves of different
bacterial strains were changed when base analogues of the aza type were
incorporated prior to UV irradiation. Generally, incorporation of UV-sensitive aza compounds enhanced the sensitivity of bacteria to UV radiation,
while incorporation of less-UV-sensitive aza compounds led to a decrease.
Therefore, if more-UV-sensitive aza compounds are incorporated into
defined genes of synchronized growing cells, UV defects should be
preferentially located in those genes. This might yield a higher degree of
selectivity in photobiological experiments. The same idea holds true for
minor bases found in DNA. Since their sensitivity toward UV radiation
does not deviate significantly from that of known major bases, careful
spectroscopic and photochemical experiments are needed to check whether
selective damage is possible. As indicated in Section 2.1.5, the photoalkylation of purines has been studied. Alkylated purines are important in the
field of cancer chemotherapy. Photoalkylation has three obvious advantages
compared to chemical dark alkylation: (1) since the photoalkylation at
purine sites takes place under UV irradiation, light can be used as a regula-

tory agent in genetic processes; (2) photoalkylation can occur with alcohols, amines, and amino acids, and there is no necessity to use chemically reactive mustard compounds; (3) photoalkylation takes place preferentially at purine sites and has a higher degree of selectivity than chemical dark alkylation.

The photobiological effects of pyrimidine photoadducts are not yet well understood. In spores, however, their occurrence is well documented. Other examples of the biological importance of photoadducts are still lacking.

A large number of physicochemical studies have been made on the photoaddition of furocoumarins to nucleic acids. Moreover, a great number of papers provide convincing evidence for biological effects related to the formation of furocoumarin–DNA adducts. Psoriasis lesions have been cleared by the combined action of furocoumarin and UV radiation. More recently, it has been demonstrated that photoaddition of furocoumarins requires B-form DNA, while A-type structures are not attacked to any great extent (Kittler and Zimmer, 1976). In addition, furocoumarin already covalently bound prevents DNA from undergoing a B → A conformational transition. This opens at least two perspectives: (1) Since DNA *in situ* may partially deviate in its conformation from the B form, photoaddition of suitable furocoumarins could give information on the distribution of B-type and non-B-type DNA. (2) DNA *in situ* probably accepts A form in the transcriptional state. The inhibition of B → A conformational transition by covalently bound furocoumarins provides a plausible hypothesis for the interference of furocoumarins with RNA synthesis.

Excited states of the complexes of fluorescent organic ligands with nucleic acids have a widespread field of application. Thus fluorescence techniques are routinely used as tools to obtain binding data on the formation of fluorescent organic complexes with nucleic acids. Since the fluorescent properties of molecules are strongly dependent on environmental conditions, it is possible to distinguish polar and nonpolar surroundings. For example, the use of bound fluorescence markers yielded information on polar and hydrophobic regions of globular proteins (Brand and Gohlke, 1972). Similar investigations for nucleic acids are rare. Mostly advanced are fluorescence measurements in which base Y of various tRNAs or ethidium bound to tRNA reflects the nature of induced conformational transitions (Beardsley *et al.*, 1970; Tao *et al.*, 1970; Yguerabide, 1972). The fluorescence changes occurring when a dye passes from its water-solvated state to the DNA-bound state may be interpreted by the altered surrounding of dye molecules. The presumably different fluorescent properties of dyes bound to GC-rich and AT-rich DNA are used as the spectroscopic bases for the production of fluorescence banding patterns on eukaryotic chromosomes. The longitudinal differentiation of chromosomes by means of fluorescent dyes is increasingly used as a tool in studying chromosome morphology. Although a transfer of excitation energy in either nucleic acids or nucleic acid–dye complexes has been discussed in the past, investigations on energy migration processes in nucleoproteins are in the early stages. Recent investigations have indicated that a transfer of excitation energy from an amino acid residue of protein to a pyrimidine dimer of DNA could be the physical basis of photoreactivation.

It was suggested by a number of authors that guanine residues were destroyed selectively during the dye-sensitized photodynamic treatment of nucleic acids. In many cases, photooxidation alters the physicochemical properties of nucleic acids. For example, prolonged photodynamic treatment causes a decrease in the viscosity and an increase in the sedimentation of DNA solutions, a fact due to the occurrence of strand scissions of the sugar–phosphate chain. It is well known that *in vivo* photodynamic action causes predominantly inactivations and mutations. The question arises whether it is possible to damage other bases than guanine during a photodynamic treatment. As shown in this chapter, it should be possible to find sensitizers possessing very positive oxidation potentials. Such sensitizers are capable of photodynamically attacking both guanine and adenine residues. However, it must be expected that, if adenine and guanine are present simultaneously, the latter will be preferentially damaged. Despite this fact, it may be assumed that sensitizer molecules bound to AT DNA might also yield adenine defects during photodynamic treatment. Such AT-rich DNA regions are to be found, for example, in the satellite DNAs, and perhaps photodynamic treatment will open a new approach to the study of their biochemical function.

6. REFERENCES

Adman, E., and Jensen, L. H., 1970, The crystal and molecular structure of the *cis-syn* photodimer of uracil, *Acta Crystallogr. Sect. B* **26**:1326–1334.

Alcantara, R., and Wang, S. Y., 1965a, Photochemistry of 1,3-dimethylthymine in aqueous solution, *Photochem. Photobiol.* **4**:465–472.

Alcantara, R., and Wang, S. Y., 1965b, Photochemistry of thymine in aqueous solution, *Photochem. Photobiol.* **4**:473–476.

Alexander, P., and Moroson, H., 1962, Cross-linking of deoxyribonucleic acid to protein following ultraviolet irradiation of different cells, *Nature (London)* **194**:882–883.

Alper, T., and Hodgins, B., 1969, "Excision repair" and dose-modification: Questions raised by radiobiological experiments with acriflavine, *Mutat. Res.* **8**:15–23.

Baden, H. P., Parrington, J. M., Delhanty, J. D. A., and Pathak, M. A., 1972, DNA synthesis in normal and XP-fibroblasts following treatment with 8-methoxypsoralen and long wave ultraviolet light, *Biochim. Biophys. Acta* **262**:247–255.

Bagchi, B., Basu, S., Misra, D. N., and Das Gupta, N. N., 1969, Conformation of UV-irradiated DNA, *Int. J. Radiat. Biol.* **16**:301–310.

Balcarová, Z., Janovská, E., Kleinwächter, V., Koudelka, J., and Löber, G., 1971, Comparison of biological activity and some physico-chemical properties of *N*-derivatives of acridine orange, *Stud. Biophys.* **27**:205–212.

Barr, H. J., and Ellison, J. R., 1971, Quinacrine staining of chromosomes and evolutionary studies in *Drosophila, Nature* (*London*) **223**:190–191.

Basu, S., and Greist, J., 1963, Resonance transfer of excitation energy between nucleosides and acridine orange. *J. Phys. Chem.* **67**:1394–1395.

Beardsley, K., Tao, T., and Cantor, C. R., 1970, Studies on the conformation of the anticodon loop of phenylalanine transfer ribonucleic acid. Effect of environment on the fluorescence of the Y base, *Biochemistry* **9**:3524–3532.

Becarevic, A., Djordjevic, B., and Sutic, D., 1963, Effect of ultra-violet light on Tobacco Mosaic Virus containing 5-fluorouracil, *Nature* (*London*) **198**:612–613.

Becker, J., Le Blanc, J. C., and Johns, H. E., 1967, The UV photochemistry of cytidylic acid. *Photochem. Photobiol.* **6**:733–743.

Bellin, J. S., 1965, Properties of pigments in the bound state: A review, *Photochem. Photobiol.* **4**:33–44.

Bellin, J. S., and Grossman, L. J., 1965, Photodynamic degradation of nucleic acids, *Photochem. Photobiol.* **4**:45–53.

Bellin, J. S., and Yankus, C. A., 1966, Effects of photodynamic degradation on the viscosity of deoxyribonucleic acid, *Biochim. Biophys. Acta* **112**:363–371.

Ben-Hur, E., and Ben-Ishai, R., 1968, *Trans-syn* thymine dimers in ultraviolet irradiated denatured DNA: Identification and photoreactivability, *Biochim. Biophys. Acta* **166**:9–15.

Ben-Hur, E., and Elkind, M. M., 1973, Psoralen plus near ultraviolet light inactivation of cultured chinese hamster cells and its relation to DNA cross-links, *Mutat. Res.* **18**:315–324.

Ben-Hur, E., and Rosenthal, J., 1970, Photosensitized splitting of pyrimidine dimers, *Photochem. Photobiol.* **11**:163–168.

Ben-Hur, E., Elad, D., and Ben-Ishai, R., 1967, The photosensitized dimerization of thymine in solution, *Biochim. Biophys. Acta* **149**:355–360.

Ben-Ishai, R., Ben-Hur, E., and Hornfeld, Y., 1968, Photosensitized dimerization of thymine and cytosine in DNA, *Isr. J. Chem.* **6**:769–775.

Ben-Ishai, R., Green, M., Graff, E., Elad, D., Steinmaus, H., and Salomon, J., 1973, Photoalkylation of purines in DNA, *Photochem. Photobiol.* **17**:155–167.

Benzer, S., and Freese, E., 1958, Induction of specific mutations with 5-bromouracil, *Proc. Natl. Acad. Sci. USA* **44**:112–119.

Berg, H., 1975, Photopolarographie und Photodynamie, in: *Sitzungsberichte der Sächsischen Akademie der Wissenschaften zu Leipzig,* pp. 5–19, Akademic Verlag, Berlin.

Berg, H., and Bär, H., 1967, Photodynamische Destabilisierung der DNS-Doppelhelix, in: *Molekulare Mechanismen photodynamischer Effekte, Stud. Biophys.* **3**:133–138.

Berg, H., and Gollmick, F. A., 1974, Photodynamic mechanisms and redoxpotentials of excited sensitizers, in: *Progress in Photobiology* (G. O. Schenck, ed.), No. 006, Deutsche Gesellschaft für Lichtforschung, Frankfurt/M.

Berg, H., and Jungstand, W., 1966, Photodynamische Wirkung auf das solide Ehrlich Karzinom, *Naturwissenschaften* **53**:481–482.

Berg, H., Gollmick, F. A., Jacob, H.-E., and Triebel, H., 1972, Sensibilisierte Photooxidation durch Methylenblau, Thiopyronin und Pyronin. II. Physikochemische Grundlagen der photodynamischen Wirkung von Thiopyronin, *Photochem. Photobiol.* **16**:125–138.

Bergstrom, D. E., and Leonard, N. J., 1972, Photoreaction of 4-thiouracil with cytosine. Relation to photoreactions in *E. coli* transfer ribonucleic acids, *Biochemistry* **11**:1–9.

Bersohn, R., and Isenberg, I., 1964, Phosphorescence in nucleotides and nucleic acids, *J. Chem. Phys.* **40**:3175–3180.

Beukers, R., 1965, The effect of proflavine on U.V.-induced dimerization of thymine in DNA, *Photochem. Photobiol.* **4**:935–937.

Beukers, R., and Berends, W., 1961, The effect of UV irradiation on nucleic acids and their components, *Biochim. Biophys. Acta* **49**:181–189.

Beukers, R., Ijlstra, J., and Berends, W., 1958, The effect of ultraviolet light on some components of the nucleic acid. II. In rapidly frozen solutions, *Rec. Trav. Chim. Pays-Bas* **77**:729–732.

Beukers, R., Ijlstra, J., and Berends, W., 1959, The effect of ultraviolet light on some components of the nucleic acid. III. Apurinic acid, *Rec. Trav. Chim. Pays-Bas* **78**:247–251.

Bevilacqua, R., and Bordin, F., 1973, Photo-C_4-cycloaddition of psoralen and pyrimidine bases: Effect of oxygen and paramagnetic ions, *Photochem. Photobiol.* **17**:191–194.

Bidet, R., Chambron, J., and Weill, G., 1970, Analyse quantitative des courbes de fusion des complexes DNA-proflavine obtenues par fluorescence, *Ann. Phys. Biol. Med.* **1**:1–30.

Bishop, J. M., Quintrell, N., and Koch, G., 1967, Poliovirus double-stranded RNA: Inactivation by ultraviolet light, *J. Mol. Biol.* **24**:125–128.

Bittmann, R., 1969, Studies of the binding of ethidium bromide to transfer ribonucleic acid: Absorption, fluorescence, ultracentrifugation and kinetic investigations. *J. Mol. Biol.* **46**:251–268.

Blackburn, G. M., and Davies, R. J. H., 1966*a*, The structure of uracil photo-dimer, *Tetrahedron Lett.* **37**:4471–4474.

Blackburn, G. M., and Davies, R. J. H., 1966*b*, The structure of thymine photo-dimer, *J. Chem. Soc. (C) Sect. A,* 2239–224.

Blum, H. F., 1941, *Photodynamic Action and Diseases Caused by Light,* Reinhard, New York.

Böhme, H., 1968, Absence of repair of photodynamically induced damage in two mutants of *Proteus mirabilis* with increased sensitivity to monofunctional alkylating agents. *Mutat. Res.* **6**:166–168.

Böhme, H., and Adler, B., 1972, Reparatur von DNA-Schäden, in: *Desoxyribonucleinsäure—Schlüssel des Lebens* (E. Geissler, ed.), pp. 134–150, Akademie Verlag, Berlin.

Böhme, H., and Wacker, A., 1963, Mutagenic activity of thiopyronine and methylene blue in combination with visible light, *Biochem. Biophys. Res. Commun.* **12**:137–139.

Bollum, F. J., and Setlow, R. B., 1963, Ultraviolet inactivation of DNA primer activity. I. Effects of different wavelengths and doses, *Biochim. Biophys. Acta* **68**:599–607.

Bonneau, R., Faure, J., and Joussot-Dubiene, J., 1967, Study of the kinetics and acid base properties of semireduced species in the photoreduction of thiazine dyes in aqueous solution, *Ber. Bunsenges. Phys. Chem.* **72**:263–266.

Bonneau, R., Pottier, R., Bagno, O., and Joussot-Dubien, J., 1975, pH dependence of singlet oxygen production in aqueous solutions using thiazine dyes as photosensitizers, *Photochem. Photobiol.* **21**:159–163.

Bordin, F., and Baccichetti, 1974, The furocoumarin photosensitizing effect on the virus-producing Graffi leukaema cells, *Z. Naturforsch.* **29c**:630–632.

Bordin, F., Baccichetti, F., and Musajo, L., 1973, Inhibition of nucleic acids synthesis in Ehrlich ascites tumor cells by irradiation *in vitro* in the presence of skin-photosensitizing furocoumarins, *Experientia* **29**:272–273.

Borisova, O. F., and Minyat, E. E., 1969, Complexes of deoxyribonucleoprotein with acridine orange dye, *Mol. Biol. (USSR)* **3**:758–767.

Borisova, O. F., and Tumerman, L. A., 1964, The luminescence of acridine orange and nucleic acid complexes, *Biophysika (USSR)* **9**:537–544.

Borisova, O. F., Potapov, A. P., Surovaya, A. N., Trubitsyn, S. N., and Volkenstein, M. V., 1973, Dependence of the fluorescence quantum yield of tRNA-acriflavine complexes on the structure of tRNA, *Mol. Biol. (USSR)* **4:**509–516.

Borisova, O. F., Razjivin, A. P., and Zaregorodzev, V. I., 1974, Evidence for the quinacrine fluorescence on three AT pairs of DNA, *FEBS Lett.* **46:**239–242.

Boyland, E., and Green, B., 1964, On the reported sedimentation of polycyclic hydrocarbons from aqueous solutions of DNA, *J. Mol. Biol.* **9:**589–597.

Boyle, R. E., Nelson, S. S., Dollish, F. R., and Olsen, M. J., 1962, The interaction of deoxyribonucleic acid and acridine orange *Arch. Biochem. Biophys.* **96:**47–50.

Brand, L., and Gohlke, J. R., 1972, Fluorescence probes for structure, *Annu. Rev. Biochem.* **41:**843–868.

Brendel, M., 1970, Different photodynamic action of proflavine and methylene blue on bacteriophage. *Mol. Gen. Genet.* **108:**303–311.

Brendel, M., and Kaplan, R. W., 1967, Photodynamische Mutations auslösung und Inaktivierung beim *Serratia* phagen K durch Methylenblau and Licht, *Mol. Gen. Genet.* **99:**181–190.

Brendel, M., and Winkler, U., 1966, Photodynamische Inaktivierung des *E. coli* phagen T4: Untersuchung der Kreuzungsreaktivierung und des Schutzeffektes von Spermin, *Z. Vererbungsl.* **98:**41–48.

Bridges, B. A., 1971, Genetic damage induced by 254 nm ultraviolet light in *Escherichia coli:* 8-Methoxypsoralen as protective agent and repair inhibitor, *Photochem. Photobiol.* **14:**659–662.

Bridges, B. A., and Munson, R. J., 1970, UV-mutagenesis, *Stud. Biophys.* **19:**49–57.

Brown, I. H., and Johns, H. E., 1967, Mathematical aspects of the ultraviolet photochemistry of pyrimidine dinucleoside phosphates, *Photochem. Photobiol.* **6:**469–483.

Brown, I. H., and Johns, H. E., 1968, Photochemistry of uracil. Intersystem crossing and dimerization in aqueous solution, *Photochem. Photobiol.* **8:**273–286.

Brown, I. H., Freeman, K. B., and Johns, H. E., 1966, Photochemistry of uridylyl-(3′ → 5′)-uridine, *J. Mol. Biol.* **15:**640–662.

Brunk, C. F., 1973, Distribution of dimers in ultraviolet-irradiated DNA, *Nature (London) New Biol.* **241:**74–76.

Bruton, C. J., and Harley, B. S., 1970, Chemical studies on methionyl-tRNA synthetase from *E. coli, J. Mol. Biol.* **52:**165–178.

Buc, M.-H., and Scott, J. F., 1966, Effects of ultraviolet light on the biological functions of transfer RNA, *Biochem. Biophys. Res. Commun.* **22:**459–465.

Burns, V. W. F., 1969, Fluorescence decay time characteristics of the complex between ethidium bromide and nucleic acid. *Arch. Biochem. Biophys.* **133:**420–424.

Burr, J. G., 1968, Advances in the photochemistry of nucleic acid derivatives, in: *Advances in Photochemistry,* Vol. 6 (W. A. Noyes, Jr., G. S. Hammond, and J. N. Pitts, Jr., eds.), pp. 193–299, Interscience, New York.

Burr, J. G., and Park, E. H., 1968a, Photochemical genetics. I. The ionic nature of uracil photohydration, *Adv. Chem. Ser.* **81:**418–434.

Burr, J. G., and Park, E. H., 1968b, Photochemical genetics. II. The kinetic role of water in the photohydration of uracil and 1,3-dimethyluracil, *Adv. Chem. Ser.* **81:**435–444.

Burr, J. G., Gordon, B. R., and Park, E. H., 1968, The mechanism of photohydration of uracil and *N*-substituted uracils, *Photochem. Photobiol.* **8:**73–78.

Burr, J. G., Park, E. H., and Chan, A., 1972, Nature of the reactive species in the photohydration of uracil and cytosine derivatives, *J. Am. Chem. Soc.* **94:**5866–5872.

Calberg-Bacq, C. M., and Van de Vorst, A., 1974, Induction of free radicals in DNA by proflavine and visible light: Influence of oxygen and ionic strength, *Photochem. Photobiol.* **20:**433–439.

Calberg-Bacq, C. M., Delmelle, M., and Duchesne, J., 1968, Inactivation and mutagenesis due to the photodynamic action of acridines and related dyes on extracellular bacteriophage T₄B, *Mutat. Res.* **6**:15–24.

Calendi, E., Di Marco, A., Reggiani, M., Scarpinato, B., and Valentini, L., 1965, On physico-chemical interactions between daunomycin and nucleic acids, *Biochim. Biophys. Acta* **103**:25–49.

Camerman, N., and Camerman, A., 1968, Photodimer of thymine in ultraviolet-irradiated DNA: Proof of structure by X-ray diffraction, *Science (Washington)* **160**:1451–1452.

Camerman, N., Weinblum, D., and Nyburg, S. C., 1969, The structure of *dl* photodimer C of 1,3-dimethylthymine, *J. Am. Chem. Soc.* **91**:982–986.

Campbell, J. M., Schulte-Frohlinde, D., and v. Sonntag, C., 1974, Quantum yields in the UV photolysis of 5-bromo-uracil in the presence of hydrogen donors, *Photochem. Photobiol.* **20**:465–467.

Caporale, G., Musajo, L., Rodighiero, G., and Baccichetti, F., 1967, Skin-photosensitizing activity of some methylpsoralens, *Experientia* **23**:985–986.

Carpenter, J. M., and Kleczkowski, A., 1969, The absence of photoreversible pyrimidine dimers in the RNA of UV-irradiated TMV, *Virology* **39**:542–548.

Carré, D. S., Thomas, G., and Favre, A., 1974, Conformation and functioning of tRNAs: Cross-linked tRNAs as substrate for tRNA nucleotidyl-transferase and aminoacyl synthetases, *Biochimie* **56**:1089–1101.

Caspersson, T., Farber, S., Foley, G. E., Kudynowski, J., Modest, E. J., Simonsson, E., Wagh, O., and Zech, L., 1968, Chemical differentiation along metaphase chromosomes, *Exp. Cell Res.* **49**:219–222.

Caspersson, T., Zech, L., and Lindsten, J., 1972, Die Identifikation menschlicher Chromosomen mit Hilfe der Fluoroeszenzmethode, *Triangel* **11**:73–80.

Cerutti, P. A., 1975, Excision repair of DNA base damage, *Life Sci.* **15**:1567–1575.

Cerutti, P., Miles, H. T., and Frazier, J., 1966, Interaction of partially reduced polyuridylic acid with polyadenylic acid, *Biochem. Biophys. Res. Commun.* **22**:466–472.

Chambers, R. W., Waits, H. P., and Freude, A., 1969, Photosensitized inactivation of alanine transfer RNA, *J. Am. Chem. Soc.* **91**:7203–7204.

Chan, L. M., and Van Winkle, Q., 1969, Interaction of acriflavine with DNA and RNA, *J. Mol. Biol.* **40**:491–495.

Chandra, P., 1972, Photodynamic action: A valuable tool in molecular biology, in: *Research Progress in Organic, Biological and Medicinal Chemistry* (U. Gallo and L. Santamaria, eds.), pp. 232–258, North-Holland, Amsterdam.

Chandra, P., and Wacker, A., 1968, Photodynamic effects on the template activity of nucleic acids, *Z. Naturforsch.* **211**:663–666.

Chandra, P., and Wacker, A., 1970, Structure and function of nucleic acids treated with dyes, furocoumarins and ketones in the presence of light, in: *Interaktionen bei Biopolymeren, Stud. Biophys.* **24/25**:437–446.

Chandra, P., Kraft, S., and Wacker, A., 1971a, Studies on the reactivation of bacteria photodamaged by psoralen, *Biophysik* **7**:251–258.

Chandra, P., Rodighiero, G., Dall'Acqua, F., Marciani, S., Kraft, S., and Wacker, A., 1971b, Studies on the reactivation of bacteria photodamaged by furocoumarins, *Stud. Biophys.* **29**:53–61.

Chandra, P., Wacker, A., Lisy, V., and Škoda, J., 1971c, Acetone-sensitized inactivation of guanylyl-uridylyl-uridine (GUU) in the binding of valyl-tRNA to ribosomes, *Biophysik* **7**:245–246.

Chandra, P., Biwas, R. K., Dall'Acqua, F., Marciani, S., Baccichetti, F., Vedaldi, D., and Rodighiero, C., 1973, Postirradiated dark recovery of photodamage to DNA induced by furocourmarins, *Biophysik* **9**:113–119.

Chandra, P., Dall'Acqua, F., Marciani, S., and Rodighiero, G., 1974, Studies on the repair of DNA photodamaged by furocoumarins, in: *Sunlight and Man. Normal and Abnormal Photobiologic Responses* (M. A. Pathak, L. C. Harber, M. Seiji, and A. Kukita, eds.), pp. 411–417, University of Tokyo Press.

Charlier, M., and Helénè, C., 1967, Photosensitized dimerization of orotic acid in aqueous solution, *Photochem. Photobiol.* **6**:501–504.

Charlier, M., and Helénè, C., 1975, Photosensitized splitting of pyrimidine dimers in DNA by indole derivatives and tryptophan-containing peptides, *Photochem. Photobiol.* **21**:31–37.

Chessin, M., 1960, Photodynamic inactivation of infectious nucleic acid, *Science* (*Washington*) **132**:1840–1841.

Cleaver, J. E., 1967, The relationship between the rate of DNA synthesis and its inhibition by ultraviolet light in mammalian cells. *Radiat. Res.* **30**:795–810.

Cleaver, J. E., 1971, Repair of damaged DNA in human and other eukaryotic cells, in: *Nucleic Acid–Protein Interactions–Nucleic Acid Synthesis in Viral Infection* (D W Ribbons, J. F. Woessner, and J. Schultz, eds.), pp. 87–112, North-Holland, Amsterdam.

Cleaver, J. E., 1973, DNA repair with purines and pyrimidines in radiation- and carcinogen-damaged normal and xeroderma pigmentosum human cells, *Cancer Res.* **33**:362–369.

Cleaver, J. E., and Trosko, J. E., 1970. Absence of excision of ultraviolet-induced cyclobutane dimers in xeroderma pigmentosum, *Photochem. Photobiol.* **11**:547–550.

Cohn, W. E., Leonard, N. J., and Wang, S. Y., 1974, Abbreviations for pyrimidine photoproducts, *Photochem. Photobiol.* **19**:89–94.

Cole, R. S., 1970, Psoralen monoadducts and interstrand cross-links in DNA, *Biochim. Biophys. Acta* **254**:30–39.

Cole, R. S., 1973, Repair of DNA containing interstrand cross-links in *Escherichia coli:* Sequential excision and recombination, *Proc. Natl. Acad. Sci. USA* **70**:1064–1068.

Colombo, G., 1967, Phosensitization of sea-urchin sperm to long-wave ultraviolet light by psoralen, *Exp. Cell Res.* **48**:167–169.

Colombo, G., Levis, A. G., and Torlone, V., 1965, Photosensitization of mammalian cells and of animal viruses by furocoumarins, *Prog. Biochem. Pharmacol.* **1**:392–399.

Comings, D. E., Avelino, E., Okada, T. A., and Wyandt, H. E., 1973, The mechanism of C- and G-banding of chromosomes, *Exp. Cell Res.* **77**:469–493.

Conolly, J. S., and Linschitz, H., 1968, Photoaddition of alcohols to purine, *Photochem. Photobiol.* **7**:791–806.

Cramer, W. A., and Uretz, R. B., 1966a, Acridine orange sensitized photoinactivation of T4 bacteriophage. I. Parameter affecting phage sensitivity to visible light, *Virology* **29**:462–468.

Cramer, W. A., and Uretz, R. B., 1966b, Acridine orange sensitized photoinactivation of T4 bacteriophage. II. Genetic studies with photoinactivated phage, *Virology* **29**:469–479.

Cusachs, L. C., and Steele, R. H., 1967, Singlet oxygen molecules and carcinogenic aromatic hydrocarbons, *Int. J. Quantum Chem. Symp.* **1**:175–183.

Dall'Acqua, F., Terbojevich, M., and Benvenuto, F., 1968, Light-scattering and viscosimetric studies on DNA after the photoreaction with some furocoumarins, *Z. Naturforsch.* **24b**:667–671.

Dall'Acqua, F., Marciani, S., and Rodighiero, G., 1969, The action of xanthotoxin and bergapten for the photoreaction with native DNA, *Z. Naturforsch.* **24b**:667–671.

Dall'Acqua, F., Marciani, S., and Rodighiero, G., 1970, Interstrand cross-linkage occuring in the photoreaction between psoralen and DNA, *FEBS Lett.* **9**:121–123.

Dall'Acqua, F., Marciani, S., Ciavatta, L., and Rodighiero, G., 1971, Formation of interstrand cross-linkings in the photoreactions between furocoumarins and DNA, *Z. Naturforsch.* **26b**:561–569.

Dall'Acqua, F., Marciani, S., Vedaldi, D., and Rodighiero, G., 1972, Formation of interstrand cross-linkings on DNA of guinea pig skin after application of psoralen and irradiation of 365 nm, *FEBS Lett.* **27**:192–194.

Dall'Acqua, F., Marciani, S., Vedaldi, D., and Rodighiero, G., 1974a, Skin photosensitization and cross-linkings formation in native DNA by furocoumarins, *Z. Naturforsch.* **29c**:635–636.

Dall'Acqua, F., Marciani, S., Vedaldi, D., and Rodighiero, G., 1974b, Studies on the photoreactions (365 nm) between DNA and some methylpsoralens, *Biochim. Biophys. Acta* **353**:267–273.

Danilov, V. J., Kruglyak, Y. A., Kuprievich, V. A., and Ogloblin, V. V., 1969, Electronic aspects of photodimerization of the pyrimidine bases and of their derivatives, *Theor. Chim. Acta (Berlin)* **14**:242–249.

Danziger, R. M., Hayon, E., and Langmuir, M. E., 1968, Pulse-radiolysis and flash-photolysis study of aqueous solutions of simple pyrimidines. Uracil and bromouracil, *J. Phys. Chem.* **72**:3842–3849.

Davis, S. L., and Tinoco, I., Jr., 1966, Ultra-violet absorption spectrum of thymine in ice, *Nature (London)* **210**:1286–1286.

Debey, P., and Douzon, P., 1970, Photodynamic action and singlet oxygen, *Isr. J. Chem.* **8**:115–123.

De Boer, G., and Johns, H. E., 1970, Hydrogen exchange in photohydrates of cytosine derivatives, *Biochim. Biophys. Acta* **204**:18–30.

De Boer, G., Pearson, M., and Johns, H. E., 1967, Ultraviolet photoproducts in ordered structures of poly U and their effects on secondary structure, *J. Mol. Biol.* **27**:131–144.

De la Chapelle, A., Schröder, J., and Selander, R. K., 1973, *In situ* localization and characterization of different classes of chromosomal DNA: Acridine orange and quinacrine mustard fluorescence, *Chromosoma (Berlin)* **40**:347–360.

Dellweg, H., and Wacker, A., 1962, Strahlenchemische Veränderungen von Thymin und Cytosin in der DNA durch UV-Licht, *Z. Naturforsch.* **17b**:827–834.

Dellweg, H., and Wacker, A., 1964, Strahlenchemische Veränderungen der 5-Halogen-Uracile in der DNS durch UV-Strahlen, *Z. Naturforsch.* **19b**:305–311.

Dellweg, H., and Wacker, A., 1965, Über die UV-Bestrahlung von Desoxyribonucleinsäure in Glykol, *Z. Naturforsch.* **20b**:141–143.

Dellweg, H., Jacherts, D., Weinblum, D., and Wacker, A., 1964, Die UV-strahlensensibilisierte Wirkung der 5-Halogenuracile, *Biophysik* **1**:391–395.

Delmelle, M., and Duchesne, J., 1967, Electron spin resonance study of photosensitization of deoxyribonucleic acid and its constituents by acridine dyes, in: *Molekulare Mechanismen photodynamischer Effekte, Stud. Biophys.* **3**:121–126.

Delmelle, M., and Duchesne, J., 1968, Effect of light on dyes and photodynamic action on biomolecules, in: *Molecular Associations in Biology* (B. Pullman, ed.), pp. 299–308, Academic Press, New York.

Delmelle, M., Depireux, J., and Duchesne, J., 1966, Aspects quantitatifs de la production de radicaux libres induits par le rayonnement visible dans quelques substances photomutagenes, *C. R. Acad. Sci. (Paris) Ser. D* **263**:1625–1627.

Distèche, C., and Bontemps, J., 1974, Chromosome regions containing DNAs of known base composition, specifically evidenced by 2,7-di-t-butyl-proflavine, *Chromosoma (Berlin)* **47**:263–281.

Dönges, K. H., and Fahr, E., 1966, Die Struktur des bei der UV-Bestrahlung von Uracil entstehenden dimeren Uracils, *Z. Naturforsch.* **21b**:87–87.

Dörhöfer, G., and Fahr, E., 1966, Die Synthese von trans-Dimeren Uracilen, *Tetrahedron Lett.* **37**:4511–4516.

Doskočil, J., 1968, Pusobeni 5-azacytidinu pri bakteriofagovych infekcich, *Biol. Listy* **33**:289–303.

Doskočil, J., and Šorm, F., 1970, The effect of 5-azacytidine and 5-azauridine on protein synthesis, *Biochem. Biophys. Res. Commun.* **38**:569–574.

Doskočil, J., and Šorm, F., 1971a, The effects of 5-azacytidine and 5-azauridine on protein synthesis in *Escherichia coli, Biochem. Biophys. Res. Commun.* **38**:569–574.

Doskočil, J., and Šorm, F., 1971b, Differential incorporation of 5-azapyrimidines into the RNA of phage f2 and of bacterial host. *Eur. J. Biochem.* **23**:253–261.

Drake, J. W., and McGuire, J., 1967, Properties of *r* mutants of bacteriophage T4 photodynamically induced in the presence of thiopyronin and psoralen, *J. Virol.* **1**:260–267.

Eisinger, J., and Lamola, A. A., 1967, The excited state precursor of the thymine dimer, *Biochem. Biophys. Res. Commun.* **28**:558–565.

Eisinger, J., and Shulman, R. G., 1967, The precursor of the thymine dimer in ice, *Proc. Natl. Acad. Sci. USA* **58**:895–900.

Elad, D., and Rosenthal, J., 1969, Photochemical alkylation of caffeine with amino-acids, *Chem. Commun.*, 905–908.

Elad, D., and Salomon, J., 1971, Ultraviolet- and radiation-induced reactions of caffeine with amines, *Tetrahedron Lett.* **50**:4783–4784.

Ellerton, N. F., and Isenberg, I., 1969, Fluorescence polarization study of DNA-proflavine complexes, *Biopolymers* **8**:767–786.

Elpiner, I. E., and Shebaldina, A. D., 1967, Action of ultrasonic waves on photodynamic effect, in: *Molekulare Mechanismen photodynamischer Effekte, Stud. Biophys.* **3**:197–203.

Erikson, R. L., and Szybalski, W., 1963, Molecular radiobiology of human cell lines. IV. Variation in ultraviolet light and X-ray sensitivity during the division cycle, *Radiat. Res.* **18**:200–212.

Evans, B., and Wolfenden, R., 1970, A potential transition state analog for adenosine deaminase, *J. Am. Chem. Soc.* **92**:4751–4752.

Faddejeva, M. D., 1969, The correlation between the increase of melting temperature of DNA through the interaction of DNA with basic dyes and their inhibitory action on the reaction of enzymatic hydrolysis of DNA by DNAse I, *Zytologija* (*USSR*) **11**:225–233.

Fahr, E., 1969, Chemische Untersuchungen über die molekularen Ursachen biologischer Strahlenschäden, *Angew. Chem.* **81**:581–597.

Fahr, E., 1970, Die molekularen Ursachen biologischer Strahlenschäden, *Stud. Biophys.* **19**:1–20.

Fahr, E., Kleber, R., and Boebinger, E., 1966, Untersuchung über die photochemische Addition von Wasser an Cytosin, Cytidin, Cytidylsäure und Thymin, *Z. Naturforsch.* **21b**:214–223.

Fahr, E., Fürst, G., Maul, P., and Wieser, H., 1972a, Die UV-Dimerisation von 1,3-Dimethyluracil in der Eismatrix, *Z. Naturforsch.* **27b**:1475–1480.

Fahr, E., Maul, P., Lehner, K.-A., and Scheutzow, D., 1972b, Die ^{1}H-NMR-spektroskopische Untersuchung der Struktur der dimeren 1,3-Dimethyluracile, *Z. Naturforsch.* **27b**:1481–1484.

Fahr, E., Pastille, R., Pelz, N., and Scheutzow, D., 1974, Die NMR-spektroskopische Strukturaufklärung des bei der UV-Bestrahlung von Thymin/Uracil-Gemischen in der Eismatrix entstehenden Thymin/Uracil-Mischdimeren, *Z. Naturforsch.* **29b**:410–413.

Favre, A., 1974, Luminescence and photochemistry of 4-thiouridine in aqueous solution, *Photochem. Photobiol.* **19**:15–19.

Favre, A., and Fourrey, J.-L., 1974, Intramolecular cross-linking of single-stranded copolymers of 4-thiouridine and cytidine, *Biochim. Biophys. Res. Commun.* **58**:507–515.

Favre, A., and Yaniv, M., 1971, Introduction of an intramolecular fluorescent probe in *E. coli* tRNA$_I^{Val}$, *FEBS Lett.* **17**:236–240.

Favre, A., Yaniv, M., and Michelson, A. M., 1969, The photochemistry of 4-thiouridine in *E. coli* tRNA$_I^{Val}$, *Biochem. Biophys. Res. Commun.* **37**:266–271.

Favre, A., Michelson, A. M., and Yaniv, M., 1971, Photochemistry of 4-thiouridine in *E. coli* transfer RNA$_I^{Val}$, *J. Mol. Biol.* **58**:367–379.

Favre, A., Roques, B., and Fourrey, J.-L., 1972, Chemical structures of the TU-C and TU-C^{red} products derived from *E. coli* tRNA, *FEBS Lett.* **24**:209–214.

Fenselau, C., and Wang, S. Y., 1969, Mass spectra of some dimeric photoproducts of pyrimidines, *Tetrahedron* **25**:2853–2863.

Fenster, A., and Johns, H. E., 1973, Temperature studies for quenching of pyrimidine triplet states, *J. Phys. Chem.* **77**:2246–2249.

Festy, B., and Daune, M., 1973, Hydroxystilbamidine. A nonintercalating drug as a probe of nucleic acid conformation, *Biochemistry* **12**:4827–4834.

Fikus, M., and Shugar, D., 1966, Alkaline transformations of the photohydrates of some 2,4-diketopyrimidines and their glycosides, *Acta Biochim. Pol.* **13**:39–56.

Fikus, M., Wierzchowski, K. L., and Shugar, D., 1965, Photochemistry of 5-fluorouracil analogues, glycosides and poly FU, *Photochem. Photobiol.* **4**:521–536.

Fisher, G. J., and Johns, H. E., 1973, Thymine hydrate formed by ultraviolet and gamma irradiation of aqueous solutions, *Photochem. Photobiol.* **18**:23–27.

Fisher, G. J., Varghese, A. J., and Johns, H. E., 1974, Ultraviolet induced reactions of thymine and uracil in the presence of cysteine, *Photochem. Photobiol.* **20**:109–120.

Flippen, J. L., Gilardi, R. D., Karle, I. L., Rhoades, D. F., and Wang, S. Y., 1971, Crystal and molecular structure of a pyrimidine phototetramer, *J. Am. Chem. Soc.* **93**:2556–2557.

Foote, C. S., 1968, Mechanisms of photosensitized oxidation, *Science (Washington)* **162**:963–976.

Foote, C. S., and Denny, R. W., 1968, Chemistry of singlet oxygen. VII. Quenching by β-carotene, *J. Am. Chem. Soc.* **90**:6233–6237.

Foote, C. S., Chang, Y. C., and Denny, R. W., 1970, Chemistry of singlet oxygen. VI. *cis-trans* Isomerization of carotenoids by singlet oxygen and a probable quenching mechanism, *J. Am. Chem. Soc.* **92**:5218–5226.

Fowlks, W. L., Griffith, D. G., and Oginsky, E. L., 1958, Photosensitization of bacteria by furocourmarins and related compounds, *Nature (London)* **181**:571–572.

Fraser, D., and Mahler, H. R., 1961, Studies in partially resolved bacteriophage-host systems. VII. Diamines, dyes, empty phage heads, and protoplast infecting agent, *Biochim. Biophys. Acta* **53**:199–213.

Fredericq, E., and Houssier, C., 1972, Study of the interaction of DNA and acridine orange by various optical methods, *Biopolymers* **11**:2281–2308.

Freeman, K. B., Hariharan, P. V., and Johns, H. E., 1965, The ultraviolet photochemistry of cytidylyl-(3′-5′)-cytidine, *J. Mol. Biol.* **13**:833–848.

Freifelder, D., and Uretz, R. B., 1966, Mechanism of photoinactivation of coliphage T7 sensitized by acridine orange, *Virology* **30**:97–103.

Freifelder, D., Davison, P. F., and Geiduschek, E. P., 1961, Damage by visible light to the acridine orange-DNA complex, *Biophys. J.* **1**:389–400.

Friedberg, E. C., 1975, DNA repair of ultraviolet-irradiated bacteriophage T4, *Photochem. Photobiol.* **21**:277–289.

Fritzsche, H., Lang, H., and Pohle, W., 1976, Evidence for B-C transition in ultraviolet-irradiated DNA. An infrared linear dichroism study, *Biochim. Biophys. Acta* **432**:409–412.

Füchtbauer, W., and Mazur, P., 1966, Kinetics of the ultraviolet-induced dimerization of thymine in frozen solutions, *Photochem. Photobiol.* **5**:323–335.

Fujita, H., and Yamazaki, H., 1970, The photosensitized reaction of deoxyguanosine in the presence of methylene blue, *Bull. Chem. Soc. Jpn.* **43**:1177–1181.

Galley, W. C., 1968, On the triplet state of polynucleotide-acridine orange-complexes. I. Triplet energy delocalization in the complex 9-aminoacridine DNA-complex, *Biopolymers* **6**:1279–1296.

Gattner, H., and Fahr, E., 1963, Darstellung des bei der UV-Bestrahlung wässriger Uracil-Lösungen eststehenden 4-hydroxy-dihydrouracils, *Liebigs Ann. Chem.* **670**:84–87.

Gauri, K. K., Pflughaupt, K. W., and Müller, R., 1969, Synthese und photochemische Eigenschaften von 1′-(2′-Desoxy-β-D-ribofuranosyl)-(4-³H)-5-äthyluracil, *Z. Naturforsch.* **24b**:833–836.

Gauri, K. K., Rüger, W., and Wacker, A., 1971, Photochemistry and photobiology of 5-ethyl- and 5-propyldeoxyuridine, *Z. Naturforsch.* **26b**:167–168.

Geissler, E., 1967, Untersuchungen über die Irreparabilität photodynamischer Schäden, *Stud. Biophys.* **2**:95–102.

Geissler, E., 1968, Reactivation of photodynamically inactivated lambda phages, *Mol. Gen. Genet.* **103**:233–237.

Geissler, E., 1970, Wirtszellenreaktivierung photodynamisch und U.V. geschädigter Bakteriophagen, *Stud. Biophys.* **19**:163–170.

Geissler, E., and Wacker, A., 1963, Untersuchungen über den Mechanismus der Induktion, VI. Die Induktion lyosgener Bakterien durch Belichtung in Gegenwart von Thiopyronin, *Acta Biol. Med. Ger.* **11**:937–942.

Giese, A. C., 1968, Ultraviolet action spectra in perspective: With special reference to mutation, *Photochem. Photobiol.* **8**:527–546.

Gilbert, E., and Cristallini, C., 1973, UV-Photolyse von 5-Bromuracil in wäβriger Lösung, *Z. Naturforsch.* **28b**:615–619.

Gilbert, E., Wagner, G., and Schulte-Frohlinde, D., 1971, Photolyse von 5-Joduracil in wäβriger, sauerstoffgesättigter Lösung in Gegenwart von Methanol, *Z. Naturforsch.* **26b**:209–213.

Gill, J. E., 1970, Fluorescence of 5-methylcytosine, *Photochem. Photobiol.* **11**:259–269.

Gill, J. E., 1971, Fluorescence of synthetic DNAs at room temperature and neutral pH, *Biochem. Biophys. Res. Commun.* **44**:779–785.

Gill, J. E., Marzimas, J. A., and Bishop, C., 1974, Physical studies on synthetic DNAs containing 5-methylcytosine, *Biochim. Biophys. Acta* **335**:330–348.

Glisin, V. R., and Doty, P., 1967, The cross-linking of DNA by ultraviolet radiation, *Biochim. Biophys. Acta* **142**:314–322.

Gollmick, F. A., and Berg, H., 1968, Photosensibilisierte Photooxydation durch Methylenblau, Thiopyronin und Pyronin. I. Mitteilung: Flash-Photooxidation von *p*-Diaminotoluol, *Photochem. Photobiol.* **7**:471–475.

Gollmick, F. A., and Berg, H., 1972, Sensibilisierte Photooxydation durch Methylenblau, Thiopyronin und Pyronin, III. Mitteilung: Uber den Mechanismus der photosensibilisierten Oxydation des Guanine durch Thiopyronin, *Photochem. Photobiol.* **16**:447–453.

Gollnick, K., 1968, Type II photooxygenation reactions in solution, *Adv. Photochem.* **6**:1–39.

Gollnick, K., and Schenck, G. O., 1964, Mechanism and stereoselectivity of photosensitized oxygen transfer reactions, *Pure Appl. Chem.* **9**:507–525.

Gorelic, L. S., Lisagor, P., and Yang, N. C., 1972, The photochemical reactions of 1,3-dimethyluracil with 1-aminopropane and poly-1-lysine, *Photochem. Photobiol.* **16**:465–480.

Grahn, D., 1972, Genetic effects of low level irradiation, *Biol. Sci.,* 535–540.

Gräslund, A., Rigler, R., and Ehrenberg, A., 1969, Light-induced free radicals in DNA-acridine complexes studied by ESR, *FEBS Lett.* **4**:227–230.

Gräslund, A., Rupprecht, A., and Ström, G., 1975, Light-induced free radicals in oriented DNA-proflavine complexes, *Photochem. Photobiol.* **21**:153–157.

Greenstock, C. L., and Johns, H. E., 1968, Photosensitized dimerization of pyrimidines, *Biochem. Biophys. Res. Commun.* **30**:21–27.

Greenstock, C. L., Brown, J. H., Hunt, J. W., and Johns, H. E., 1967, Photodimerization of pyrimidine nucleic acid derivatives in aqueous solution and the effect of oxygen, *Biochem. Biophys. Res. Commun.* **27**:431–436.

Grossman, L., 1963, The effects of UV-irradiated poly U in cell-free protein synthesis in *E. coli* II, *Proc. Natl. Acad. Sci. USA* **50**:657–664.

Grossweiner, L. J., 1969, Molecular mechanisms in photodynamic action, *Photochem. Photobiol.* **10**:183–191.

Grossweiner, L. J., and Kepka, A. C., 1972, Photosensitization in biopolymers, *Photochem. Photobiol.* **16**:305–314.

Grünberger, D., and Sorm, F., 1963, Relationship between 8-azaguanine-containing ribonucleic acid and protein synthesis in *B. cereus, Collect. Czech. Chem. Commun.* **28**:1044–1051.

Günther, H., and Prusoff, W. H., 1962, Decrease of sensitivity to ultraviolet radiation of *Streptococcus faecalis* grown in media supplemented with 6-azathymine, an analog of thymine, *Biochim. Biophys. Acta* **55**:778–780.

Günther, H. L., and Prusoff, W. H., 1967, Protective effect of 6-azathymine and 6-azauracil against ultraviolet irradiation, *Biochim. Biophys. Acta* **142**:304–312.

Guschelbauer, W., Favre, A., and Michelson, A. M., 1965, Photochemistry of polynucleotides. I. Ultraviolet photolysis of substituted pyrimidines, *Z. Naturforsch.* **20b**:1141–1145.

Hackett, A. J., 1962, The photodynamic effects of acridine orange on a RNA virus (vesicular exanthema), *Photochem. Photobiol.* **1**:147–154.

Hahn, F. E., and Krey, A. K., 1971, Interactions of alkaloids with DNA, in: *Progress in Molecular and Subcellular Biology,* Vol. 2 (F. E. Hahn, ed.), pp. 134–151, Springer Verlag, Berlin.

Hanawalt, P. C., 1972, Repair of genetic material in living cells, *Endeavour* **31**:83–87.

Hariharan, P. V., and Johns, H. E., 1967, Photochemical cross sections in cytidylyl-(3′-5′)-cytidine, *Can J. Biochem.* **46**:911–918.

Hariharan, P. V., and Johns, H. E., 1968a, Rate constants for the dehydration of single and double hydrates of cytidylyl-(3′-5′)-cytidine, *Photochem. Photobiol.* **7**:239–252.

Hariharan, P. V., and Johns, H. E., 1968b, Dimer photoproducts in cytidylyl-(3′-5′)-cytidine, *Photochem. Photobiol.* **8**:11–22.

Harm, W., 1970, Reparatur von Ultraviolett-Schäden in der Erbsubstanz, *Umsch. Wiss. Tech.* **70**:469–472.

Harriman, P. D., and Zachau, H. G., 1966, Ultraviolet inactivation of transfer ribonucleic acid functions, *J. Mol. Biol.* **16**:387–403.

Haug, A., 1964, Photochemical decomposition of TpBU, *Z. Naturforsch.* **19b**:143–147.

Haug, A., and Douzou, P., 1965, Electron paramagnetic resonance and phosphorescence measurements of the triplet state of orotic acid and related pyrimidines, *Z. Naturforsch.* **20b**:509–512.

Hauswirth, W., and Wang, S. Y., 1973, Pyrimidine adduct fluorescence in UV-irradiated nucleic acids, *Biochem. Biophys. Res. Commun.* **51**:819–826.

Hauswirth, W., Hahn, B. S., and Wang, S. Y., 1972, Spontaneous and light induced hydration of pyrimidines, *Biochem. Biophys. Res. Commun.* **48**:1614–1621.

Hayes, F. N., Williams, D. L., Ratliff, R. L., Varghese, A. J., and Rupert, C. S., 1971, Effect of a single thymine photodimer on the oligodeoxythymidylate-polydeoxyadenylate interaction, *J. Am. Chem. Soc.* **93**:4940–4942.

Heléne, C., and Charlier, M., 1971*a*, Photosensitized splitting of pyrimidine dimers by indole derivatives, *Biochem. Biophys. Res. Commun.* **43**:252–257.

Heléne, C., and Charlier, M., 1971*b*, Photosensitized reactions in nucleic acids. Photosensitized formation and splitting of pyrimidine dimers, *Biochimie* **53**:1175–1180.

Heléne, C., Santus, R., and Douzou, P., 1966*a*, Photoionisation and biphotonic processes of nucleic acids derivatives in frozen solutions, *Photochem. Photobiol.* **5**:127–133.

Heléne, C., Santus, R., and Michelson, A. M., 1966*b*, Energy transfer in dinucleotides, *Proc. Natl. Acad. Sci. USA* **55**:376–381.

Heléne, C., Brun, F., and Yaniv, M., 1969, Fluorescence study of interactions between valyl-tRNA synthetase and valine-specific tRNAs from *E. coli, Biochim. Biophys. Res. Commun.* **37**:393–398.

Heléne, C., Montenay-Garestier, T., and Dimicoli, J. L., 1971, Interactions of tyrosine and tyramine with nucleic acids and their components. Fluorescence, nuclear magnetic resonance and circular dichroism studies, *Biochim. Biophys. Acta* **254**:349–365.

Helleiner, C. W., Pearson, M. L., and Johns, H. E., 1963, The ultraviolet photochemistry of deoxyuridylyl-(3′ → 5′) deoxyuridine, *Proc. Natl. Acad. Sci. USA* **50**:761–767.

Henry, B. R., and Hunt, R. V., 1971, Triplet–triplet absorption studies on coumarin and related molecules, *J. Mol. Spectros.* **39**:466–470.

Herbert, M. A., and Johns, H. E., 1971, Flash photolysis studies of orotic acid, *Photochem. Photobiol.* **14**:693–704.

Herbert, M. A., Hunt, J. W., and Johns, H. E., 1968, Detection of the triplet state in orotic acid by flash photolysis, *Biochem. Biophys. Res. Commun.* **33**:643–648.

Herbert, M. A., Le Blanc, J. C., Weinblum, D., and Johns, H. E., 1969, Properties of thymine dimers, *Photochem. Photobiol.* **9**:33–43.

Hessler, A. Y., 1965, Acridine resistance in bacteriophage T2H as a function of dye penetration measured by mutagenesis and photoinactivation, *Genetics* **52**:711–722.

Hewitt, R., Billen, D., and Jorgensen, G., 1967, Radiation-induced reorientation of chromosome replication sequence: Generality in *E. coli.* Independence of prophage or 5-bromouracil toxity, *Radiat. Res.* **32**:214–226.

Hiatt, C. W., 1967*a*, Inactivation of viruses by photodynamic action, in: *Molekulare Mechanismen photodynamischer Effekte, Stud. Biophys.* **3**:157–164.

Hiatt, C. W., 1967*b*, Kinetics of virus inactivation by photodynamic action, in: *Radiation Research* (G. Silini, ed.), pp. 857–868, North-Holland, Amsterdam.

Hill, R. F., 1965, Ultraviolet-induced lethality and reversion to prototrophy in *E. coli* strains with normal and reduced dark repair ability, *Photochem. Photobiol.* **4**:563–568.

Hilwig, I., and Gropp, A., 1972, Staining of constitutive heterochromatin in mammalian chromosomes with a new fluorochrome, *Exp. Cell Res.* **75**:122–126.

Hollander, D. H., and Borgaonkar, D. S., 1971, The quinacrine fluorescence method of Y-chromosome identification, *Acta Cytol.* **15**:452–454.

Horrii, Z. I., and Suzuki, K., 1968, Degradation of the DNA of *E. coli K 12 rec⁻* (JC 1569b) after irradiation with ultraviolet light, *Photochem. Photobiol.* **8**:93–105.

Horii, Z. I., and Suzuki, K., 1970, Degradation of the DNA of rec A mutants of *E. coli K 12* after irradiation with ultraviolet light. II. Further studies including a *rec A UVr A* double mutant, *Photochem. Photobiol.* **11**:99–107.

Horvath, L., Matolcsy, G., and Pozsar, B. J., 1969, Incorporation of radiocarbon labelled uracil- and thymine-analogues into the DNA of bean leaf tissues, *Acta Bot. Acad. Sci. Hung.* **15**:79–80.

Hosszu, J. L., and Rahn, R. O., 1967, Thymine dimer formation in DNA between 25° and 100°C, *Biochim. Biophys. Res. Commun.* **29**:327–330.

Howard, B. D., and Tessman, I., 1964, Identification of the altered bases in mutated single-stranded DNA. III. Mutagenesis by ultraviolet light, *J. Mol. Biol.* **9**:372–375.

Howard-Flanders, P., and Boyce, R. B., 1966, DNA repair and genetic recombination: Studies on mutants of *Escherichia coli* defective in these processes, *Radiat. Res. Suppl.* **6**:156–184.

Hunter, J., Burk, D., and Woods, M., 1967, Effects of light with acridine and thiazine dyes on aerobic and anaerobic glucose metabolism of *Ehrlich* cells, in: *Molekulare Mechanismen photodynamischer Effekte, Stud. Biophys.* **3**:211–224.

Hutchison, F., 1973, The lesions produced by ultraviolet light in DNA containing 5-bromouracil, *Q. Rev. Biophys.* **6**:201–246.

Ichimura, S., Zama, M., and Fujita, H., 1971, Quantitative determination of single-stranded sections in DNA using the fluorescent probe acridine orange, *Biochim. Biophys. Acta* **240**:485–495.

Igali, S., Bridges, B. A., Ashwood-Smith, M. J., and Scott, B. R., 1970, Mutagenesis in *Escherichia coli*. Photosensitization to near ultraviolet light by 8-methoxypsoralen, *Mutat. Res.* **9**:21–30.

Isenberg, I., Leslie, R. B., Baird, S. L., Jr., Rosenbluth, R., and Bersohn, R., 1964, Delayed fluorescence in DNA-acridine dye-complexes, *Proc. Natl. Acad. Sci. USA* **52**:379–387.

Ishihara, H., and Wang, S. Y., 1966*a*, Photochemistry of 5-bromouracil in aqueous solution, *Biochemistry* **5**:2307–2313.

Ishihara, H., and Wang, S. Y., 1966*b*, Photochemistry of 5-bromouracil: Isolation of 5,5′-diuracil, *Nature (London)* **210**:1222–1225.

Ito, T., and Kobayashi, K., 1974, *In vivo* evidence for the participation of singlet excited oxygen molecules in the photodynamic inactivation, *Sci. Pap. Coll. Gen. Educ. Univ. Tokyo* **24**:33–36.

Ito, T., Yamasaki, T., and Ischizaka, S., 1967, Photoinactivation of acridine-sensitized yeast cells, *Sci. Pap. Coll. Gen. Educ. Univ. Tokyo* **17**:35–42.

Jacob, H.-E., 1971, *In vivo* production of DNA single-strand breaks by photodynamic action, *Photochem. Photobiol.* **14**:743–745.

Jacob, H.-E., and Kittler, L., 1970, Kultivierung von Bakterien mit 8-Azaadenine und sein Einfluß auf die UV-Empfindlichkeit, *Stud. Biophys.* **19**:123–129.

Jacob, H.-E., Berg, H., and Fliess, F.-R., 1967, Die Wirkung von Photodynamika-Mischungen auf Mikroorganismen, in: *Molekulare Mechanismen photodynamischer Effekte, Stud. Biophys.* **3**:189–196.

Jagger, J., and Stafford, R. S., 1965, Evidence for two mechanisms of photoreactivation in *E. coli B., Biophys. J.* **5**:75–88.

Jagger, J., Takebe, H., and Snow, J. M., 1970, Photoreactivation of killing in streptomyces: Action spectra and kinetic studies, *Photochem. Photobiol.* **12**:185–196.

Jakubetz, W., Lischka, H., and Polansky, O. E., 1973, personal communication.

Janovská, E., and Pillich, J., 1968, Inactivation of the phage of *Staphylococcus aureus* with acridine orange, *Int. J. Radiat. Biol.* **14**:59–65.

Jellinek, T., and Johns, R. B., 1970, The mechanism of photochemical addition of cysteine to uracil and formation of dihydrouracil, *Photochem. Photobiol.* **11**:349–359.

Jennings, B. H., Pastra, S.-C., and Wellington, J. L., 1970, Photosensitized dimerization of thymine, *Photochem. Photobiol.* **11**:215–226.

Jennings, B. H., Pastra-Landis, L., and Lerman, J. W., 1972, Photosensitized dimerization of uracil, *Photochem. Photobiol.* **15**:479–491.

Johns, H. E., 1966, Photoproducts produced in nucleic acids by ultraviolet light, *Radiat. Res.* 733–755.

Johns, H. E., 1968, Intersystem crossing and dimerization in aqueous solutions of uracil and orotic acid, *Photochem. Photobiol.* **7**:633–636.

Johns, H. E., 1971, in: *Creation and Detection of Excited State,* Vol. I (A. Lamola, ed.), pp. 123–172, Marcel Dekker, New York.

Johns, H. E., Pearson, M. L., Le Blanc, J. C., and Helleiner, C. W., 1964, The ultraviolet photochemistry of thymidylyl-(3′-5′) thymidine, *J. Mol. Biol.* **9**:503–524.

Johns, H. E., Le Blanc, J. C., and Freeman, D. K., 1965, Reversal and deamination rates of the main ultraviolet photoproduct of cytidylic acid, *J. Mol. Biol.* **13**:849–861.

Johns, H. E., Pearson, M., and Brown, I. H., 1966, Mathematical aspects of the ultraviolet photochemistry of poly U, *J. Mol. Biol.* **20**:231–243.

Jones, T. C., and Dove, W. F., 1972, Photosensitization of transcription by bromodeoxyuridine substitution, *J. Mol. Biol.* **64**:409–416.

Jungstand, W., and Berg, H., 1967, In-vivo-Versuche sur Cytostase durch photodynamische Effekte von Redoxfarbstoffen, in: *Molekulare Mechanismen photodynamischer Effekte, Stud. Biophys.* **3**:225–231.

Kahn, M., 1974, The effect of thymine dimers on DNA: DNA hybridization, *Biopolymers* **13**:669–675.

Kalab, D., 1967, Photodynamic effect on *Bacillus subtilis* bacteriophage, in: *Molekulare Mechanismen photodynamischer Effekte, Stud. Biophys.* **3**:181–186.

Kalousek, F., Raška, K., Jurovčik, M., and Šorm, F., 1966, Effect of 5-azacytidine on the acceptor activity of sRNA, *Collect. Czech. Chem. Commun.* **31**:1421–1424.

Kaplan, R. W., 1950, Auslösung von Phagenresistenzmutationen bei *B. coli* durch Erythrosyn mit und ohne Belichtung, *Naturwissenschaften* **37**:308–308.

Karle, I. L., Wang, S. Y., and Varghese, A. J., 1969, Crystal and molecular structure of a thymine-thymine adduct, *Science (Washington)* **164**:183–184.

Kearns, D. R., 1971, Physical and chemical properties of singlet molecular oxygen, *Chem. Rev.* **71**:395–427.

Kepka, A. C., and Grossweiner, L. J., 1971, Photodynamic oxidation of iodide ion and aromatic amino acids by eosin, *Photochem. Photobiol.* **14**:621–639.

Khan, A. V., and Kasha, M., 1970, An optical-residue singlet-oxygen theory of photocarcinogenicity, *Ann. N.Y. Acad. Sci.* **171**:24–32.

Khattak, M. N., and Wang, S. Y., 1969, Uracil photoproducts from uracil irradiated in ice, *Science (Washington)* **163**:1341–1342.

Khattak, M. N., and Wang, S. Y., 1972, The photochemical mechanism of pyrimidine cyclobutyl dimerization, *Tetrahedron* **28**:945–957.

Khattak, M. N., Hauswirth, W., and Wang, S. Y., 1972, Photohydration of pyrimidines in "acid puddles," *Biochem. Biophys. Res. Commun.* **48**:1622–1629.

Killander, D., and Rigler, R., 1969, Activation of deoxyribonucleoprotein in human leucocytes stimulated by phytohemagglutinin. I. Kinetics of the binding of acridine orange to deoxyribonucleoprotein, *Exp. Cell Res.* **54**:163–170.

Kittler, L., 1968, Photochemisches und polarographisches Verhalten molekularbiologisch wichtiger Azapyrimidine und Azapurine, Thesis, Friedrich-Schiller-Universität Jena.

Kittler, L., 1969, Photochemie anomaler Nucleinsäurebausteine. V. Photoreaktionen der 5-Halogenderivate des 6-Azauracils, *Biophysik* **5**:310–314.

Kittler, L., 1970, Photochemie anomaler Nucleinsäurebausteine. VII. UV-Photoreaktion des 6-Azacytosins und 6-Azacytidins, *Stud. Biophys.* **19**:21–29.

Kittler, L., 1972, Photochemie anomaler Nucleinsäurebausteine. VIII. Zur Photochemie des 6-Azacytosins und 6-Azacytidins, *Photochem. Photobiol.* **16**:39–49.

Kittler, L., and Berg, H., 1967, Photodegradation von Azapyrimidinen und Azapurinen, *Photochem. Photobiol.* **6**:199–204.

Kittler, L., and Berg, H., 1968, Photochemie anomaler Nucleinsäurebausteine. II. Elektronenakzeptor—Eigenschaften von Azaanalogen der Pyrimidin- und Purinreihe aus polarographischen Messungen, *J. Electroanal. Chem.* **16**:251–260.

Kittler, L., and Löber, G., 1968, Photochemie anomaler Nucleinsäurebausteine. IV. Der Einfluß paramagnetischer Metallionen auf die Photoreaktion des 6-Azauracils, *Stud. Biophys.* **6:**41–48.

Kittler, L., and Löber, G., 1969, Zum photochemischen Reaktions mechanismus des 6-Azauracils, *Photochem. Photobiol.* **10:**35–44.

Kittler, L., and Löber, G., 1971, Photochemie anomaler Nucleinsäurebausteine. IX. Zur photochemischen Hydratation des 6-Azauracils, *Monatsber. Dtsch. Akad. Wiss. Berlin* **13:**216–221.

Kittler, L., and Löber, G., 1973, Photochemistry of some anomalous nucleic bases. Deviations from the photochemical behaviour of the normal constituents, *Stud. Biophys.* **36/37:**5–19.

Kittler, L., and Löber, G., 1974, On relationship between polarographic oxidation potentials and photodynamic activity of dyes, *Stud. Biophys.* **45:**175–182.

Kittler, L., and Zimmer, C., 1976, Conformational changes of nucleic acids and poly d(A-T)·d(A-T) caused by photoreaction with furocoumarins, *Nucleic Acids Res.* **3:**191–203.

Kittler, L., Hradečna, Z., Jacob, H.-E., and Löber, G., 1975, Photobiological behaviour of bacteria and phages supplemented with aza-analogues of nucleic acid bases, *Z. Allg. Mikrobiol.* **15:**323–331.

Kleber, R., Fahr, E., and Boebinger, E., 1965, Die Struktur der bei der UV-Bestrahlung von Cytosin, Cytidin und Cytidylsäure entstehenden reversiblen Bestrahlungsprodukte, *Naturwissenschaften* **52:**513–514.

Kleinwächter, V., 1972, Luminescence spectra of polynucleotides, *Stud. Biophys.* **33:**1–50.

Kleinwächter, V., and Koudelka, J., 1964, Thermal denaturation of deoxyribonucleic acid-acridine orange complexes, *Biochim. Biophys. Acta* **91:**539–540.

Kleinwächter, V., Balcarová, Z., and Bohaček, J., 1969, Thermal stability of complexes of amino acridines with deoxyribonucleic acids of varying base content, *Biochim. Biophys. Acta* **174:**188–201.

Kleopfer, R., and Morrison, H., 1972, Organic photochemistry. XVII. The solution-phase photodimerization of dimethylthymine, *J. Am. Chem. Soc.* **94:**255–264.

Klimek, M., 1966, Thymine dimerization in *l*-strain mammalian cells after irradiation with ultraviolet light and the search for repair mechanisms, *Photochem. Photobiol.* **5:**603–607.

Klimek, M., and Sevčikova, P., 1973, Comparison of the effect of acridine derivatives and similar substances on the dimerization of thymine in mammalian DNA *in situ* and isolated (Russ.), *Stud. Biophys.* **36/37:**205–210.

Klimek, M., and Vasuček, J., 1970, The role of pyrimidine dimers in the inhibition of DNA synthesis in mammalian cells after ultraviolet irradiation in the mathematical interpretation of results, *Math. Biosci.* **9:**165–177.

Knowles, A., 1967, Dye sensitization of nucleotides, in: *Molekulare Mechanismen photodynamischer Effekte, Stud. Biophys.* **3:**97–104.

Knowles, A., 1971, A mechanism for the methylene blue sensitized oxidation of nucleotides, *Photochem. Photobiol.* **13:**473–487.

Knowles, A., and Gurnani, S., 1972, A study of the methylene blue sensitized oxidation of aminoacids, *Photochem. Photobiol.* **16:**95–108.

Kochetkov, N. K., Budowski, E. J., Swerdlov, E. D., Simukova, N. A., Turtschinski, M. F., and Schibaev, B. N., 1970, in: *Organic Chemistry of Nucleic Acids* (N. K. Kochetkov and E. J. Budowski, eds.), pp. 615–697, *Chemistry (Moscow)* (in Russian).

Köhnlein, W., and Hutchinson, F., 1969, ESR-studies of normal and 5-bromouracil-substituted DNA of *B. subtilis* after irradiation with ultraviolet light, *Radiat. Res.* **39:**745–757.

Koizumi, M., Obata, H., and Hayashi, S., 1964, Studies of the photoreduction of thiazine dyes in aqueous solutions, *Bull. Chem. Soc. Jpn.* **37:**108–117.

Kondo, S., and Kato, T., 1966, Action spectra for photoreactivation of killing and mutation to phototrophy in UV sensitive strains of *E. coli* possessing and lacking photoreactivating enzyme, *Photochem. Photobiol.* **5**:827–837.

Kondo, Y., and Witkop, B., 1969, Selective photoreductions of nucleic acids and their building stones. VIII. Photoreduction and dimerization of 1,3-dimethyluracil, *J. Am. Chem. Soc.* **91**:5264–5270.

Kornhauser, A., Krinsky, N. J., Huang, P.-K. C., and Clagett, D. C., 1973, A comparative study of photodynamic oxidation and radiofrequency-discharge-generated $^1O_2$ oxidation of guanosine, *Photochem. Photobiol.* **18**:63–69.

Krajewska, E., and Shugar, D., 1971, Photochemical transformation of 5-alkyluracils and their nucleosides, *Science (Washington)* **173**:435–437.

Krämer, D. M., and Pathak, M. A., 1970, Photoaddition of psoralen and of 4,5′,8-trimethyl-psoralen to DNA, *Photochem. Photobiol.* **12**:333–337.

Kramer, H. E. A., and Maute, A., 1972a, Sensitized photooxygenation according to type I mechanism (radical mechanism). I. Flash photolysis experiments, *Photochem. Photobiol.* **15**:7–24.

Kramer, H. E. A., and Maute, A., 1972b, Sensitized photooxygenation according to type I mechanism (radical mechanism). II. Flash photolysis experiments, *Photochem. Photobiol.* **15**:25–32.

Kramer, H. E. A., and Maute, A., 1973, Sensitized photooxygenation: Change from type I (radical) to type II (singlet oxygen) mechanism, *Photochem. Photobiol.* **17**:413–423.

Krauch, C. H., Krämer, D. M., and Wacker, A., 1967a, Zum Wirkungsmechanismus photodynamischer Furocoumarine. Photoreaktion von Psoralen (4-^{14}C) mit DNS, RNS, Homopolynucleotiden und Nucleosiden, *Photochem. Photobiol.* **6**:341–354.

Krauch, C. H., Krämer, D. M., Chandra, P., Mildner, P., Feller, H., and Wacker, A., 1967b, Durch Aceton sensibilisierte Photodimerisation von Urazil, *Angew. Chem.* **79**:944–945.

Kubota, Y., 1970a, Luminescence in DNA-acridine dye complexes. I. Phosphorescence and delayed fluorescence due to triplet–triplet annihilation of acridine dyes in the complexes, *Bull. Chem. Soc. Jpn.* **43**:3121–3125.

Kubota, Y., 1970b, Luminescence in DNA-acridine dye complexes. II. Sensitized delayed fluorescence, *Bull. Chem. Soc. Jpn.* **43**:3126–3130.

Kubota, Y., 1973, Fluorescence lifetimes and quantum yields of acridine dyes bound to DNA, *Chem. Lett. (Jpn.)*, 299–304.

Kubota, Y., Fujisaki, Y., and Miura, M., 1969, Delayed fluorescence of the DNA-acridine dye complexes in a frozen aqueous solution, *Bull. Chem. Soc. Jpn.* **42**:853–853.

Lamola, A. A., 1966, Solution photochemistry of thymine and uracil, *Science (Washington)* **154**:1560–1561.

Lamola, A. A., 1968, Excited state precursors of thymine photodimers, *Photochem. Photobiol.* **7**:619–632.

Lamola, A. A., 1969, Electronic energy transfer in solution: Theory and Application, in: *Technique of Organic Chemistry*, Vol. XIV (P. A. Leermakers and A. Weissberger, eds.), pp. 17–132, Interscience, New York.

Lamola, A. A., 1970, Triplet photosensitization and the photobiology of thymine dimers in DNA, *Pure Appl. Chem.* **24**:599–610.

Lamola, A. A., 1972, Photosensitization in biological systems and the mechanism of photoreactivation, *Mol. Photochem.* **4**:107–133.

Lamola, A. A., and Eisinger, J., 1968, On the mechanism of thymine photodimerization, *Proc. Natl. Acad. Sci. USA* **59**:46–51.

Lamola, A. A., and Mittal, J. P., 1966, Solution photochemistry of thymine and uracil, *Science (Washington)* **154**:1560–1561.

Lamola, A. A., and Yamane, T., 1967, Sensitized photodimerization of thymine in DNA, *Proc. Natl. Acad. Sci. USA* **58**:44–446.

Lang, H., 1974*a*, CD studies of conformational changes of DNA upon photosensitized UV-irradiation at 313 nm, *Nucleic Acid Res.* **2**:179–183.

Lang, H., 1974*b*, On the structure of UV-irradiated DNA, *Stud. Biophys.* **42**:157–159.

Lang, H., and Luck, G., 1973, Ultraviolet-light-induced conformational changes in DNA, *Photochem. Photobiol.* **17**:387–393.

Langmuir, M. E., and Hayon, E., 1969, Transient species produced in the photochemistry of 5-bromouracil and its *N*-methyl derivatives, *J. Chem. Phys.* **51**:4893–4899.

Lawley, P. D., 1966, in: *Progress in Nucleic Acid Research and Molecular Biology,* Vol. 5 (Davidson and Cohn, eds.), pp. 89–131, Academic Press, New York.

Lawrence, J. J., and Louis, M., 1972, Étude du rôle des histones dans l'interaction de la proflavine avec le DNA de la chromatine, *Biochim. Biophys. Acta* **272**:231–237.

Leonard, N. J., Bergstrom, D. E., and Tolman, G. L., 1971, Photoproducts from 4-thiouracil and cytosine and from 4-thiouridine and cytidine: Refinement of tertiary tRNA structure, *Biochem. Biophys. Res. Commun.* **44**:1524–1530.

Leonov, D., Salomon, J., Sasson, S., and Elad, D., 1973, Ultraviolet- and x-ray-induced reactions of nucleic acid constituents with alcohols. On the selectivity of these reactions for purines, *Photochem. Photobiol.* **17**:465–468.

Le Pecq, J. B., and Paoletti, C., 1967, A fluorescent complex between ethidium bromide and nucleic acids, *J. Mol. Biol.* **27**:87–106.

Le Pecq, J. B., Yot, P., and Paoletti, C., 1964, Interaction du bromhydrate d'ethidium (BET) avec les acides nucleiques (A.N.). Étude spectrofluorimetrique, *C. R. Acad. Sci.* **259**:1786–1789.

Lerman, L. S., 1961, Structural considerations in the interaction of DNA and acridines, *J. Mol. Biol.* **3**:18–30.

Leutzen, D., and Walker, J. R., 1970, Bromodeoxyuridine sensitization of the ultraviolet-sensitive *E. coli* ras-mutant to ultraviolet irradiation, *Mol. Gen. Genet.* **108**:218–224.

Linschitz, H., and Conolly, J. S., 1968, The photochemical addition of alcohols to purine, *J. Am. Chem. Soc.* **90**:2979–2980.

Lion, M. B., and Köhnlein, W., 1974, Effect of DNA conformation on the UV damage in 5-bromouracil substituted DNA of T3 coliphage, in: *Progress in Photobiology* (G. O. Schenck, ed.), Abst. 107, Deutsche Gesellschaft für Lichtforschung, Frankfurt.

Liquori, A. M., De Lerma, B., Ascoli, F., Botre, C., and Frasciatti, M., 1962, Interaction between DNA and polycyclic aromatic hydrocarbons, *J. Mol. Biol.* **5**:521–526.

Lisewski, R., and Wierzchowski, K. L., 1969, Photodimerization and van der Waals stacking of dimethylthymine in water, *Chem. Commun.,* 348–349.

Lisewski, R., and Wierzchowski, K. L., 1970, Solid state photochemistry of thymine, its *n*-methylated derivatives and orotic acids in KBr matrices, *Photochem. Photobiol.* **11**:327–347.

Litwin, J., and Riesterer, Z., 1973, The effect of photosensitizing dyes on the ³H-thymidine incorporation of cells grown *in vitro, Exp. Cell Res.* **79**:191–198.

Löber, G., 1965, On the fluorescence of acridine derivatives in the presence of DNA, *Photochem. Photobiol.* **4**:607–612.

Löber, G., 1968, On the complex formation of acridine dyes with DNA. IV. The equilibrium constants of substituted proflavine and acridine orange derivatives, *Photochem. Photobiol.* **8**:23–30.

Löber, G., 1971, Acridine, ihre physikochemische und biochemische Bedeutung: Eine Betrachtung anlässlich der Entdeckung des Acridins vor 100 Jahren, *Z. Chem.* **11**:135–145.

Löber, G., 1975, On the spectroscopic basis of acridine-induced fluoresence banding patterns in chromosomes, *Stud. Biophys.* **48**:109–123.

Löber, G., and Achtert, G., 1969, On the complex formation of acridine dyes with DNA. VII. Dependence of the binding on the dye structure, *Biopolymers* **8**:595–608.

Löber, G., and Kittler, L., 1973, Photochemie und Photobiologie von Nucleinsäuren und Nucleinsäurebausteinen, *Stud. Biophys.* **41**:81–153.

Löber, G., and Kittler, L., 1977, Selected topics in photochemistry of nucleic acids. Recent results and perspectives, *Photochem. Photobiol.* **25**:215–233.

Löber, G., Fleck, W., Jacob, H.-E., and Rost, K., 1970, Beziehungen zwischen der Komplexbindung mit DNS und einigen biologischen Wirkungen von Acridinfarbstoffen, in: *Wirkungsmechanismen von Fungiziden, Antibiotika und Cytostatika* (H. Lyr and W. Rawald, eds.), pp. 39–49, Akademie-Verlag, Berlin.

Löber, G., Schütz, H., and Kleinwächter, V., 1972, Effect of organic solvents on the properties of the complexes of DNA with proflavine and similar compounds, *Biopolymers* **11**:2439–2459.

Löber, G., Koudelka, J., and Smékal, E., 1974a, Stacking interactions of ethidium bromide bound to a polyphosphate and phage DNA *in situ, Biophys. Chem.* **2**:158–163.

Löber, G., Kleinwächter, V., Koudelka, J., and Smékal, E., 1974b, On spectral properties of type-I complexes of dyes with deoxyribonucleic acid and human serum albumin, *Stud. Biophys.* **45**:91–103.

Löber, G., Kleinwächter, V., Koudelka, J., Balcarová, Z., Filkuka, J., Krejči, P., Döbel, P., Beensen, V., and Rieger, R., 1976, Molecular and spectroscopic aspects of chromosome banding, *Biol. Zentralbl.,* **95**:169–191.

Lochmann, E. R., and Michler, A., 1973, Binding of organic dyes to nucleic acids and the photodynamic effect, in: *Physicochemical Properties of Nucleic Acids,* Vol. 1 (J. Duchesne, ed.), pp. 223–267, Academic Press, New York.

Lochmann, E. R., and Stein, W., 1964, Zur Inaktivierung durch Thiopyronin mit und ohne Licht, *Naturwissenschaften* **51**:59–61.

Lochmann, E. R., and Stein, W., 1967, Hemmung der RNS-Synthese bei *Saccharomyceszellen* verschiedenen Ploidiegrades durch Farbstoffe in Gegenwart und Abwesenheit von sichtbarem Licht, *Z. Naturforsch.* **22b**:196–200.

Lochmann, E. R., and Stein, W., 1968, Die Wirkung von Thiopyronin auf die Dunkelreaktivierung von UV-, Röntgen- und photodynamischen Schäden bei *Saccharomyces, Biophysik* **5**:78–84.

Loeser, C. N., West, S. S., and Schoenberg, M. D., 1960, Absorption and fluorescence studies on biological systems: Nucleic acid-dye complexes, *Anat. Rec.* 163–178.

Logan, D. M., and Whitmore, G. F., 1966, Dehydration of UV irradiated uridine and its derivatives, *Photochem. Photobiol.* **5**:143–156.

Lomant, A. J., and Fresco, J. R., 1972a, Polynucleotides. X. Influences of polynucleotide conformation on susceptibility of pyrimidine residues to photochemical attack, *J. Mol. Biol.* **66**:49–64.

Lomant, A. J., and Fresco, J. R., 1972b, Ultraviolet photochemistry as a probe of polynucleotide conformation, in: *Progress in Nucleic Acid Research,* Vol. 12 (J. N. Davidson and W. E. Cohn, eds.), pp. 1–27, Academic Press, New York.

Lomant, A. J., and Fresco, J. R., 1973, Polynucleotides. XIV. Photochemical evidence for an extrahelical solvent-accesible environment of non-complementary residues in polynucleotide helices, *J. Mol. Biol.* **77**:345–354.

Lozeron, H. A., and Gordon, M. P., 1964, Ultraviolet sensitization and photoreactivation of tobacco mosaic virus ribonucleic acid containing 5-fluorouracil, *Biochemistry* **3**:507–510.

Lozeron, H. A., Gordon, M. P., Gabriel, T., Tautz, W., and Duschinsky, R., 1964, The photochemistry of 5-fluor-uracil, *Biochemistry* **3:**1844–1850.

Maelicke, A., 1970, Interaction of ethidium with specific transfer ribonucleic acids and influence on the aminoacylation, in: *Interaktionen bei Biopolymeren, Stud. Biophys.* **24/25:**343–350.

Mantulin, W. W., and Song, P. S., 1973, Excited states of skin-sensitizing coumarins and psoralens: Spectroscopic studies, *J. Am. Chem. Soc.* **95:**5122–5129.

Marciani, S., Dall'Acqua, F., Gielfi, L., and Vedaldi, D., 1971. Photoreactivity (365 nm) of some coumarins and 4′,5′-dihydro-furocoumarins with nucleic acids, *Z. Naturforsch.* **26b:**1129–1136.

Marciani, S., Terbojevich, M., Dall'Acqua, F., and Rodighiero, G., 1973a, Light scattering and flow dichroism studies on DNA after the photoreaction with psoralen, *Z. Naturforsch.* **27b:**196–200.

Marciani, S., Terbojevich, M., Dall'Acqua, F., and Rodighiero, G., 1973b, Bifuncional photobinding of psoralen to single stranded nucleic acids, *Z. Naturforsch.* **28c:**370–375.

Matheson, J. B. C., Etheridge, R. D., Kratovich, N. R., and Lee, J., 1975, The quenching of singlet oxygen by amino acids and proteins, *Photochem. Photobiol.* **21:**165–171.

Matolscy, G., Pinter, J., and Pozsar, B. J., 1969, Incorporation of radiocarbon labelled uracil- and thymine-analogoues into the RNA of bean leaf tissues, *Acta Bot. Acad. Sci. Hung.* **15:**119–121.

Mattern, M., Binder, R., and Cerutti, P., 1972, Cytidine photohydration in R 17 RNA, *J. Mol. Biol.* **66:**201–204.

Matthews, M. M., 1963, Comparative study of lethal photosensitizations of *S. lutea* by 8-methoxypsoralene, *J. Bacteriol.* **85:**322–328.

Matthews, R. E., and Smith, J. E., 1956, Distribution of 8-azaguanine in the nucleic acids of *Bacillus cereus, Nature* (*London*) **177:**271–272.

McLaren, A. D., and Shugar, D., 1964, *Photochemistry of Proteins and Nucleic Acids,* Pergamon Press, Oxford.

McLaren, A. D., and Takahashi, W. N., 1970, Inactivation of infectious nucleic acid from tobacco mosaic virus by ultraviolet light (2537Å), *Radiat. Res.* **6:**532–542.

Meistrich, M. L., Lamola, A. A., and Gabbay, E. J., 1970, Sensitized photoinactivation of bacteriophage T4, *Photochem. Photobiol.* **11:**169–178.

Mennigmann, H.-D., and Wacker, A., 1970, Photoreactivation of *Escherichia coli* B$_{S-3}$ after inactivation by 313 nm radiation in the presence of acetone, *Photochem. Photobiol.* **11:**291–196.

Michaelis, L., 1947, The structure of the interaction of nucleic acids and nuclei with basic dye-stuffs, *Cold Spring Harbor Symp. Quant. Biol.* **12:**131–142.

Micheler, A., Pietsch, J., and Lochmann, E.-R., 1973, Über die RNS-Synthese bei *Saccharomyces*-Zellen nach photodynamischer Behandlung und nach Röntgenbestrahlung, *Biophysik* **10:**249–256.

Michelson, A. M., Moony, C., and Kovoor, A., 1972, Action of quinacrine mustard on polynucleotides, *Biochimie* **54:**1129–1136.

Miller, N., and Cerutti, P., 1968, Structure of the photohydration products of cytidine and uridine, *Proc. Natl. Acad. Sci. USA* **59:**34–38.

Minyat, E. E., Borisova, O. F., Volkenstein, M. V., and Georgiev, G. P., 1970, On the deoxyribonucleoprotein structure. I. Studies of base content of protein-free regions of DNA, *Mol. Biol.* **4:**291–301.

Modak, S. P., and Setlow, J. K., 1969, Synthesis of deoxyribonucleic acid after ultraviolet irradiation of sensitive and resistant *Haemophilus influenzae, J. Bacteriol.* **98:**1195–1198.

Mönkehaus, F., 1973, Einfluss von Cysteamin auf die UV-Empfindlichkeit von λ-Phagen mit variablem Bromurazil-Gehalt, *Z. Naturforsch.* **29c**:286–288.

Mönkehaus, F., 1974, UV-Empfindlichkeit von λ-Phagen mit variablem Bromuracil-Gehalt, *Z. Naturforsch.* **29c**:289–293.

Mönkehaus, F., and Köhnlein, W., 1972, Experimente zur intramolekularen Energieleitung in BU-DNA des Phagen PBSH aus *B. subtilis* nach Bestrahlung mit langwelligem UV, *Z. Naturforsch.* **27b**:833–839.

Mönkehaus, F., and Köhnlein, W., 1973, Single- and double-strand breaks in 5-bromouracil-substituted DNA of *B. subtilis* and phage PBSH after irradiation with longwave length UV and their correlation to intramolecular energy transfer, *Biopolymers* **12**:329–340.

Mönkehaus, F., and Köhnlein, W., 1974, Intramolecular energy transfer in BU-DNA after irradiation: Single and double strand breakage rates, in: *Progress in Photobiology, 1972* (G. O. Schenck, ed.), No. 106, Deutsche Gesellschaft für Lichtforschung, Frankfurt.

Montenay-Garestier, T., and Hélénè, C., 1968, Molecular interaction between tryptophan and nucleic acid components in frozen aqueous solutions, *Nature (London)* **217**:844–845.

Moore, A. M., 1958, Ultraviolet irradiation of pyrimidine derivatives. II. Synthesis of the product of reversible photolysis of uracil, *Can. J. Chem.* **36**:281–295.

Moore, A. M., and Thomson, C. H., 1955, Ultraviolet irradiation of pyrimidine derivatives, *Science (Washington)* **122**:594–595.

Moore, A. M., and Thomson, C. H., 1957, Ultraviolet irradiation of pyrimidine derivatives. I. 1,3-dimethyluracil, *Can J. Chem.* **35**:163–174.

Moore, T. A., Mantulin, W. W., and Song, P. S., 1973, Excited states and reactivity of carcinogenic benzpyrene; A comparison with skin-sensitizing coumarins, *Photochem. Photobiol.* **18**:185–194.

Morita, M., and Kato, S., 1969, Studies of the transient intermediates of a thiopyronine aqueous solution under flash excitation, *Bull. Chem. Soc. Jpn.* **42**:25–35.

Morrison, H., Feeley, A., and Kleopfer, R., 1968, Solution-phase photodimerization of dimethylthymine, *Chem. Commun.,* 358–367.

Müller, W., Crothers, D. M., and Waring, M. J., 1973, A non-intercalating proflavine derivative, *Eur. J. Biochem.* **39**:223–234.

Mund, C., and Venner, H., 1967, Spektrophotometrische Untersuchungen über die Sekundärstruktur UV-bestrahlter DNA, *Stud. Biophys.* **3**:57–64.

Murcia, D., Kleopfer, R., Maleski, R., and Morrison, H., 1972, Formation of a new 1,3-dimethylthymine photoproduct in the presence of carbon tetrachloride, *Mol. Photochem.,* 61–65.

Musajo, L., and Rodighiero, G., 1970, Studies on the photo-C_4-cyclo-addition reactions between skin-photosensitizing furocoumarins and nucleic acids, *Photochem. Photobiol.* **11**:27–35.

Musajo, L., and Rodighiero, G., 1972, in: *Photophysiology, Current Topics in Photobiology and Photochemistry* (A. C. Giese, ed.), pp. 115–147, Academic Press, New York.

Musajo, L., Rodighiero, G., Colombo, G., Torlone, V., and Dall'Acqua, F., 1965, Photosensitizing furocoumarins: Interactions with DNA and photo-inactivation of DNA containing viruses, *Experientia* **21**:22–24.

Musajo, L., Rodighiero, G., Breccia, A., Dall'Acqua, F., and Malesami, G., 1966, Skin-photosensitizing furocoumarins: Photochemical interaction between DNA and $O^{14}CH_3$ bergapten (5-methoxypsoralen), *Photochem. Photobiol.* **5**:739–745.

Musajo, L., Rodighiero, G., Caporale, G., Dall'Acqua, F., Marciani, S., Bordin, F., Baccichetti, F., and Bevilacqua, R., 1974, in: *Sunlight and Man—Normal and Abnormal Photobiologic Response* (M. A. Pathak, L. C. Harber, M. Seiji, and A. Kukita, eds.), pp. 369–387, University of Tokyo Press, Tokyo.

Nakai, S., and Saeki, T., 1964, Induction of mutation by photodynamic action in *Escherichia coli, Genet. Res.* **5**:158–161.

Nicolau, C., 1972, Short-lived free radicals in aqueous solution of purine, in: *The Purines— Theory and Experiment. The Jerusalem Symposia on Quantitative Chemistry and Biochemistry, IV,* pp. 519–527, The Israel Academy of Sciences and Humanities, Jerusalem.

Nicoletti, B., and Trippa, G., 1967, Sull'azione mutagena del psoralene irradiato con luce ultravioletta in *Drosophila melanogaster, Rend. Accad. Naz. Lincei (Roma)* **43**:259–263.

Nilsson, R., Merkel, P. B., and Kearns, D. R., 1972, Unambiguous evidence for the participation of singlet oxygen ($^1\Delta$) in photodynamic oxidation of amino acids, *Photochem. Photobiol.* **16**:117–124.

Nino, J., Favre, A., and Yaniv, M., 1969, Molecular model for transfer RNA, *Nature (London)* **223**:1333–1335.

Nirmala, J., and Sastry, K. S., 1971, Some factors influencing photodynamic degradation of guanosine, *Ind. J. Biochem. Biophys.* **8**:263–265.

Noble, M.-C., and Bradley, S. G., 1972, Photosensitization of actinophages by crystal violet or proflavine, in: *Dev. Ind. Microbiol. (Washington)* **13**:412–420.

Ofengard, J., and Bierbaum, J., 1973, Use of photochemically induced cross-linking as a conformational probe of the tertiary structure of certain regions in transfer ribonucleic acid, *Biochemistry* **12**:1977–1989.

Oginsky, E. L., Green, G. S., Griffith, D. G., and Fowlks, W. L., 1959, Lethal photosensitization of bacteria with 8-methoxypsoralen to long wavelength ultraviolet radiation, *J. Bacteriol.* **78**:821–833.

Ono, J., and Shimazu, Y., 1967, *In vivo* cleavage of a circular, single-stranded DNA of bacteriophage ϕR irradiated with ultraviolet light, *J. Mol. Biol.* **24**:491–495.

Ono, J., Wilson, R. G., and Grossman, L., 1965a, Effects of ultraviolet light on the template properties of polycytidylic acid, *J. Mol. Biol.* **11**:600–612.

Ono, J., Wilson, R. G., and Grossman, L., 1965b, Continuity of DNA synthesis in *E. coli, J. Mol. Biol.* **11**:650–653.

Orbob, G. B., 1963, Some effects of photosensitizing dyes on three plant viruses, *Virology* **21**:291–299.

Ottensmeyer, F. P., and Whitmore, G. F., 1968, Coding properties of ultraviolet photoproducts of uracil. I. Binding studies and polypeptide synthesis, *J. Mol. Biol.* **38**:1–16.

Pačes, V., Doskočil, J., and Šorm, F., 1968, Incorporation of 5-azacytidine into nucleic acids of *E. coli, Biochim. Biophys. Acta* **161**:352–360.

Parker, C. A., and Joyce, T. A., 1973, Prompt and delayed fluorescence of some DNA adsorbates, *Photochem. Photobiol.* **18**:467–474.

Parrish, J. A., Fitzpatrick, T. B., Tanenbaum, L., and Pathak, M. A., 1974, Photochemotherapy of psoriasis with oral methoxsalen and longwave ultraviolet light, *N. Engl. J. Med.* **291**:1207–1222.

Pathak, M. A., and Krämer, D. M., 1969, Photosensitization of skin *in vivo* by furocoumarins (psoralens), *Biochim. Biophys. Acta* **195**:197–206.

Pathak, M. A., Fellman, J. H., and Kaufmann, K. D., 1960, The effect of structural alterations on the erythemal activity of furocoumarins: psoralens, *J. Invest. Dermatol.* **35**:165–183.

Pathak, M. A., Worden, L. R., and Kaufmann, K. D., 1967, Effect of structural alterations on the photosensitizing potency of furocoumarins (psoralens) and related compounds, *J. Invest. Dermatol.* **48**:103–111.

Patrick, M. H., Haynes, R. H., and Uretz, R. B., 1964, Dark recovery phenomena in yeast. I. Comparative effects with various inactivating agents, *Radiat. Res.* **21**:144–163.

Peacocke, A. R., and Skerrett, J. N. H., 1956, The interaction of aminoacridines with nucleic acids, *Trans. Faraday Soc.* **52**:261–279.

Pearson, M., and Johns, H. E., 1966*a*, Excision of dimers and hydrates from ultraviolet-irradiated poly U by pancreatic ribonuclease, *J. Mol. Biol.* **19**:303–319.

Pearson, M., and Johns, H. E., 1966*b*, Suppression of hydrate and dimer formation in ultraviolet-irradiated poly (A + U) relative to poly U, *J. Mol. Biol.* **20**:215–229.

Pearson, M. L., Ottensmeyer, F. P., and Johns, H. E., 1965, Properties of an unusual photoproduct of UV-irradiated thymidylyl–thymidine, *Photochem. Photobiol.* **4**:739–747.

Pearson, M., Whillans, D. W., Le Blanc, J. C., and Johns, H. E., 1966, Dependence on wavelength of photoproduct yields in ultraviolet-irradiated poly U., *J. Mol. Biol.* **20**:245–261.

Peter, H. H., and Drewer, R. J., 1970, Photoproducts of bromouracil-labelled DNA and the structure of 5-bromodeoxyuridylyl-(3′ → 5′)-thymine photoproduct, *Photochem. Photobiol.* **12**:269–282.

Peter, H. H., and Drewer, R. J., 1971, The photochemistry of ^{14}C-5-bromo-2′-deoxyuridylyl-(3′ → 5′)-thymidine determination of quantum yields as a function of pH, *Photochem. Photobiol.* **14**:561–567.

Petrissant, G., and Favre, A., 1972, Separation and characterization of intramolecular cross-linked form of tRNA^{Met} from *E. coli, FEBS Lett.* **23**:191–194.

Phillips, S. L., Person, S., and Jagger, J., 1967, Division delay induced in *E. coli* by near ultraviolet radiation, *J. Bacteriol.* **94**:165–170.

Pichowska, M., and Shugar, D., 1965, Replacement of 5-methyluracil (thymine) by 5-ethyluracil in bacteria DNA, *Biochim. Biophys. Res. Commun.* **20**:768–773.

Pietrzykowska, I., 1973, On the mechanism of bromouracil induced mutagenesis, *Mutat. Res.* **19**:1–9.

Pietrzykowska, I., and Shugar, D., 1966, Replacement of thymine by 5-ethyluracil in bacteriophage DNA, *Biochim. Biophys. Res. Commun.* **25**:567–572.

Pietrzykowska, I., and Shugar, D., 1968, 5-ethyldeoxyuridine, thymidine analog: Photochemical transformation, *Science (Washington)* **161**:1248–1249.

Pietrzykowska, I., and Shugar, D., 1970, Photochemistry of 5-ethyluracil and its glycosides, *Acta Biochim. Polon.* **17**:361–384.

Pilet, J., and Brahms, J., 1973, Investigation of DNA structural changes by infrared spectroscopy, *Biopolymers* **12**:387–403.

Pochon, F., Pascal, Y., Pitha, P., and Michelson, A. M., 1970, Photochimie des polynucleotides. IV. Photochimie de quelques nucleosides puriques, *Acta Biochim. Biophys.* **213**:273–281.

Pochon, F., Balny, C., Scheit, K. H., and Michelson, A. M., 1971, Photochimie des polynucleotides. V. Études sur des polymeres contenant de la 4-thio-uridine, *Biochim. Biophys. Acta* **228**:49–56.

Pohl, D., and Kaplan, R. W., 1968, Einfluß von Bromuracil auf die Mutationsauslösung und Inaktivierung durch UV und Röntgenstrahlen beim Phagen, *Biophysik* **4**:196–213.

Pörschke, D., 1973*a*, A specific photoreaction in polydeoxyadenylic acid, *Proc. Natl. Acad. Sci. USA* **70**:2683–2686.

Pörschke, D., 1973*b*, Analysis of a specific photoreaction in oligo- and polydeoxyadenosine acids, *J. Am. Chem. Soc.* **95**:8840–8846.

Portocală, R., Sorodoc, G., Peiulescu, P., Surdan, G., and Stoian, N., 1972, Photodynamic action of toluidine blue on a strain of *Clamidia psittaci, Rev. Roum. Virol.* **9**:251–252.

Pritchard, N. J., Blake, A., and Peacocke, A. R., 1966, Modified interaction model for the interaction of amino acridines and DNA, *Nature (London)* **212**:1360–1361.

Pullman, A., and Pullman, B., 1955, *Cancerisation par les Substances Chimiques Moleculaires*, Masson, Paris.

Pullman, B., 1968, Electronic factors in the photodimerization of thymine, *Photochem. Photobiol.* 7:525–530.

Rada, B., and Zavada, J., 1962, Screening-test for cytostatic and virostatic substances, *Neoplasma* 9:57–65.

Rahn, R. O., 1970, Physical and environmental factors influencing the photochemistry of DNA, in: *Photochemistry of Macromolecules* (R. F. Reinisch, ed.), pp. 15–29, Plenum Press, New York.

Rahn, R. O., 1973, Denaturation in ultraviolet-irradiated DNA, in: *Photophysiology,* Vol. VIII (A. C. Giese, ed.), pp. 231–255, Academic Press, New York.

Rahn, R. O., and Hosszu, J. L., 1968*a*, Photochemistry of polynucleotides, a summary of temperature effects, *Photochem. Photobiol.* 7:637–642.

Rahn, R. O., and Hosszu, J. L., 1968*b*, Photoproduct formation in DNA at low temperatures, *Photochem. Photobiol.* 8:53–63.

Rahn, R. O., and Hosszu, J. L., 1969*a*, Influence of relative humidity on the photochemistry of DNA films, *Biochim. Biophys. Acta* 190:126–131.

Rahn, R. O., and Hosszu, J. L., 1969*b*, Photochemical studies of thymine in ice, *Photochem. Photobiol.* 10:131–137.

Rahn, R. O., and Landry, L. C., 1971, Pyrimidine dimer formation in poly(dT) and apurinic acid, *Biochim. Biophys. Acta* 247:197–206.

Rahn, R. O., and Landry, L. C., 1973, Ultraviolet irradiation of nucleic acids complexed with heavy atoms. II. Phosphorescence and photodimerization of DNA complexed with Ag, *Photochem. Photobiol.* 18:29–38.

Rahn, R. O., and Schleich, T., 1974, Proton magnetic resonance studies of ultraviolet-irradiated apurinic acid, *Nucleic Acid Res.* 1:999–1005.

Rahn, R. O., and Stafford, R. S., 1974, Measurement of defects in ultraviolet-irradiated DNA by the kinetic formaldehyde method, *Nature (London)* 248:52–54.

Rahn, R. O., Setlow, J. K., and Hosszu, J. L., 1969, Ultraviolet inactivation and photoproducts of transforming DNA irradiated at low temperatures, *Biophys. J.* 9:510–517.

Rahn, R. O., Battista, M. D. C., and Landry, L. C., 1970, Influence of mercuric ions on the phosphorescence and photochemistry of DNA, *Proc. Natl. Acad. Sci. USA* 67:1390–1397.

Ramenda, K. P., and Sinsheimer, R. L., 1971, Nature of the complementary strands synthesized *in vitro* upon the single-stranded circular DNA of bacteriophage ϕX174 after ultraviolet irradiation, *Biophys. J.* 11:355–369.

Ramstein, J., and Leng, M., 1975, Effect of DNA base composition on the intercalation of proflavine: A kinetic study, *Biophys. Chem.* 3:234–240.

Rauth, A. M., and Whitmore, G. F., 1966, The survival of synchronized L cells after ultraviolet irradiation, *Radiat. Res.* 28:84–95.

Rawls, H. R., and van Santen, R. J., 1970, Singlet oxygen: A possible source of the original hydroperoxides in fatty acids, *Ann. N.Y. Acad. Sci.* 171:135–142.

Regan, J. D., Setlow, R. B., and Ley, R. D., 1971, Normal and defective repair of damaged DNA in human cells: A sensitive assay utilizing the photolysis of bromodeoxyuridine, *Proc. Natl. Acad. Sci. USA* 68:708–712.

Remsen, J. F., Miller, N., and Cerruti, P. A., 1970, Photohydration of uridine in the RNA of coliphage R17. II. The relationship between ultraviolet inactivation and uridine photohydration, *Proc. Natl. Acad. Sci. USA* 65:460–469.

Rhoades, D. F., and Wang, S. Y., 1970, Uracil-thymine adduct from a mixture of uracil and thymine irradiated with ultraviolet light, *Biochemistry* 9:4416–4420.

Rhoades, D. F., and Wang, S. Y., 1971*a*, Photochemistry of polycytidylic acid, deoxycytidine, and cytidine, *Biochemistry* **10**:4603–4611.

Rhoades, D. F., and Wang, S. Y., 1971*b*, A new photoproduct of cytosine. Structure and mechanism studies, *J. Am. Chem. Soc.* **93**:3779–3781.

Rigler, R., 1966, Microfluorometric characterization of intracellular nucleic acids and nucleoproteins by acridine orange, *Acta Physiol. Scand.* **67**:1–122 (Suppl. 267).

Rigler, R., Cronvall, E., Hirsch, R., Pachmann, U., and Zachau, H., 1970, Interactions of seryl-tRNA synthetase with serine and phenylalanine specific tRNA, *FEBS Lett.* **11**:320–323.

Ritchie, D. A., 1964, Mutagenesis with light and proflavine in phage T4, *Genet. Res.* **5**:168–169.

Ritchie, D. A., 1965, Mutagenesis with light and proflavine in phage T4, II. Properties of the mutants, *Genet. Res.* **6**:474–478.

Rodighiero, G., Chandra, P., and Wacker, A., 1970*a*, Structural specifity for the photoinactivation of nucleic acids by furocoumarins, *FEBS Lett.* **10**:29–32.

Rodighiero, G., Musajo, L., Dall'Acqua, F., Marciani, S., Carporale, G., and Ciavatta, L., 1970*b*, Mechanism of skin photosensitization by furocoumarins: Photoreactivity of various furocoumarins with native DNA and with ribosomal RNA, *Biochim. Biophys. Acta* **217**:40–49.

Rodighiero, G., Dall'Acqua, F., Marciani, S., Chandra, P., Feller, H., Götz, A., and Wacker, A., 1971, Studies on the reactivation of bacteria photodamaged by an angular furocoumarin: Angelicin, *Biophysik* **8**:1–8.

Romanovskaja, L. N., Kulba, A. M., and Gabrilovich, I. M., 1972, On the mechanism of interaction between acridine dyes and nucleic acids, *Biophysika* (*USSR*) **17**:313–316.

Rosen, B., Rothman, F., and Weigert, M. G., 1969, Miscoding caused by 5-fluorouracil, *J. Mol. Biol.* **44**:363–375.

Roth, D., 1973, Effect of ultraviolet irradiation of DNA on the dissociation transition of the strong DNA-acriflavine complex, *Photochem. Photobiol.* **18**:437–439.

Roth, J. A., and McGormick, D. B., 1967, Complexing of riboflavin and its 2-substituted analogs with adenosine and other 6-substituted purine derivatives, *Photochem. Photobiol.* **6**:657–664.

Rothman, W., and Kearns, D. R., 1967, Triplet states of bromouracil and iodouracil, *Photochem. Photobiol.* **6**:775–778.

Rupp, W. D., and Prusoff, W. H., 1964, Incorporation of 5-iodo-2-deoxyuridine into bacteriophage T1 as related to ultraviolet sensitization or protection, *Nature* (*London*) **202**:1288–1290.

Rupp, W. D., and Prusoff, W. H., 1965*a*, Photochemistry of iodouracil. I. Photoproducts obtained in water, *Biochem. Biophys. Res. Commun.* **18**:145–151.

Rupp, W. D., and Prusoff, W. H., 1965*b*, Photochemistry of iodouracil. II. Effects of sulfur compounds, ethanol, and oxygen, *Biochem. Biophys. Res. Commun.* **18**:158–164.

Salomon, J., and Elad, D., 1974, Selective photochemical alkylation of purines in DNA, *Biochim. Biophys. Res. Commun.* **58**:890–895.

Samejima, R., Kita, M., Saneyoski, M., and Sawada, F., 1969, Optical rotatory dispersion and circular dichroism of sulfur-containing nucleosides and nucleotides and of the ribonuclease-thionucleotide complex, *Biochem. Biophys. Acta* **179**:1–9.

Santamaria, L., 1967, Natural photodynamic sensitivity in retina and cancer cells, *Stud. Biophys.* **3**:269–275.

Santus, R., Helénè, C., Ovadia, J., and Grossweiner, L. J., 1972, Splitting of thymine dimer by hydrated electrons, *Photochem. Photobiol.* **16**:65–67.

Sarin, P. S., and Johns, H. E., 1968, UV induced conformational changes in transfer RNA, *Photochem. Photobiol.* **7:**203–210.

Sastry, K. S., and Gordon, M. P., 1966, The photodynamic inactivation of *Tobacco Mosaic Virus* and its ribonucleic acid by acridine orange, *Biochim. Biophys. Acta* **129:**42–48.

Sawada, F., and Kanbayashi, N., 1973, Fractionation of ribonuclease A photosensitized with 4-thiouridylic acid, *J. Biochem.* **74:**459–471.

Sawada, F., 1974, Kinetics of 4-thiouridylate-sensitized photoinactivation of ribonuclease A, *Photochem. Photobiol.* **20:**523–526.

Scaife, J. F., 1970, RNA synthesis and uridine pools in normal and BUdR-containing human kidney T-cells after UV-irradiation, *Int. J. Radial. Biol.* **18:**189–192.

Schenck, G. O., 1970, Mechanism of formation of singlet oxygen in photosensitized oxygenation, *Ann. N.Y. Acad. Sci.* **171:**67–78.

Schoentjes, M., and Fredericq, E., 1972, Proflavine binding of yeast rRNA and ribosomes as related to structure, *Biopolymers* **11:**361–374.

Scholes, C. P., Hutchinson, F., and Hales, H. B., 1967, Ultraviolet-induced damage to DNA independent of molecular weight, *J. Mol. Biol.* **24:**471–474.

Schott, H. N., and Shetlar, M. D., 1974, Photochemical addition of amino acids to thymine, *Biochem. Biophys. Res. Commun.* **59:**1112–1116.

Schreiber, J. P., and Daune, M. P., 1974, Fluorescence of complexes of acridine dyes with synthetic polydeoxyribonucleotides: A physical model of frameshift mutations, *J. Mol. Biol.* **83:**487–501.

Schuster, H., 1964, Photochemie von Ribonucleinsäuren, *Z. Naturforsch.* **19b:**815–830.

Secrist, J. A., Barrio, J. R., and Leonard, N. J., 1971, Attachement of a fluorescent label to 4-thiouracil and 4-thiouridine, *Biochim. Biophys. Res. Commun.* **45:**1262–1270.

Sehgal, V. N., 1971, Oral application of trimethylpsoralen in *vitiligo* in children: Preliminary report, *Biol. J. Dermatol.* **85:**454–456.

Sekely, L., and Prusoff, W. H., 1966, Anti-viral activity of azathymidine and uracil methyl sulphone, *Nature (London)* **211:**1260–1260.

Sela, J., 1969, Fluorescence of nucleic acids with ethidium bromide: An indication of the configurative state of nucleic acids, *Biochim. Biophys. Acta* **190:**216–219.

Selander, R. K., 1974*a*, The binding of quinacrine mustard to nucleic acids, *Acta Chem. Scand. B* **28:**45–55.

Selander, R. K., 1974*b*, Interaction of quinacrine mustard with whole and partially deproteinized calf thymus deoxynucleoproteins, *Acta Chem. Scand. B* **28:**937–948.

Seliskar, C. J., and Brand, L., 1971, Electronic spectra of L-aminonaphthalene-6-sulfonate and related molecules, *J. Am. Chem. Soc.* **93:**5414–5420.

Setlow, J. K., 1964, Effects of UV on DNA: Correlations among biological changes, physical changes and repair mechanisms, *Photochem. Photobiol.* **3:**405–413.

Setlow, J. K., 1966, The molecular basis of biological effects of ultraviolet radiation and photoreactivation, in: *Current Topics in Radiation Research,* Vol. II (M. Ebert and A. Howard, eds.), pp. 195–248, North-Holland, Amsterdam.

Setlow, R. B., 1964, Physical changes and mutagenesis, *J. Cell. Comp. Physiol. Suppl.* **64:**51–68.

Setlow, R. B., 1966, Cyclobutane-type pyrimidine dimers in polynucleotides, *Science (Washington)* **153:**379–386.

Setlow, R. B., 1968, Photoproducts in DNA irradiated *in vivo, Photochem. Photobiol.* **7:**643–649.

Setlow, R. B., and Carrier, W. L., 1963, Identification of ultraviolet-induced thymine dimers in DNA by absorbance measurements, *Photochem. Photobiol.* **2:**49–57.

Setlow, R. B., and Carrier, W. L., 1966, Pyrimidine dimers in ultraviolet irradiated DNAs, *J. Mol. Biol.* **17**:237–254.

Setlow, R. B., and Carrier, W. L., 1967, Formation and destruction of pyrimidine dimers in polynucleotides by ultraviolet irradiation in presence of proflavine, *Nature* (*London*) **213**:906–909.

Setlow, R. B., and Setlow, J. K., 1962, Evidence that ultraviolet-induced thymine dimers in DNA cause biological damage, *Proc. Natl. Acad. Sci. USA* **48**:1250–1257.

Setlow, R. B., and Setlow, J. K., 1970, Macromolecular synthesis in irradiated bacteria, *Mutat. Res.* **9**:434–436.

Setlow, R. B., Swenson, P. A., and Carrier, W. L., 1963, Thymine dimers and inhibition of DNA synthesis by ultraviolet irradiation of cells, *Science* (*Washington*) **142**:1464–1465.

Setlow, R. B., Carrier, W. L., and Bollum, F. J., 1965, Pyrimidine dimers in UV-irradiated poly dI:dC, *Proc. Natl. Acad. Sci. USA* **5**:1111–1118.

Shafranovskaya, N. N., Trifonov, E. N., Lazurkin, Y. S., and Frank-Kamenetskii, 1973, Clustering of thymine dimers in ultraviolet-irradiated DNA and the long range transfer of electronic excitation along the molecule, *Nature* (*London*) *New Biol.* **241**:58–60.

Simon, M. J., and Van Vunakis, H., 1962, The photodynamic reaction of methylene blue with deoxyribonucleic acid, *J. Mol. Biol.* **4**:488–499.

Simon, M. J., Grossman, L., and Van Vunakis, H., 1965, Photosensitized reaction of polyribonucleotides. I. Effects on their susceptibility to enzyme digestion and their ability to act as synthetic messengers, *J. Mol. Biol.* **12**:50–58.

Sinclair, W. K., and Morton, R. A., 1965, X-ray and ultraviolet sensitivity of synchronized chinese hamster cells at various stages of the cell cycle, *Biophys. J.* **5**:1–25.

Singer, B., and Fraenkel-Conrat, H., 1966, Dye-catalyzed photoinactivation of tobacco mosaic virus ribonucleic acid, *Biochemistry* **5**:2446–2450.

Sinsheimer, R. L., 1954, The photochemistry of uridylic acid, *Radiat. Res.* **1**:505–513.

Sinsheimer, R. L., 1957, The photochemistry of cytidylic acid, *Radiat. Res.* **6**:121–125.

Sinsheimer, R. L., and Hastings, R., 1949, A reversible photochemical alteration of uracil and uridine, *Science* (*Washington*) **110**:525–527.

Škoda, J., 1963, Mechanism of action and application of azapyrimidines, in: *Progress in Nucleic Acid Research*, Vol. II (J. N. Davidson and W. E. Cohn, eds.), pp. 197–219, Academic Press, New York.

Škoda, J., 1968, Dead code triplets, in: *Biochemistry of Ribosomes and Messenger-RNA* (R. Lindigkeit, P. Langen, and J. Richter, eds.), pp. 499–507, Akademie Verlag, Berlin.

Škoda, J., 1969, The role of pharmacologically active nucleoside derivatives in RNA translation, *FEBS Symp.* **16**:23–30.

Škoda, J., and Šorm, F., 1964, Biosynthesis of co-polymers of uridylic and cytidylic acids with 6-azacytidylic acid, *Biochim. Biophys. Acta* **91**:352–354.

Smets, L. A., and Cornelis, J. J., 1971, Repairable and irrepairable damage in 5-bromouracil-substituted DNA exposed to ultraviolet radiation, *Int. J. Radiat. Biol.* **19**:445–457.

Smith, B. J., 1966, Some effects of bromouracil on the kinetics of thymineless death in *E. coli*, *J. Mol. Biol.* **20**:21–28.

Smith, K. C., 1961/1962, A chemical basis for the sensitization of bacteria to ultraviolet light by incorporated bromouracil, *Biochem. Biophys. Res. Commun.* **6**:458–463.

Smith, K. C. 1962, Dose dependent decrease in extractability of DNP from bacteria following irradiation with ultraviolet light or with visible light plus dye, *Biochem. Biophys. Res. Commun.* **8**:157–163.

Smith, K. C., 1963, Photochemical reactions of thymine, uracil, uridine, cytosine and bromouracil in frozen solutions and in dried films, *Photochem. Photobiol.* **2**:503–517.

Smith, K. C., 1964*a*, The photochemical interaction of deoxyribonucleic acid and protein *in vivo* and its biological importance, *Photochem. Photobiol.* **3**:415–427.

Smith, K. C., 1964*b*, Photochemistry of nucleic acids, in: *Photophysiology,* Vol. II (A. C. Giese, ed.), pp. 329–388, Academic Press, New York.

Smith, K. C., 1964*c*, The photochemistry of thymine and bromouracil *in vivo, Photochem. Photobiol.* **3**:1–10.

Smith, K. C., 1966*a*, Physical and chemical changes induced in nucleic acids by ultraviolet light, *Radiat. Res. Suppl.* **6**:54–79.

Smith, K. C., 1966*b*, An isomer of the cyclobutane type thymine dimer, *Biochem. Biophys. Res. Commun.* **25**:426–433.

Smith, K. C., 1968, The biological importance of UV-induced DNA–protein cross-linking *in vivo* and its probable chemical mechanism, *Photochem. Photobiol.* **7**:651–660.

Smith, K. C., 1969, Photochemical addition of amino acids to ^{14}C-uracil, *Biochem. Biophys. Res. Commun.* **34**:354–357.

Smith, K. C., 1970, A mixed photoproduct of thymine and cysteine: S-S-cysteine, 6-hydrothymine, *Biochem. Biophys. Res. Commun.* **39**:1011–1016.

Smith, K. C., 1974, Photoaddition of proteins and other molecules to nucleic acids, in: *Progress in Photobiology, 1972* (G. O. Schenck, ed.), Abst. 017, Deutsche Gesellschaft für Lichtforschung, Frankfurt.

Smith, K. C., 1976, The radiation-induced addition of proteins and other molecules to nucleic acids, in: *Photochemistry and Photobiology of Nucleic Acids,* Vol. 2 (S. Y. Wang, ed.), pp. 187–218, Academic Press, New York.

Smith, K. C., and Aplin, R. T., 1966, A mixed photoproduct of uracil and cysteine (5-*s*-cysteine-6-hydrouracil): A possible model for the *in vivo* cross-linking of DNA and protein by UV-light, *Biochemistry* **5**:2125–2130.

Smith, K. C., and Hanawalt, P. C., 1969, *Molecular Photobiology—Inactivation and Recovery,* Academic Press, New York.

Smith, K. C., and Meun, H. C., 1968, Kinetics of the photochemical addition of ^{35}S-cysteine to polynucleotides and nucleic acids, *Biochemistry* **7**:1033–1037.

Smith, K. C., Hodgkins, B., and O'Leary, M. E., 1966, The biological importance of ultraviolet light induced DNA–protein cross links in *E. coli* 15 TAU, *Biochim. Biophys. Acta* **114**:1–15.

Snyder, L. C., Shulman, R. C., and Neuman, D. B., 1970, Electronic structure of thymine, *J. Chem. Phys.* **53**:256–267.

Song, P. S., and Gordon, W. H., 1970, A spectroscopic study of the excited states of coumarins, *J. Phys. Chem.* **74**:4234–4240.

Song, P. S., Harter, M. L., Moore, T. A., and Herndon, W. C., 1971, Luminescence spectra and photocycloaddition of the excited coumarins to DNA bases, *Photochem. Photobiol.* **14**:521–530.

Song, P. S., Mantulin, W. W., Mc.Inturff, D., Felkner, I. C., and Harter, M. L., 1975, Photoreactivity of hydroxypsoralens and their photobiological effects in *Bacillus subtilis, Photochem. Photobiol.* **21**:317–324.

Šorm, F., and Škoda, J., 1964, The mechanism of action of cancerostatically important azapyrimidines, *Acta Unio Int. Contra Cancrum* **20**:37–38.

Šorm, F., Šormova, Z., Raška, K., and Jurovcik, M., 1966, Comparison of the metabolism and inhibitory effects of 5-azacytidine and 5-aza-2′-deoxycytidine in mammalian tissues, *Rev. Roum. Biochim.* **3**:139–147.

Spectra-Physics Laser Review, 1975, **2**(2):3–4.

Spikes, J. D., 1967, Sensitized photochemical processes in biological systems, *Ann. Rev. Phys. Chem.* **18**:409–436.

Spikes, J. D., and Livingston, R., 1969, The molecular biology of photodynamic action: Sensitized photoautoxidations in biological systems, in: *Advances in Radiation Biology,* Vol. 3 (L. G. Augenstein, R. Mason, and M. Zelle, ed.), pp. 29–121, Academic Press, New York.

Spikes, J. D., and MacKnight, M. L., 1970, The dye-sensitized photooxidation of biological macromolecules, in: *Photochemistry of Macromolecules* (R. F. Reinisch, ed.), pp. 67–83, Plenum Press, New York.

Stankunas, A., Rosenthal, I., and Pitts, J. N., 1971, Photochemical and radiochemical alkylation of caffeine by alkyl amines, *Tetrahedron Lett.* 4479–4782.

Steele, P. H., and Cusachs, L. C., 1967, Energy terms of oxygen and riboflavin—a biological quantum ladder? *Nature (London)* 213:800–801.

Steinmaus, H., Rosenthal, I., and Elad, D., 1969, Photochemical and γ-ray-induced reactions of purines and purine nucleosides with 2-propanol, *J. Am. Chem. Soc.* 91:4921–4923.

Steinmaus, H., Elad, D., and Ben-Ishai, R., 1971, Ultraviolet light-induced purine modified DNA, *Biochim. Biophys. Res. Commun.* 40:1021–1025.

Stephan, G., Miltenburger, H. G., and Hotz, G., 1970, UV-induzierte Brüche in 5-Bromuracil-substituierter DNA des Phagen T1, *Z. Naturforsch.* 25b:1037–1042.

Stepien, E., Lisewski, R., and Wierzchowski, K. L., 1973a, Cyclobutane dimers of 1-methylthymine: Isolation, identification and properties, *Acta Biochim. Pol.* 20:297–311.

Stepien, E., Lisewski, R., and Wierzchowski, K. L., 1973b, Photochemistry of 2,4-diketopyrimidines: Photodimerization, photohydration and stacking association of 1,3-dimethyluracil in aqueous solution, *Acta Biochim. Pol.* 20:313–323.

Stermitz, F. R., Wei, C. C., and O'Donell, C. M., 1970, Photochemistry of qinoline and some substituted quinoline derivatives, *J. Am. Chem. Soc.* 9:2745–2752.

Stockert, J. C., and Lisanti, J. A., 1972, Acridine orange differential fluorescence of fast- and slow-reassociating chromosomal DNA after *in situ* denaturation and reassociation, *Chromosoma (Berlin)* 37:117–130.

Strauss, B. S., 1975, Repair of DNA in mammalian cells, *Life Sci.* 15:1685–1693.

Summers, W. A., and Burr, J. G., 1972, Viscosity effects on the photohydration of pyrimidines, *J. Am. Chem. Soc.* 76:3137–3141.

Summers, W. A., Enwall, C., Burr, J. G., and Letsinger, R. L., 1973, The photoaddition of nucleophiles to uracil, *Photochem. Photobiol.* 17:295–301.

Surovaya, A., and Trubitsyn, S., 1972, Binding isotherms of tRNA—acriflavine complexes, *FEBS Lett.* 25:349–352.

Sussenbach, J. S., and Berends, W., 1963, Photosensitized inactivation of deoxyribonucleic acid, *Biochim. Biophys. Acta* 76:154–156.

Sussenbach, J. S., and Berends, W., 1964, Photodynamic degradation of guanine, *Biochim. Biophys. Res. Commun.* 16:263–266.

Sussenbach, J. S., and Berends, W., 1965, Photodynamic degradation of guanine, *Biochim. Biophys. Acta* 95:184–185.

Sutherland, B. M., and Sutherland, J. C., 1969a, Mechanisms of inhibition of pyrimidine dimer formation in deoxyribonucleic acid by acridine dyes, *Biophys. J.* 9:292–302.

Sutherland, J. C., and Sutherland, B. M., 1969b, Energy transfer in the DNA-chloroquine complex, *Biochim. Biophys. Acta* 190:545–548.

Sutherland, J. C., and Sutherland, B. M., 1970, Ethidium bromide-DNA complex: Wavelength dependence of pyrimidine dimer inhibition and sensitized fluorescence as probes of excited state, *Biopolymers* 9:639–653.

Swenson, P. A., and Setlow, R. B., 1966, Effects of ultraviolet radiation on macromolecular syntheses in *Escherichia coli, J. Mol. Biol.* 15:201–219.

Swierkowski, M., and Shugar, D., 1969, A new thymine base analogue, 5-ethyluracil: 5-Ethyluridine-5′-pyrophosphate and poly-5-ethyluridylic acid, *Acta Biochim. Pol.* **16**:263–277.

Szabo, A. G., Riddell, W. D., and Yip., R. W., 1970, Detection and chemistry of the triplet state in acetonitril, *Can. J. Chem.* **48**:694–696.

Sztumpf, E., and Shugar, D., 1965, Preparation and properties of photoproducts of orotic acid analogues, *Photochem. Photobiol.* **4**:719–733.

Sztumpf-Kulikowska, E., Shugar, D., and Boag, J. W., 1967, Kinetics of photodimerization of orotic acid in aqueous medium, *Photochem. Photobiol.* **6**:41–54.

Szybalski, W., 1967, Molecular events resulting in radiation injury, repair and sensitization of DNA, *Radiat. Res. Suppl.* **7**:147–159.

Tamm, C., Shappiro, H. S., Lipschitz, H., and Chargaff, E., 1953, Distribution density of nucleotides within a deoxyribonucleic acid chain, *J. Biol. Chem.* **203**:673–688.

Tao, T., Nelson, J. H., and Cantor, C. R., 1970, Conformational studies on transfer ribonucleic acid: Fluorescence lifetime and nanosecond depolarization measurements on bound ethidium bromide, *Biochemistry* **9**:3514–3524.

Taylor, E. C., Maki, Y., and Evans, A. C., 1969, Photochemical addition of alcoholes to an amidine —C=N—bond, *J. Am. Chem. Soc.* **91**:5181–5182.

Thomas, J. C., Weill, G., and Daune, M., 1969, Fluorescence of proflavine–DNA complexes: Heterogeneity of binding sites, *Biopolymers* **8**:647–659.

Tomasz, M., and Chambers, R. W., 1966, The chemistry of pseudouridine. VII. Selective cleavage of polynucleotides containing pseudouridylic acid residues by a unique photochemical reaction, *Biochemistry* **5**:773–781.

Tomita, G., 1968, Absorption and fluorescence properties of some basic dyes complexing with nucleosides or nucleic acids, *Z. Naturforsch.* **23b**:922–925.

Toulme, J. T., Charlier, M., and Helénè, C., 1974, Specific recognition of single-stranded regions in ultraviolet-irradiated and heat-denatured DNA by tryptophan-containing peptides, *Proc. Natl. Acad. Sci. USA* **71**:3185–3188.

Träger, L., Türck, G., Ishimoto, M., and Wacker, A., 1964, Strahlenchemische Reaktionen zur Aufklärung molekulargenetischer Vorgänge, *Biophysik* **1**:403–406.

Tramer, Z., Wierzchowski, K. L., and Shugar, D., 1969, Influence of polynucleotide secondary structure on thymine photodimerization, *Acta Biochim. Pol.* **16**:83–107.

Trosko, J. E., and Isoun, M., 1971, Photosensitizing effect of tripsoralen on DNA synthesis in human cell grown *in vitro, Int. J. Radial. Biol.* **19**:87–92.

Trosko, J. E., Krause, D., and Isoun, M., 1970, Sunlight-induced pyrimidine dimers in human cells *in vitro, Nature* (*London*) **228**:358–360.

Tsugita, A., Okada, Y., and Uehara, K., 1965, Photosensitized inavitation of ribonucleic acids in the presence of riboflavin, *Biochem. Biophys. Acta* **103**:360–363.

Tubbs, K. R., Ditmars, W. E., Jr., and Van Winkle, Q., 1964, Heterogeneity of the interaction of DNA with acriflavine, *J. Mol. Biol.* **9**:545–557.

Uliana, R., Creach, P. V., and Ducastaing, A., 1971, Ouelques aspects de la radiodenaturation de l'acide desoxyribonucleique (ADN), *Biochimie* **53**:461–468.

Uretz, R. B., 1964, Sensitivity to acridine sensitized photoinactivation in *E. coli* B, B/r, and B_{s-1}, *Radiat. Res.* **22**:245–253.

Van de Vorst, A., and Lion, Y., 1971*a*, Formation de radicaux libres dans les constituants du DNA photosensibilises par la proflavine, *Biochim. Biophys. Acta* **238**:417–428.

Van de Vorst, A., and Lion, Y., 1971*b*, Mécanisme de la photosensibilisation des constituents du DNA par la proflavine: Une étude par resonance paramagnétique electronique, *Biochim. Biophys. Acta* **246**:421–429.

Varghese, A. J., 1970a, Photochemistry of thymidine in ice, *Biochemistry* 9:4781–4787.

Varghese, A. J., 1970b, Photochemistry of thymidine as a thin solid film, *Photochem. Photobiol.* 13:357–364.

Varghese, A. J., 1970c, 5-thyminyl-5,6-dihydrothymine from DNA irradiated with ultraviolet light, *Biochim. Biophys. Res. Commun.* 38:484–490.

Varghese, A. J., 1971a, Photochemical reactions of cytosine nucleosides in frozen aqueous solution and in deoxyribonucleic acid, *Biochemistry* 10:2194–2199.

Varghese, A. J., 1971b, Photochemistry of uracil and uridine, *Biochemistry* 10:4283–4289.

Varghese, A. J., 1972a, Photochemistry of nucleic acids and their constituents, in: *Photophysiology*, Vol. VII (A. C. Giese, ed.), pp. 207–274, Academic Press, New York.

Varghese, A. J., 1972b, Acetone-sensitized dimerization of cytosine derivatives, *Photochem. Photobiol.* 15:113–118.

Varghese, A. J., 1973, Properties of photoaddition products of thymine and cysteine, *Biochemistry* 12:2725–2730.

Varghese, A. J., 1974a, Photoaddition products of uracil and cysteine, *Biochim. Biophys. Acta* 374:109–114.

Varghese, A. J., 1974b, Photochemical addition of glutathione to uracil and thymine, *Photochem. Photobiol.* 20:339–343.

Varghese, A. J., 1974c, Photoreactions of 5-bromouracil in the presence of cysteine and glutathione, *Photochem. Photobiol.* 20:461–464.

Varghese, A. J., and Day, R. S., 1970, Excision of cytosine–thymine adduct from the DNA of ultraviolet-irradiated *Micrococcus radiodurans*, *Photochem. Photobiol.* 11:511–517.

Varghese, A. J., and Rupert, C. S., 1971, Ultraviolet irradiation of cytosine nucleosides in frozen solution products cyclobutane-type dimeric products, *Photochem. Photobiol.* 13:365–368.

Varghese, A. J., and Wang, S. Y., 1967a, Cis-syn thymine homodimer from ultraviolet-irradiated calf thymus DNA, *Nature (London)* 213:909–910.

Varghese, A. J., and Wang, S. Y., 1967b, Ultraviolet irradiation of DNA *in vitro* and *in vivo* produced a third thymine-derived product, *Science (Washington)* 156:955–957.

Varghese, A. J., and Wang, S. Y., 1968a, Thymine–thymine adduct as a photoproduct of thymine, *Science (Washington)* 160:186–187.

Varghese, A. J., and Wang, S. Y., 1968b, Photoreversible photoproduct of thymine, *Biochim. Biophys. Res. Commun.* 33:102–107.

Venner, H., and Zimmer, C., 1964, Zum Mechanismus der durch UV-Bestrahlung hervorgerufenen Veränderungen an DNA, in: *Physikalische Chemie biogener Makromoleküle* (H. Berg, ed.), pp. 341–347, Akademie Verlag, Berlin.

Vorličkova, M., and Paleček, E., 1974, A study of changes in DNA conformation caused by ionizing and ultra-violet radiation by means of pulse polarography and circular dichroism, *Int. J. Radial. Biol.* 26:363–372.

Vosa, C. G., 1971, The quinacrine-fluorescence patterns of chromosomes of *Allium carinatum*, *Chromosoma (Berlin)* 33:382–385.

Wacker, A., and Jacherts, D., 1962, UV-Resistenz Azathyminhaltiger Bakterienzellen, *J. Mol. Biol.* 4:413–418.

Wacker, A., and Lodemann, E., 1965, Einfluss der Grenzflächenenergie organischer Lösungsmittel auf die photochemische Dimerisierung von Thymidylyl-(3′ → 5′)-thymidin, *Angew. Chem.* 77:133–134.

Wacker, A., Dellweg, H., and Weinblum, D., 1960, Strahlenchemische Veränderung der Bakteriendesoxyribonucleinsäure *in vivo*, *Naturwissenschaften* 47:477–477.

Wacker, A., Dellweg, H., and Lodemann, E., 1961, Strahlenchemische Veränderungen der Nukleinsäuren *in vivo* und *in vitro*. 2. Mitteilung, *Angew. Chem.* 73:64–65.

Wacker, A., Dellweg, H., and Jacherts, D., 1962, Thymine dimerization and survival of bacteria, *J. Mol. Biol.* **4**:410–412.

Wacker, A., Türck, I. G., and Gerstenberger, A., 1963, Zum Wirkungsmechanismus photodynamischer Farbstoffe, *Naturwissenschaften* **50**:377–377.

Wacker, A., Ishimoto, M., Chandra, P., and Selzer, 1964*a*, Photoreaktivierung von UV-inaktivierter Polyuridylsäure, *Z. Naturforsch.* **19b**:406–408.

Wacker, A., Dellweg, H., Träger, L., Kornhauser, A., Lodemann, E., Türck, G., Selzer, R., Chandra, P., and Ishimoto, M., 1964*b*, Organic photochemistry of nucleic acids, *Photochem. Photobiol.* **3**:369–394.

Wacker, A., Kornhauser, A., and Träger, L., 1965, Isotopeneffekte bei der photochemischen Umwandlung von Tritium markiertem Uracil, *Z. Naturforsch.* **20b**:1043–1047.

Wacker, A., Chandra, P., Mildner, P., and Feller, H., 1968, Photodynamic action of acetone on the template activity of nucleic acids, *Biophysik* **4**:283–288.

Wagner, P. J., and Bucheck, D. J., 1968, Causes for the low efficiency of thymine and uracil photodimerization in solution, *J. Am. Chem. Soc.* **90**:6530–6532.

Wallnöfer, P., and Bukatsch, F., 1960, Untersuchungen über den photodynamischen Effekt von Acridinfarbstoffen an *Escherichia coli* und *Bacillus subtilis, Naturwissenschaften* **47**:282–283.

Walter, J. F., and Voorhees, J. J., 1973, Psoriasis improved by psoralen plus black light, *Acta Dermatol. Venereol. (Stockholm)* **53**:469–472.

Wang, S. Y., 1958, Photochemistry of nucleic acids and related compounds. I. The first step in the ultraviolet irradiation of 1,3-dimethyluracil, *J. Am. Chem. Soc.* **80**:6196–6198.

Wang, S. Y., 1959, Phototautomerization of cytosine derivatives by ultraviolet irradiation, *Nature (London)* **184**:184–186 (Suppl. 4).

Wang, S. Y., 1961, Photochemical reactions in frozen solution, *Nature (London)* **190**:690–694.

Wang, S. Y., 1964, The mechanism for frozen aqueous solution irradiation of pyrimidines, *Photochem. Photobiol.* **3**:395–398.

Wang, S. Y., 1965, Photochemical reactions of nucleic acid components in frozen solutions, *Fed. Proc.* **24**:71–79.

Wang, S. Y., 1971, Thymine phototrimer, *J. Am. Chem. Soc.* **93**:2768–2771.

Wang, S. Y., and Nnadi, J. C., 1968, Mechanism for the photohydration of pyrimidines, *Chem. Commun.* 1160–1162.

Wang, S. Y., and Rhoades, D. F., 1971, Pyrimidine phototetramer, *J. Am. Chem. Soc.* **93**:2554–2556.

Wang, S. Y., and Varghese, A. J., 1967, Cytosine–thymine addition product from DNA irradiated with ultraviolet light, *Biochem. Biophys. Res. Commun.* **29**:543–549.

Wang, S. Y., Patrick, M. H., Varghese, A. J., and Rupert, C. S., 1967, Concerning the mechanism of formation of UV-induced thymine photoproducts in DNA, *Proc. Natl. Acad. Sci. USA* **57**:465–472.

Waskell, L. A., Sastry, K. S., and Gordon, M. P., 1966, Studies on the photosensitized breakdown of guanosine by methylene blue, *Biochim. Biophys. Acta* **129**:49–53.

Webb, R. B., and Kubitschek, H. E., 1963, Mutagenic and antimutagenic effects of acridine orange in *E. coli, Biochem. Biophys. Res. Commun.* **13**:90–94.

Webb, R. B., and Petrusek, R. L. 1966, Oxygen effect in the protection of *E. coli* against U.V. inactivation and mutagenesis by acridine orange, *Photochem. Photobiol.* **5**:645–654.

Weill, G., and Calvin, M., 1963, Optical properties of chromophore-macromolecule complexes: Absorption and fluorescence of acridine dyes bound to polyphosphates and DNA, *Biopolymers* **1**:401–417.

Weinblum, D., 1967, Characterization of the photodimers from DNA, *Biochim. Biophys. Res. Commun.* **27**:384–390.

Weinblum, D., and Johns, H. E., 1966, Isolation and properties of isomeric thymine dimers, *Biochim. Biophys. Acta* **114**:450–459.

Weinblum, D., Ottensmeyer, F. P., and Wright, G. F., 1968, The structures of the isomeric thymine dimers as deduced from their dipole moments, *Biochim. Biophys. Acta* **155**:24–31

Weisblum, B., 1973, Fluorescent probes of chromosomal DNA structure: Three classes of acridines, *Cold Spring Harbor Symp. Quant. Biol.* **38**:441–449.

Weisblum, B., and Haenssler, E., 1974, Fluorometric properties of the bibenzimidazole derivative Hoechst 33258, a fluorescent probe specific for AT concentration in chromosomal DNA, *Chromosoma (Berlin)* **46**:255–260.

Weisblum, B., and De Haseth, P. L., 1972, Quinacrine, a chromosome stain specific for deoxyadenylate–deoxythymidylate-rich regions in DNA, *Proc. Natl. Acad. Sci. USA* **69**:629–632.

Weisblum, B., and De Haseth, P. L., 1973, Nucleotide specifity of the quinacrine staining reaction for chromosomes, *Chromosomes Today* **4**:35–51.

Whillans, D. W., and Johns, H. E., 1969, Dependence of intersystem crossing on excitation energy in orotic acid, *Photochem. Photobiol.* **9**:323–330.

Whillans, D. W., and Johns, H. E., 1971, Properties of the triplet states of thymine and uracil in aqueous solution, *J. Am. Chem. Soc.* **93**:1358–1362.

Whillans, D. W., and Johns, H. E., 1972a, Photoreactions in aqueous solutions of thymine, pH 12, *J. Phys. Chem.* **76**:489–493.

Whillans, D. W., and Johns, H. E., 1972b, Triplet state studies of the nucleoside and nucleotide derivatives of uracil and thymine, *Biochim. Biophys. Acta* **277**:1–6.

Whillans, D. W., Herbert, M. A., Hunt, J. W., and Johns, H. E., 1969, Optical detection of the triplet state of uracil, *Biochem. Biophys. Res. Commun.* **36**:912–918.

Wierzchowski, K. L., and Shugar, D., 1957, Photochemistry of cytosine nucleosides and nucleotides, *Biochim. Biophys. Acta* **25**:355–369.

Wierzchowski, K. L., and Shugar, D., 1959, Studies of reversible photolysis in oligo- and polyuridylic acids, *Acta Biochim. Pol.* **6**:313–334.

Wierzchowski, K. L., and Shugar, D., 1961a, Photochemistry of model oligo- and polynucleotides. IV. Hetero-oligo-nucleotides and high molecular weight single and double-stranded polymer chains, *Photochem. Photobiol.* **1**:21–36.

Wierzchowski, K. L., and Shugar, D., 1961b, Photochemistry of cytosine nucleosides and nucleotides. II. *Acta Biochim. Pol.* **8**:219–234.

Witkin, E. M., 1969, Ultraviolet-induced mutation and DNA repair, *Annu. Rev. Genet.* **3**:525–552.

Witkop, B., 1968, Mechanisms of photoreductions and hydrogenolysis of pyrimidine nucleosides and their photodimers, *Photochem. Photobiol.* **7**:813–827.

Witmer, H., and Fraser, D., 1970, Photodynamic action of proflavine on *Coliphage T3*. I. Kinetics of inactivation, *J. Virol.* **7**:314–318.

Witmer, H., and Fraser, D., 1971a, Photodynamic action of proflavine on *Coliphage T3*. II. Protection by L-cysteine. *J. Virol.* **7**:319–322.

Witmer, H., and Fraser, D., 1971b, Photodynamic action of proflavine on *Coliphage T3*. III. Damages to the deoxyribonucleic acid associated with RX1 and RX2, *J. Virol.* **7**:323–331.

Wulff, D. L., and Fraenkel, G., 1961, On the nature of thymine photoproduct, *Biochim. Biophys. Acta* **51**:332–339.

Yamabe, S., 1969, A fluorospectrophotometric study on the binding of acridine orange with DNA and its bases, *Arch. Biochem. Biophys.* **130**:148–155.

Yamamoto, N., 1958, Photodynamic inactivation of bacteriophage and its inhibition, *J. Bacteriol.* **75**:443–448.

Yamamoto, N., 1967, Photodynamic action on bacteriophage genome: Inactivation and genetic recombination of bacteriophage, in: *Molekulare Mechanismen photodynamischer Effekte, Stud. Biophys.* **3**:175–180.

Yamasaki, N., 1973, Differentielle Darstellung der Metaphasechromosomen von *Cypropedium debile* mit Chinacrin- und Giemsafärbung, *Chromosoma (Berlin)* **41**:403–412.

Yan, Y., 1969, Effect of UV-irradiated DNA containing 5-bromuracil on reactivation of UV damage in phage λ. *Int. J. Radial. Biol.* **16**:367–376.

Yang, N. C., Gorelic, L. G., and Kim, B., 1971, A new photochemical reaction of purine, photochemical alkylation of purine by 1-propylamine, *Photochem. Photobiol.* **13**:275–278.

Yaniv, M., Chestier, A., Gros, F., and Favre, A., 1971, Biological activity of irradiated tRNA^{Val} containing 4-thiouridine-cytosine dimer, *J. Mol. Biol.* **58**:381–388.

Yasuda, K., and Fukutome, H., 1970, Inactivation of *E. Coli* ribosomes by ultraviolet irradiation. III. The activity of poly U-directed binding of phenylalanyl-tRNA, *Biochim. Biophys. Acta* **217**:142–147.

Yguerabide, J., 1972, Nanosecond fluorescence spectroscopy of macromolecules, in: *Methods in Enzymology*, Vol. XXVI, Enzyme Structure Part C (C. H. W. Hirs and S. N. Timasheff, eds.), pp. 498–578, Academic Press, New York.

Yielding, K. L., and Sternglanz, H., 1971, Comments on the interaction of LSD with DNA, in: *Progress in Molecular and Subcellular Biology*, Vol. 2 (F. E. Hahn, ed.), pp. 163–165, Springer Verlag, Berlin.

Yip, R. W., Riddell, W. D., and Szabo, A. G., 1970, Triplet state of orotic acid and orotic acid methyl ester in solution, *Can. J. Chem.* **48**:987–999.

Zampieri, A., and Greenberg, J., 1965, Mutagenesis by acridine *orange and proflavine in Escherichia coli* strain S, *Mutat. Res.* **2**:552–556.

Zavilgelskij, G. B., Iljasenko, B. W., Minjat, E. E., and Rudoenko, C. N., 1964, *Dokl. Akad. Nauk (USSR)* **155**:937–939.

Zenda, K., Saneyoshi, M., and Chihara, G., 1965, Biological photochemistry. I. The correlation between the photodynamical behaviours and the chemical structures of nucleic acid bases, nucleosides, and related compounds in the presence of methylene blue, *Chem. Pharm. Bull. (Tokyo)* **13**:1108–1117.

Zierenberg, B. E., Krämer, D. M., Geisert, M. G., and Kirste, R. G., 1971, Effect of sensitized and unsensitized long-wave U.V.-irradiation on the solution properties of DNA, *Photochem. Photobiol.* **14**:515–520.

Zügel, M., Förster, T., and Kramer, H. E. A., 1972, Sensitized photooxygenation according to type-I mechanism (radical mechanism). III. Experiments with continuous illumination, *Photochem. Photobiol.* **15**:33–42.

3

Ultraviolet Radiation Effects on the Human Eye

Richard B. Kurzel, Myron L. Wolbarsht, and Bill S. Yamanashi

Harvard University School of Medicine, Boston, Massachusetts 02115

1. INTRODUCTION

Many papers and reviews in the past have focused their attention on the question of ultraviolet (UV) radiation damage to ocular tissues. Unfortunately, all of these works have dealt with damage to specific tissues, and to get a comprehensive view of the field, the reader has been required to consult many references. In this chapter, we will comprehensively cover what is known about UV radiation damage to all ocular tissues. For clarity and for completeness, some repetition of older work is necessary, but the study of UV radiation effects on the eye has only begun to unfold in recent years, and therefore the emphasis here is on recent work. While UV radiation effects on the cornea have been established for many years, the question of UV damage to the retina has only recently been seriously entertained and established. The literature on the effects on the retina, therefore, is still scant. The question of UV radiation-induced aging or damage to the lens is a topic of much controversy and speculation. As the latest example of possible solar radiation-induced aging of a human tissue, this is perhaps one of the most exciting new directions in ophthalmology and human biology research. We will therefore study this question in depth.

An appreciation for the processes involved in the interaction of UV radiation with ocular tissues requires a firm understanding of photochemistry and photobiology. Many excellent reviews of these subjects exist, and some of these have been referenced. An outline or mention of fundamental photochemistry has been made only in those situations where it is needed for clarity or for amplification of a particular point.

2. TRANSMISSION AND EXPECTATIONS

Research dealing with the interaction of UV radiation on ocular media has traditionally taken two directions: morphological studies of the radiation damage to particular tissues in the eye, and measurement of the spectral transmission characteristics of the ocular tissues. Damage to the cornea and conjunctiva has been recorded since ancient times, and is well accepted. UV radiation effects on the corea may take two forms: radiation-induced tumors and photokeratitis. Freeman and Knox (1964) have shown that tumors of the cornea result from UV irradiation, and wavelengths from 220 to 400 nm are effective. Photokeratitis has a more limited range ($\simeq$220–340 nm, Figs. 1 and 2), with more distinctive sensitivity to UV radiation. This problem is the subject of the next section. The subject of UV radiation damage to the lens and retina has come of age only within the last few years, and is a topic of considerable speculation and controversy.

Although the literature on transmission characteristics of the eye is rather extensive, there is much contradiction, and much work has yet to be done. For example, Cogan (1950) reported that for the cornea the percent transmittance decreases abruptly at 400 nm and is zero at 350 nm. He also pointed out that wavelengths shorter than 320 nm never reach the lens. These results are incorrect, and today it is generally accepted that the absorption cutoff due to the cornea occurs in the range of 285–295 nm. This point is not insignificant when considering the possibility of radiation

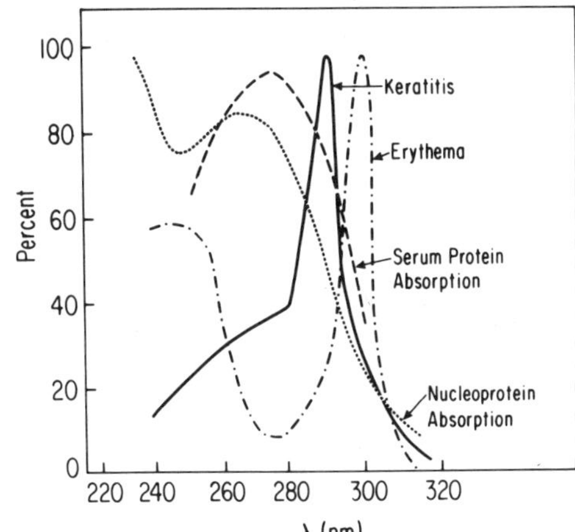

Fig. 1. Action spectra for photo-keratitis and erythema, and absorption spectra for serum protein and nucleo-protein. Adapted from Cogan and Kinsey (1946).

damage to the lens and retina. A set of reasonable transmission data was obtained by Kinsey (1948), and is presented in Table 1. Notice should be taken that the filtration effect of the various tissues is important in the eye. Much has been made of this line of reasoning to exclude the possibility of radiation damage to any tissue in the eye other than the cornea. While at first this may seem reasonable, fine points such as the essential differences in the nature of the tissues are generally ignored.

The action spectrum of each of the different types of effects of UV radiation on the eye, however, is not as well worked out. Two important points in this regard are discussed in detail by Sliney (1972). These investigations are hampered by the limited number of optical radiation sources

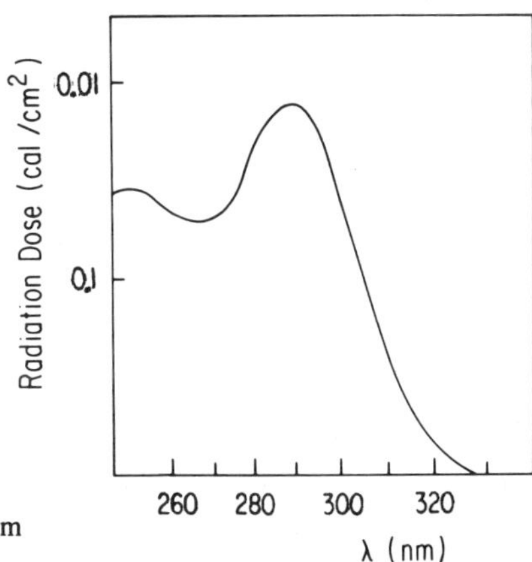

Fig. 2. Action spectrum for photokeratitis. From Sherashov (1970).

TABLE 1. Percent of Energy Incident on the Corneal Epithelium That Impinges on the Anterior Surface of the Various Ocular Media[a]

λ (nm)	Corneal stroma	Aqueous humor	Lens	Vitreous humor	Retina
230	2.7	—	—	—	—
235	10.8	—	—	—	—
240	19.0	—	—	—	—
245	26.0	—	—	—	—
250	26.0	—	—	—	—
260	26.0	—	—	—	—
265	26.5	—	—	—	—
270	28.7	—	—	—	—
275	30.7	—	—	—	—
280	33.2	—	—	—	—
285	41.5	—	—	—	—
290	51.5	2.0	0.4	—	—
295	63.0	8.7	3.2	—	—
300	70.0	27.0	14.3	—	—
305	75.0	50.0	37.0	—	—
310	78.0	64.0	50.5	—	—
320	81.0	78.0	74.0	0.33	0.29
330	84.0	80.0	76.6	0.52	0.45
350	86.5	86.0	82.5	1.82	1.62
360	88.5	88.0	84.5	4.2	3.8
370	89.5	90.0	87.0	12.1	11.0
380	93.0	91.2	88.2	28.2	25.8
390	94.0	92.8	91.4	48.5	44.6
400	94.5	94.0	93.0	68.5	63.5
450	—	96.0	96.0	84.0	80.5
500	—	96.0	96.0	86.5	84.0

[a] From Kinsey (1948).

that emit sufficiently high levels of UV radition, and by the difficulty in performing spectral irradiance measurements in this spectral region. In the past, the most practical light sources have generally been limited to the intense mercury arc wavelengths, which are not well spaced throughout the spectrum. For this reason, large segments of the shapes of most action spectra are extrapolated. There may be hidden peaks and valleys that could have considerable theoretical importance but are as yet unknown.

While the amount of UV radiation reaching the lens and retina is small, some radiation does reach these tissues. Additional points dealing with transmission characteristics will be dealt with in the respective sections on ocular tissue damage.

3. THE CORNEA AND CONJUNCTIVA: PHOTOKERATITIS

3.1. Symptoms of Photokeratitis

The mechanism of photokeratitis (the inflammation of the cornea due to exposure to light), like many other effects of radiation on cells, is poorly understood. Clinically, after the initial exposure to UV radiation, a period of latency occurs (30 min–24 h), varying in length inversely with the severity of the exposure. A sensation of sand in the eyes is common, with varying degrees of photophobia (intolerance to light), lacrimation (tearing), and blepharospasm (tonic spasm of the lid muscles). Conjunctivitis is observed, accompanied by erythema of the surrounding skin of the eyelids and face. These acute symptoms last for 6–24 h and disappear by 48 h. Rarely does permanent injury result, and the ocular system does not develop tolerance to repeated exposure. A study of the cornea on a gross scale shows, in addition to conjunctivitis, stippling or visible granulation of the corneal epithelium (mosaiclike), which is presumably due to the swelling or shrinking of some cells. The stippling may be positive (small protrusions) or negative (small indentations). With higher doses of UV radiation, clouding of the stroma occurs, along with a threshold reaction in the endothelium (Cogan and Kinsey, 1946), epithelial exfoliation *en masse*, and a duplication of layers in which mitosis is normally seldom seen.

On a finer scale, four effects of UV radiation are observed in the corneal epithelium (Duke-Elder, 1954):

1. Inhibition of mitosis. This effect occurs early and with small doses of UV radiation. The effect is maximized in early prophase.

2. Nuclear fragmentation. This is observed with higher doses of UV radiation, and occurs in four stages:
    a. Nuclear swelling.
    b. Aggregation of chromatin reticulum.
    c. Bursting of the nuclear membrane.
    d. Swelling of the cell body, after which the cell dies and is desquamated.

3. Vacuole formation. Increased dye binding to proteins in the nucleus and cytoplasm is observed in this stage. The dye globules resemble inclusion bodies in the cell body, and are associated with the vacuoles.

4. Loosening of the epithelial layer. Along with the changes in the cornea, the conjunctiva shows inflammation, with swelling and edema. It shows similar desquamation of the outer layers of the epithelial cells, while the basal cells are degenerated and spread out. Small hemorrhages are

observed, with congestion and plasma cell infiltration. The subepithelial connective tissue shows hyaline degeneration.

From the above description, two points become immediately obvious: the role of filtration is important when considering individual tissues in the eye, and also when considering the layers comprising each tissue. At a threshold dose of radiation, the stroma and endothelium are rarely affected, due to filtration by the epithelium. The progression of filtering layers acts to reduce the dose of radiation from particular wavelengths. The accumulation of exfoliated cells, or stippling, and the buildup of dead cell layers effectively provide a filtering layer to prevent further damage to the tissue. This has been observed by Hemmingsen and Douglas (1970), who noted that the increased resistance of the cornea to UV radiation varies directly with the degree of stippling. It is also worthwhile to note that each of the four characteristics of damage observed histologically is observed spatially in different parts of the tissue due to this filtration (and therefore radiation dosage) effect (Friedenwald *et al.,* 1948). Nuclear fragmentation occurs almost exclusively in the superficial layers, while loosening of the corneal epithelium takes place at its boundary with the underlying connective tissue. When an effect is observed in the basal layers, it is usually the inhibition of mitosis. Nuclear fragmentation, therefore, requires high radiation doses. It is questionable whether the clinically exhibited symptoms (e.g., inflammation) would be expressed if the dosage were sufficiently low such that the cells did not pass the mitosis inhibition stage. Vacuole formation probably is produced under high dosage. In order to make a macroscopic measurement of sensitivity (i.e., action spectrum), one must note an observable parameter such as the number of vacuoles per unit cell volume. Working at high radiation dosages not only introduces widespread damage but also may release a factor, probably near cell rupture, which may trigger the observed clinical symptoms. Therefore, the various processes induced by radiation in the cell may have different thresholds. The release of the factor signaling inflammation is probably radiation dosage and wavelength dependent, and therefore results only from the most superficial layers of the corneal tissue. We will return to this point in Section 3.7.

3.2. The Chemical Model for Photokeratitis

Presently, no data exist to support any particular model for photokeratitis. Until recently, the only information available has been in the form of action spectra, and these spectra have served only to confuse researchers further. In the discussion that follows, frequent reference will be made to research on UV radiation-induced erythema of the skin, since definite parallels exist in the two problems. In fact, workers studying photokeratitis have often taken note of the erythema problem, since the action spectra for both effects are similar in shape, and their maxima are separated by only 8 nm.

One of the first, classical works on photokeratitis was that of Cogan and Kinsey (1946), who established the sensitivity, or action spectrum, for this process (Fig. 1). The maximum efficiency for keratitis production peaks sharply at 288 nm, and has a long wavelength limit of 306–326 nm. The band shapes for keratitis and erythema are similar, but the latter is shifted so that its maximum is situated at 297 nm. The major difficulty in assigning the effect due to absorption by a particular chromophore has been the unusual shape of the action spectrum. Ideally, the action spectrum should correspond to the absorption spectrum of the responsible chromophore. Not only is the action spectrum embarrassingly "spikelike" (few if any biological macromolecules have an absorption spectrum whose width at half-height is less than 10 nm), but the maximum sits too far to the red to be conveniently assigned even to protein absorption. The various attempts to explain the absorption process have therefore relied on tricks to change the action spectrum into a somewhat more appealing form. Sherashov (1970) has remeasured the action spectrum for keratitis using a huge animal sample number and treating his data statistically (Fig. 2). Although the shape of the curve is more pleasing, the positions of the maxima have not changed (253.7 and 289.4 nm), and a log scale was used for the ordinate instead of Kinsey's linear scale. The results are still essentially the same, and no better explanation has been forwarded. Notice, however, that this more careful work has detected a small maximum at 254 nm. This minor peak was also found by Bachem (1956), along with the standard maximum for keratitis occurring at 288 nm. The work of Cogan and Kinsey (1946) must therefore stand, because of its reproducibility.

One worker has recently produced results that are in disagreement with those mentioned above. These are given Fig. 3. It is not immediately obvious how the experimental technique of Pitts and Tredici (1971) is so different from that of all the other workers to yield such a different result. Their work implicates either a single molecule with a complicated spectrum or perhaps multiple chromophores involved in the damage in a multitarget model. Such a model may be more acceptable than the one-target model of Cogan (1950), who suggested nucleoprotein as being the absorbing chromophore. In addition to the three maxima observed for primates (220, 240, and 270 nm), careful inspection of Fig. 3 shows a shoulder at ~289 nm, indicating that the curve given may consist of at least two overlapping curves, one of which is the standard curve for keratitis. Is the second curve with its

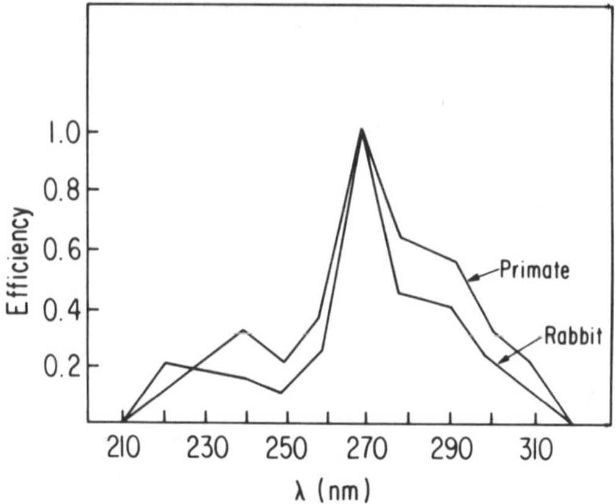

Fig. 3. Action spectra for photo-keratitis in primate and rabbit. After Pitts and Tredici (1971).

maximum at 270 nm an artifact? Considering our lack of knowledge of the processes involved and the difficulty in measuring these spectra, one hesitates to label this work artifactual. In subsequent work, Pitts was able to duplicate his action spectrum using a different technique, the measurement of light backscatter from the damaged epithelial cells and vacuoles, which is wavelength dependent (Pitts and Gibbons, 1973).

An action spectrum, in the ideal situation where there are no interfering factors, mimics the absorption spectrum of the responsible chromophore(s). Since the action spectrum is measured on intact tissues or cells, showing the cumulative effect of all chromophores, it does not necessarily identify the single specific chromophore responsible for the effect. Also, rapid cellular changes due to high radiation levels may complicate the picture. The total action spectrum for photokeratitis is, therefore, potentially nonlinear, with the involvement of more than one chromophore. This makes the identification of a chromophore on the basis of the overall (possibly composite) action spectrum tenuous. These points were demonstrated by Friedenwald *et al.* (1948), who showed that the three processes—mitosis inhibition, epithelial loosening, and nuclear fragmentation-exhibited different action spectra. (These processes are listed in order of increasing radiation dosage required to obtain maximum response. Each differs from the previous process by dose × 10.) Although each of the three spectra had its maximum at ~289 nm, the spectrum for nuclear fragmentation had a broad band shape, different from the other two processes, and, significantly, the curves crossed. This implies that, for the higher levels required for nuclear fragmentation, additional or different chromophores are active in the process.

While the underlying objective in the photokeratitis action spectra determinations was to identify the critical chromophores, almost 40 years of such spectral determinations have been relatively unrevealing, and the traditional question of the photobiologist of whether it is protein or nucleic acid that acts as the critical absorber is still unanswered. If the maximum in the keratitis action spectrum had been situated at $\lambda \sim 260$ nm, one could have made a strong argument for the primary role of nucleic acids in the cell damage, although the maximum at $\lambda \sim 288$ nm does not completely rule out this argument. In the wavelength range of 240–290 nm, nucleic acids absorb 10–20 times as much as an equal weight of protein, due to the fact that all nucleic acids are aromatic while only a few of the amino acids are. However, McLaren and colleagues (McLaren *et al.*, 1953; McLaren and Takahashi, 1957) have shown that in the middle UV (260 nm) the product of the molar absorptivity (ϵ) and the quantum yield for damage (ϕ) is comparable for proteins and nucleic acids. Setlow (1960) has confirmed this by showing that the absorption cross sections were comparable for these two classes of compounds in the middle UV and that for $\lambda < 290$ nm the absorption cross section for protein actually was larger than that for the nucleic acids. This, coupled with the fact that the nuclear/cytoplasmic ratio of the corneal epithelium cells is not large, could serve as an argument to support a role for protein absorption in cell injury. Unfortunately, the action spectrum neither confirms nor rejects the role of protein. Answers to this question will have to come from biochemical or histochemical work.

3.3. Filtration and the True Action Spectrum

In the ideal situation, where there are no interfering factors, the action spectrum corresponds to the absorption spectrum of the responsible chromophores. The most important and common of the interfering factors is absorption or filtration by layers anterior to cells of the tissue studied. From the action spectra of both erythema and keratitis, it is obvious that a filter effect is governing and distorting the true band shape of the chromophore(s). The mode of action of the filter in producing the modified absorption spectrum is outlined in Fig. 4. The action spectrum, therefore, is a difference spectrum of the absorption of the chromophore and that of a dominant anterior absorber. The resultant product is characteristically spike shaped. Attempts to implicate a role of a particular chromophore based on the band shape and position of maximum absorption of the action spectrum, as have been done in the past, are invalid in the case of UV-induced photokeratitis. In the present case, because of the effect of filtra-

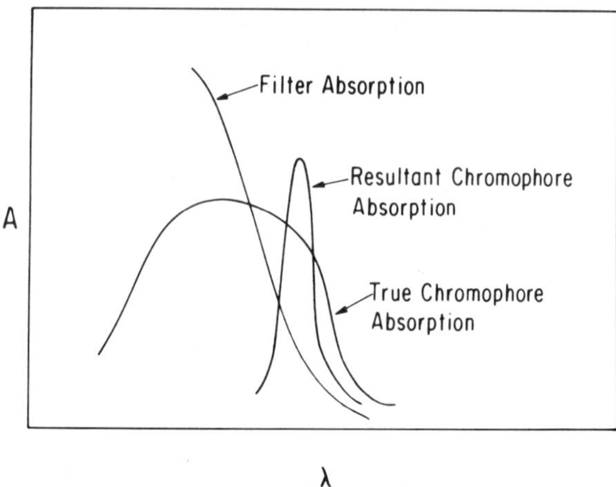

Fig. 4. Factors controlling action spectrum analysis. The combined effect (difference spectrum) of the filter absorption and the true chromophore absorption results in a typical "spike-shaped" action spectrum; hence, caution must be excercised not to interpret the resultant spectral graph as a true action spectrum.

tion, the resultant band shape bears no resemblance to the true band shape of the responsible chromophore, and the true absorption maximum has been shifted to the red. However, a few things can be said about the identity of the chromophore from just a knowledge of its higher wavelength limit of absorption. Not only aromatic amino acids but also certain nucleotide bases can be considered as potential absorbers, even though traditionally only proteins are considered seriously for $\lambda \sim 280$ nm. That the higher wavelength limit of the action spectrum approaches 326 nm implies the greatest likelihood for the involvement of tryptophan (Trp). However, to be more accurate, one should also consider the quantum yield. Less can be said about the low wavelength part of the curve, due to ignorance of the shape of the filter and chromophore absorption curves. However, it seems probable that the 254-nm absorption peak may be a second chromophore involved in the damage. It is likely that different chromophores play different (or dominant) roles in mediating the damage. The fact that the filter cuts off sharply at ~ 285 nm may imply the action of proteins rich in tyrosine (Tyr) and phenylalanine (Phe). The absorption of both Tyr and Phe drops off abruptly near 285 nm. That no damage results to the filter (or at least the damage does not manifest itself in the symptoms with which we are concerned) is implied in its role and in the action spectrum. Otherwise, it would not be a filter, but rather another chromophore. The verification of the role of filtration on the action spectrum for erythema has been obtained by Claesson *et al.* (1959), who demonstrated that the susceptibility of the skin increased and shifted to shorter wavelengths on removal of layers of the stratum corneum, which filtered out all wavelengths below 293 nm. Since the cornea has no cells or tissues lying anterior to the corneal epithelium that can act to filter radiation, perhaps the filtering absorbers are constituents of the epithelial cell membrane itself, such as glycoproteins.

3.4. Scattering

In an attempt to identify the absorbing chromophore from the action spectrum, Cogan and Kinsey (1946) noted that the absorption was not likely to be due to nucleoproteins (λ = 265 nm), and suggested that it more closely approximated cytoplasmic proteins and globulins (an invalid argument as already discussed above). This is shown in Fig. 1. Cogan (1950) suggested that the same model used to explain the band position for erythema be used to explain photokeratitis: differential scattering by superficial layers. It was estimated that by correcting for differential scattering the position of both action spectra maxima could be shifted farther to the blue to give better coincidence with typical protein absorption near 280 nm. However, this reasoning is wrong for both erythema and keratitis. The main effect of scattering is to increase the effective path length for absorption. The shift in the absorption maximum due to scattering will be minimal.

3.5. Some Recent Histochemical Results

In the last several years, three papers were published in Poland by Hamerski (1969, 1971; Hamerski and Zajaczkowska, 1969) dealing with an analysis of the chemical changes resulting during the postirradiation latency period. Clinical signs of radiation damage usually appear 6 h after irradiation. Analysis of the protein content was made for α_1-, α_2-, β-, and γ-globulins of the serum 4 h after irradiation (Hamerski, 1969). It was found that β-globulin fell in amount while γ-globulin rose, the disparity reaching a maximum at the period of most intense inflammation. The disparity regressed along with the symptoms. Histochemical tests detected changes in sulfhydryl content during the latency period (Hamerski and Zajaczkowska, 1969). Protein and free sulfhydryl groups were observed to decrease in quantity, while disulfide groups proportionately increased. The activity of alkaline phosphotase, acid phosphotase, ATPase, and 5-nucleotidase was observed to decrease. The decrease may be due to enzyme inactivation by UV absorption by cysteine (Cys), resulting from two sulfhydryl groups forming an S—S bond, as proposed by Hamerski and Zajaczkowska (1969). However, it seems more likely that the S—S bonds would form SH groups following UV irradiation. Addition of free Cys in the conjunctival sac before irradiation delayed the keratitic symptoms with decreased severity. Presumably, the free Cys acted as a filter for the radiation. The addition of glycerol to the conjunctival sac completely prevented the onset of keratitis. Hamerski and Zajaczkowska (1969) assumed that the glycerol acted to limit the role of water. The mode of action of glycerol in offering protection may be to more effectively filter the radiation, or to act as a free radical scavenger.

Hamerski's work (1969; Hamerski and Zajaczkowska, 1969) is significant because it is the first time that this problem has been approached biochemically, and it demonstrates that the damage is bound to be both complex and widespread. It is possible that the critical damage may be centered in the cytoplasm or in nucleoproteins, with widespread enzyme inactivation and structural protein alterations taking place. The choice to investigate sulfhydryl involvement was arbitrary, and similar studies on proteins involving other amino acid changes, e.g., to the aromatic amino acids, would probably have been positive for damage also. That protein damage is involved is therefore conclusive. These results do not indicate whether these changes are the critical ones for the cell, however. The role of nucleic acids is more obscure. Most recently, Hamerski (1971) reported a weakening in the histochemical tests for RNA and DNA following corneal irradiation, which is presumbly due to decreased DNA synthesis. Whether this decrease is a primary or secondary effect of the UV radiation cannot as yet be specified. At this point, Hamerski's results are consistent with the older histochemical work, which points to absorption of radiation by both the nucleoproteins and cytoplasmic proteins. Damage to the nucleoproteins could result in the observed inhibition of mitosis. This inhibition, which takes place in early prophase, could conceivably be explained in terms of spindle destruction, which has been reported to involve protein dependence (Carlson and Hollaender, 1948). Additionally, the observed binding of dyes to irradiated cytoplasmic and nucleoproteins suggests the formation of proteins with new reactive sites.

3.6. Repair Processes

The fact that patients recover from photokeratitis indicates the presence of repair mechanisms. Only one study of the repair processes in the cornea has been noted in the literature. Following irradiation of the cornea with middle UV, Genter (1971) noted that the basal cells of the corneal epithelium were several times more resistant to UV radiation than fibroblasts. He assumed that this resistance was due to both dark repair and photoreactivation processes. He did not, however, specifically test for photoreactivation.

The observation that the increasing resistance to UV radiation of the corneal epithelial cells varies directly with the degree of stippling (Hemmingsen and Douglas, 1970), as mentioned earlier, can therefore be explained in terms not only of radiation shielding of the basal layers by those lying anterior to it but also of "photoprotection" mechanisms, as suggested by Genter's (1971) results.

This is a promising field for future work, and it would be very useful to carry out tissue culture experiments to determine which repair processes really are utilized in the cornea; photoreactivation, indirect photoreactivation, dark repair, or postreplication repair. Knowledge of the presence of a particular repair mechanism would indirectly inform us of the nature of the particular lesion, which is still unknown. For example, if Genter (1971) is correct that photoreactivation is active in the cornea, this would point out the presence of pyrimidine dimers. At this point, one can say with some certainty that the damage to the cornea in photokeratitis involves the genetic apparatus (DNA or nucleoprotein), but the direct involvement of DNA is not yet established.

3.7. Release of Chemical Mediators

The idea that a specific chemical is released under UV stimulation and is responsible for mediating the inflammatory response is an old one. Although this is an idea that has not yet been verified after much research on the problem of UV radiation-induced erythema, it may be applicable for the cornea. In the case of the irradiated skin, it has been adequately shown by Grof and Kovacs (1967) that UV irradiation causes the release of histamine from the tissue mast cells, and this release coincides with the vascular response in the inflammatory reaction. Hirabayashi and Graham (1969) also showed that UV irradiation caused the released of serotonin and heparin into the tissue, in addition to histamine. Since the cornea is avascular and contains no tissue mast cells, the nature of the substance released from the injured corneal epithelial cells is still unknown, and this problem has never been approached before for the cornea. That this messenger may still be histamine is doubtful, although Ellinger (1951) has suggested that the radiation may act to generate histamine from histidine (His) via decarboxylation. It is more likely that the injured cells suffer increased permeability of their membranes, and that the messenger is actually cytoplasmic contents that have leaked out. These cytoplasmic contents may then diffuse across the corneal surface to the limbus where they encounter the vasculature and signal the process leading to the inflammatory reaction. The irradiated cell may suffer increased membrane permeability either through direct damage or more likely by radiation damage to the respiratory apparatus, thereby resulting in a failure of the membrane pump, which causes a leaky membrane. The irradiation may give rise to absorption and destruction of some of the mitochondrial quinones. Kashet and Brodie (1963) have suggested that UV radiation can destroy some of the enzymes required for oxidative phosphorylation, e.g., succinate cytochrome *c* reductase. Although these latter changes have been reported for the near-UV interruption of the electron transport chain in *Escherichia coli,* conflicting evidence has been presented by Jagger (1967).

4. ULTRAVIOLET RADIATION EFFECTS ON THE LENS

4.1. Normal Lens Pigmentation

In contrast to many other mammals which have pigmented corneas, or colored oil droplets in the cones as well as colored lenses, the human eye has pigment only in the lens. This pigment is present from birth. That the normal lens fluoresces in the blue has been known for many years, and the fluorescence arises primarily from the lens pigments. Until 1972, the number and identity of these pigments were unknown. Oguchi *et al.* (1973) and van Heyningen (1970) independently suggested that more than one pigment was responsible for the lens absorption and fluorescence, with the fluorescence ranging from purple-blue to green-yellow, depending on the pigment. Van Heyningen (1973*a*) was the first to identify the pigments in the human lens. There are at least five of these pigments, with the three main components being oxidation products of Trp. The hydroxylated products are conjugated to form glucosides. That the main pigments in the lens are oxidation products of Trp may have significance in their physiological role as pigments, and, as will be discussed in Section 4.2, was important in deducing the nature of the changes resulting in excessive pigmentation in the aged lens, and in brunescent cataractous lenses (brown-colored cataract of the nucleus of the eye). The absorption and emission characteristics of normal pigment lenses using numerous excitation lines are shown in Fig. 5. The fluorescence of Trp residues (curve 1) matches closely with the kynurenine (Kyn) residues (curve 2); however, the true identity of this chromophore and its mode of attachment to the peptide still remain to be determined. Stein *et al.* (1976) reported that fluorescence resembling that

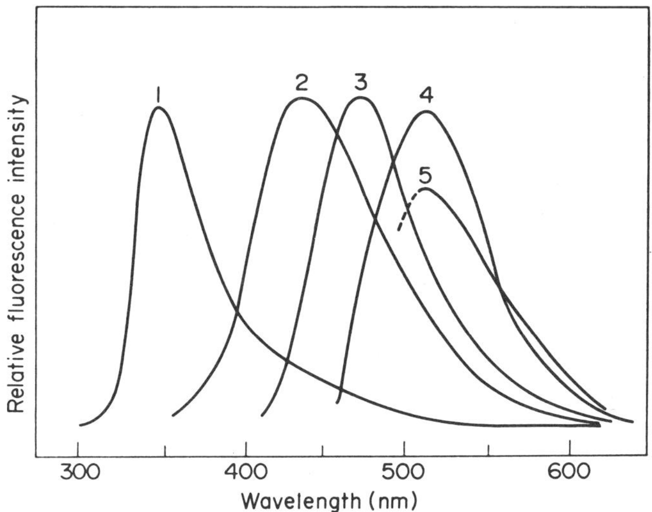

Fig. 5. Total emission spectra (fluorescence) of the healthy intact human lens. The wavelengths of exciting light are as follows: curve 1, 300 nm; curve 2, 350 nm; curve 3, 400 nm; curve 4, 450 nm; curve 5, 500 nm. Curves were not recorded at the same intensity. Adapted from Kurzel *et al.* (1973*a*).

of Kyn residues (curve 2), with a λ_{max} of 410–430 nm, is seen in both cataractous (senile and brunescent) and normal lenses for 340- or 350-nm excitation. They observed that both water-soluble and guanadine-HCl-soluble fractions fluoresce, although the brunescent cataractous lens has a much higher intensity in the guanadine-HCl-soluble fraction than a normal lens. The molecules responsible for curves 3, 4, and 5 are unknown, although the hydrolysis products of Kyn have a fluorescence spectrum similar to curve 5.

One of the earliest theories to explain the physiological role of the lens pigments was suggested by Wald (1952), who felt that the pigments served to correct for chromatic aberration in the lens. However, today it is generally accepted that the lens pigments also serve a protective role in filtering out near-UV radiation from reaching the retina. Many forms of indirect evidence support this theory. Wald (1952) has noted that the retina of the aphakic eye (i.e., with the lens removed) is 1000 times more sensitive to radiation of 365 nm wavelength, and the sensitivity of the rods and cones of the retina is extended to wavelengths down to 300 nm. The harmful effects of these wavelengths on the retina are uncertain and will be mentioned later in the discussion on the retina. Since rhodopsin has a second absorption maximum at 350 nm (Dartnall, 1957) (β band), and the cone photopigments have maxima at 315, 357, and 374 nm, near-UV radiation can result in visual stimulation (Wolbarsht, 1976; Wolbarsht and Yamanashi, 1976). It has been noted that aphakic patients can see 365-nm radiation as "violet" light (Wald, 1952). Normally, these wavelengths are excluded from the eye by the lens pigments. The filtration effect on the retina has also been noted by Said and Weale (1959), who observed that the log scotopic sensitivity (dark-adapted vision) decreased linearly with increasing concentration of lens pigment. An important and very interesting piece of indirect evidence pointing to the protective role of the lens pigments was presented by van Heyningen (1973a), who demonstrated a variation in the composition of the lens pigments with respect to age.

Since the retina of the newborn is immature, one might expect the effects of UV radiation to be especially damaging in the newborn. Although the lens pigments present in the child are the same as those present in the adult, van Heyningen (1973b) showed that the amount of o-β-D glucoside of 3-hydroxykynurenine was increased for ages 0–7 years, and steadily decreased as a function of age. The other pigments showed no appreciable concentration variation with respect to age. Such an age variation would have the effect of shifting the absorption farther to the blue in the newborn, to more effectively exclude UV radiation.

One might wonder whether there was further significance in the lens pigments being oxidation products of Trp insofar as the radiation exclusion model was concerned. Tomicic *et al.* (1973) have shown that indole derivatives, including Trp, have a unique ability to offer protection from UV radiation to DNA, as measured from absorption loss to DNA with indoles present. The mechanism of this protective effect was shown by Helénè (1973) to involve triplet–triplet energy transfer from the nucleic acid bases to Trp, which is highly favored since Trp has a triplet energy level lower than those of the nucleic acid bases. Interestingly enough, Helénè (1973) has shown that this protective transfer of energy can occur for Trp either in the free amino acid form or in its protein-bound form. Whether such a role is played by free or bound Trp *in vivo* or whether a similar effect could take place with the compounds derived from Kyn is unknown.

4.2. Pigmentary Changes with Aging

As the human lens ages, there is an increase in the amount of insoluble protein; nuclear sclerosis develops, and there is an increase in pigmentation with a change in the absorption and emission characteristics of the pigments. In 1943, Trendelenberg noted that the fluorescence of the aged lens changes from blue to blue-green. Analyzing the emissions in somewhat greater detail, Sato (1973) showed that the emission of the aging lens has both a purple (290/340) and a blue (340/420) component. The purple emission was shown to be the major component, and was relatively constant during the aging process. The blue emission component, however, increased in time, and this emission was found to be greatest for the insoluble protein fraction of the lens and least for γ-crystallin, with the emission from α- and β-crystallin lying in between. The increase in blue emission with aging was found to be of the order of 4–5 times in the aged lens ($\sim$60 yr) relative to that of a child. Progress in understanding this age-dependent increase in pigmentation was made when Yamamoto (1973) noted that the aging process was accompanied by a decrease in the 280/350 character and an increase in the 350/450 character of the lens emission. Based on spectral studies of deeply pigmented human lenses, especially brunescent cataractous lenses, Kurzel *et al.* (1973*b*) explained this as a depletion of the protein-bound Trp and its replacement by Trp oxidation products, the Kyn family of compounds, perhaps as a consequence of absorbed radiation. It was seen from the spectra that the characteristic Trp absorption was diminished while the absorption to the red was increased, although no new features were added to the spectra. This implied that the pigments produced were the same as the soluble protein-free pigments present in the normal lens, and were shown by van Heyningen (1973*a*) to be in the Kyn family of compounds. van Heyningen (1973*b*) subsequently showed that there was no buildup in protein-free pigments in the aged lens or in cataract. Lerman (1972*a*) and Bando (1973) independently demonstrated that this pigment in the aged lens is associated with the lens protein. Tryptic digestion of the separated and purified lens protein also produced the fluorescent compound. Bando (1973) showed that although this compound was associated with both the soluble and insoluble protein fractions it was greater in concentration in the insoluble fraction.

The earlier qualitative findings of absorption and fluorescence changes in the aging lens can now be easily understood. The intact normal human lens fluoresces at $\sim$440 nm, due to the soluble protein-free pigments described in the preceding section, but the lens protein itself fluoresces only in the UV region from 300 to 350 nm. The emission is primarily due to protein-bound Trp. The Tyr residues of the lens protein absorb and emit essentially in the UV region from 278 to 305 nm (Lerman, 1972*a*). The Trp residues become oxidized as the tissue ages, and are converted to the Kyn family of compounds with the emission characteristics 350/440. The α-, β-, and γ-crystallins, whose residues have been altered, tend to aggregate to form albuminoid, and therefore the pigment is highest in concentration in the albuminoid. Although the Trp and Tyr residue number in γ-crystallin is twice that in α-crystallin, the γ-crystallin fraction diminishes when adulthood is reached (Pirie, 1972*b*), and therefore less pigment is associated with this fraction.

Other amino acid residues also change in aging. Lerman (1972*b*) has shown that the concentration of free sulfhydryl groups decreases in the aged lens protein, and is accompanied by an increase in the number of disulfide bonds. These changes contribute to the aggregation of the soluble proteins and tend to increase the insoluble protein fraction. These changes are important in trying to understand the progression from lens aging to cataract formation. Lerman (1972*b*) has pointed out that, in all likelihood, brunescent cataract probably represents the end stage of the normal aging process in the lens. In this state, the excessive pigmentation causes the lens to be brown-black in color.

4.3. UV Radiation-Induced Cataract

4.3.1. Brunescent Cataract

The question of the role of UV radiation in causing cataractous changes in the lens has remained controversial over the years. Arguments have been made in favor of such a UV effect, but they are based mainly on epidemiological data (Pirie, 1972*b*) correlating an increased incidence of the dark brown or brunescent cataract with regions of the world where there is high incident radiation density, such as India. However, in general, two arguments have usually been made to rule out such an effect: the component of UV radiation in normal daylight is small, less than 8%; and absorption by the cornea and aqueous humor makes the amount of UV reaching the lens negligible.

UV radiation effects have not been considered in lens damage because of the above two arguments. This lack of serious consideration has persisted even though such effects have often been observed, such as histological changes in the anterior cortex (Duke-Elder, 1954). Kinsey (1948) remarked that "doses to which the eye would have to be exposed to produce minimal damage to the lens is the amount necessary to injure the cornea multiplied by the ratio of the total effectivities (3.02)." Duke-Elder (1954) perpetuates this argument, and assumes that the damage is induced in one exposure. The argument, unfortunately, ignores both the striking differences in repair capabilities of the two tissues and the differences between the chromophores that are acting in the two tissues. Today, the cortex and nucleus of the lens are no longer considered to be dead tissues, although their metabolic activity is very low. Cogan (1950) admitted that, because of its very slow rate of repair, the lens was more vulnerable to repetitive irradiations, although any single dose in the series may not produce visible clinical symptoms in the cornea. The second argument stressing the lack of penetration of UV radiation to the lens is now realized to be invalid. The old literature stated that the transmittance limit to the lens was 310 nm. Today, the commonly accepted limit is between 285 and 295 nm. Reference to Table 1 shows that, at 300 nm, 14.3% of the incident light at the cornea reaches the lens. It will be mentioned shortly that wavelengths up to 313 nm may be effective in causing lens damage.

Several papers have been published reporting the induction of cataract in laboratory animals by UV irradiation (Bachem, 1956; Kulczycka, 1961; Rohrschneider, 1936; Pitts, 1976; Zuclich and Connolly, 1976). Rohrschneider (1936) first showed that he could produce clouding in the anterior part of the lens of guinea pigs by irradiating them with a mercury vapor lamp emitting wavelengths between 293 and 303 nm. These results were confirmed by Bachem (1956), who produced cataract in a large number of both guinea pigs and rabbits, and who determined the action spectrum for UV radiation-induced cataract (Fig. 6). Pitts (1976) mapped the action spectrum for lens damage with a xenon arc and a monochromator, and found cataract formation at 295, 315, 337, and 355 nm. The damage at 337 nm appeared to be deep in the nucleus and the posterior cortical surface; at 315 nm, the damage was largely in the anterior portion of the lens. It should be noted that, at both wavelengths, immediate corneal damage occurred before any lenticular damage could be observed, even in cases where first signs of the cataract appeared several weeks after exposure

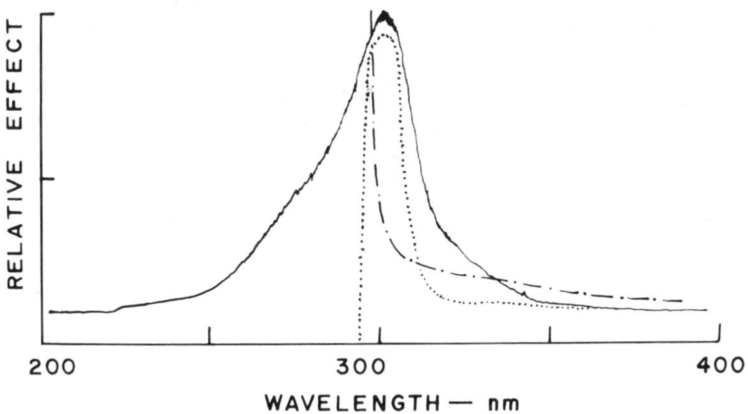

Fig. 6. Comparison of ultraviolet cataractogenesis with phosphorescence excitation and corneal absorption. The phosphorescence excitation spectrum of a healthy intact human lens (77°K) (—), λ_{max} ~300 nm, superimposed on the excitation spectrum for tryptophan. The excitation has monitored at the phosphorescence maximum 440 nm. Action spectrum for near-UV radiation-induced cataract (···), and the corneal absorption spectrum (-··-). From Kurzel *et al.* (1973*b*) and Bachem (1956).

to the UV radiation. In a similar set of experiments, Zuclich and Connolly (1976) used a series of laser exposures. The major surprise in their work appeared to be in the greatly increased effectiveness of very short pulses (a single 10-ns pulse in one case) from a nitrogen laser at 337 nm as compared with Pitts' data for much longer exposures at a similar wavelength. They found lenticular damage after a single 10-ns pulse, with a total energy of 1 J, whereas Pitts required at least 12 J over a 6-h period for comparable injury. Both investigators used rhesus monkeys.

Because of absorption by the cornea, the action spectrum again is seen to take a characteristic spike shape, with a maximum at 300 nm. This spectrum has been shown by Kurzel *et al.* (1973*b*) to be a difference spectrum of the excitation spectrum of protein Trp residues and the characteristic corneal absorption curve as given by Langham (1967), which drops off sharply at 293 nm. It should be noted that the spectrum has a very long tail, extending to ~365 nm. Protein residues absorb only to a minor extent this far into the visible spectrum, and according to Kashet and Brodie (1963) it is possible that the absorption at 340–360 nm may be due to cytochromes in the cellular respiratory apparatus. The possibility that this tail in the curve may have a contribution from photochemical changes in the Trp residues will be discussed below. Trace absorption by Tyr, Phe, Cys, and cystine is possible at 300 nm. Tryptophan dominates the absorption and is probably the key species for long-term, low-level radiation damage to the lens. It has already been discussed above that the brunescent change in the lens is accompanied by a depletion of the Trp absorption, in agreement with the action spectrum. This Trp depletion is accompanied by an increase in Trp oxidation products as determined by spectral analysis. That the pigmentation (i.e., oxidized amino acids) is protein bound has been demonstrated adequately (Lerman, 1972*a*; Bando, 1973). The current direction of research on this problem has been to isolate and identify the amino acid alterations causing the lens protein to become pigmented and aggregated. The first such determination was made by Pirie (1972*b*), who noted that, in addition to a decrease in amount of Trp, the brunescent protein had a diminished histidine (His) content. Subsequently, Lerman (1972*b*) demonstrated that the brunescent protein was deficient in sulfhydryl groups, these groups being replaced by disulfide linkages, as described earlier. We have recently reinvestigated this latter effect, and, to the contrary, have shown that the number of Cys residues in the insoluble protein fraction actually increases in number (Zigler *et al.,* 1976; Stein *et al.,* 1976). However, this determination involved reduction with thioglycolic acid, and we are studying the nature of the changes in the sulfhydryl-containing amino acids. The first comprehensive amino acid analysis on the protein of the brunescent lens was performed by Dilley and Pirie (1974), and confirmed later by Zigler *et al.* (1976) and Stein *et al.* (1976). Following proteolytic digestion by acid hydrolysis, it was biochemically confirmed for the first time that Kyn was present as a replacement for Trp. This task was difficult because of the normally low concentrations of Trp in the adult lens protein ($\sim$0–2%). However, other widespread alterations in the amino acid residues were also determined. Considering both the soluble and insoluble nuclear proteins relative to normal human nuclear lens proteins, the most marked reduction in residues occurs for Trp, with reductions also in the amount of His, and in the amount of Phe to a smaller extent. As noted before, the number of free sulfhydryl groups (Cys) may be increased in the insoluble protein fraction. Interestingly, the number of Tyr residues was relatively unchanged in the nuclear protein of the cataract. The other nonaromatic amino acids (including methionine) showed little change in the nuclear portion of the cataractous lens. The observation that Trp, Cys, His, and Phe are the only amino acids appreciably changed in the nuclear protein of brunescent cataract is interesting when considering the validity of the radiation model for lens damage. These are the amino acids with the greatest product $\phi \times \epsilon$ (300 nm). Although Tyr has a good extinction at 300 nm (ϵ: Trp > Tyr > Cys > Phe), its quantum yield for damage is low relative to the other amino acids, and therefore we see no appreciable depletion of Tyr in the nuclear protein (ϕ: Cys > His > Phe > Trp > Tyr). Thus far, the only identified residue alterations are Trp $\rightarrow$ Kyn, and perhaps —S—S— $\rightarrow$ 2 —SH (Cys). The nature of the other amino acid alterations has not been elucidated yet. Dilley and Pirie (1974) have suggested that cross-linkages other than disulfide bonds may exist in the insoluble nuclear protein of the brunescent cataract, but the nature of these linkages has not been determined. There has been some speculation that cross-linkage to protein may result from a photosensitizing substance, whereas the absorber in the lens would be a compound such as the protein-free lens pigments (van Heyningen, 1973b), or Trp in the free amino acid form (Zigman et al., 1973). Neither of these theories has been proved, and, since the amino acid residues of the proteins themselves can absorb the radiation to form reactive species, it is questionable as to why it is necessary to postulate the need for photosensitizers. However, photosensitizers may be present in certain individuals, and could act to accelerate the normal UV damage mechanisms.

An apparently unrelated change was reported by Noguchi (1973), who detected decreases in glutathione levels in the lenses of irradiated rabbits, but only when very high doses were used.

4.3.2. A Proposed Chemical Model

Sellers and Ghiron (1973) have pointed out that, at 295 nm, 97% of the radiation absorbed by protein is by Trp residues, while Tyr absorbs only 2%

and Cys 1% of the radiation. Equally important, Trp also acts as the pivotal species to which excitation energy is "funneled," and the highly reactive species that results is instrumental in mediating further change in the protein. The unique property of Trp, in its possession of the lowest-lying triplet states, results in the funneling of the excitation energy to it via energy transfer. In fact, although the number of Tyr and Phe residues greatly outnumbers Trp in the human lens protein, we have shown (Kurzel *et al.*, 1973*b*) that the phosphorescence of the human lens is totally characteristic of the emission of Trp. Any energy absorbed by Tyr is thus transferred to Trp. Depending on the protein configuration, estimates of energy transfer efficiency from Tyr to Trp of as high as 100% have been cited (Cassen and Kearns, 1969). Once Trp has been elevated to an excited state, whether by direct absorption or by this funneling effect, the energy is dissipated through the various routes that are available to it, as outlined by Feitelson (1971). These processes compete with each other, and rate constants can be determined. The nature of the excited state of Trp responsible for photochemical damage has been a point of considerable interest. Since no phosphorescence is observed from the lens at room temperature (Kurzel *et al.*, 1973*b*), the triplet-state excitation must be dissipated via internal conversion or intersystem crossing to the first excited singlet state (S_1^*). The S_1^* state is also populated by direct absorption. Weiter and Finch (1975) have studied the irradiated human lens with electron spin resonance (ESR) spectroscopy and have confirmed the presence of only the first excited triplet state of Trp, with no signals being observed from Phe or Tyr. The formation of the reactive intermediate probably follows population of the first excited singlet state. That the S_1^* state was the photoreactive state in lens protein damage was suggested by Kurzel *et al.* (1973*b*) on the basis of fluorescence quenching data. Other evidence to support this theory has been presented by a number of groups (Feitelson, 1971; Hopkins and Lumry, 1972; Konev and Volotovskii, 1966), and it seems that the reactive intermediate results from the S_1^* state primarily via electron ejection. Photoionization is the primary process resulting in the decomposition of Trp, and the probability of the process increases with increasing vibrational energy of the excited singlet state (Steen, 1974) and with the population of T_1^*. The S_1^* state (Tyr → energy transfer) therefore enhances Trp photoionization efficiency (Santus *et al.*, 1972). The ultimate end product in this decomposition involves the cleavage of the C(2—3) bond of the indole ring to form *N*-formylkynurenine (NFK). That NFK was the final product of Trp photodestruction has been proposed by Pirie (1972*b*) by *in vitro* spectral studies. It has been shown that one of the degraded residues of Trp in the protein of the human brunescent cataractous lens is either NFK or Kyn. Presumably the *N*-formyl group is cleaved off metabolically *in vivo*, according to Zigler *et al.* (1976).

Pailthorpe and Nicholls (1971) have demonstrated that, in addition to the photoejection of electrons, the absorption by Trp of radiation of wavelengths 300–365 nm results in the cleavage of the N—H bond, which ultimately results in indole ring cleavage. Since the N—H bond is the weakest bond in the indole ring, with a bond strength of only 75–80 kcal/mol, this bond can be cleaved by the absorption of radiation of wavelengths 300–365 nm (95–80 kcal/mol). This results in the process of electron abstraction from the singlet state as shown in Fig. 7. The electron rearrangement of the indole nucleus, with the free electron at the C(3) position, has been confirmed by ESR spectroscopy, and this free radical has a lifetime of 50 μs at 25°C. This reactive species also results in C(2—3) bond cleavage with NFK formation. It is important that the long tail in the action spectrum for UV-induced cataract (Fig. 6) can be explained in terms of this Trp N—H bond cleavage.

The action spectrum for UV-induced cataract formation therefore tells us that damage is probably mediated via two photochemical processes, both proceeding via the first excited singlet state of Trp. The primary process is that of electron photoejection, and the wavelength dependence for this process centers at 293–313 nm. The process of electron (or H atom) abstraction is of secondary importance, contributing to the action spectrum in the region of 300–365 nm, and is revealed by the long tail in the action spectrum for cataractogenesis. From the maximum in the action spectrum, it is possible to deduce whether lens protein damage requires Trp degradation photochemically, or whether the major role of Trp is to mediate damage by transferring its excitation via energy transfer to other residues, with the Trp remaining unchanged in the process. A region of the action spectrum corresponding to indole ring N—H bond cleavage attests to the role of Trp degradation itself in the damage process, to some extent. The question of the relative importance of energy transfer vs. photochemical production of excited species (i.e., solvated electrons or H atoms which interact with His, disulfide bridges, etc.) in mediating damage will be decided only by measurements of the quantum yield for Kyn formation relative to other residue alterations (e.g., disulfide reduction, His oxidation). Thus far, the results of our amino acid analysis on the brunescent protein

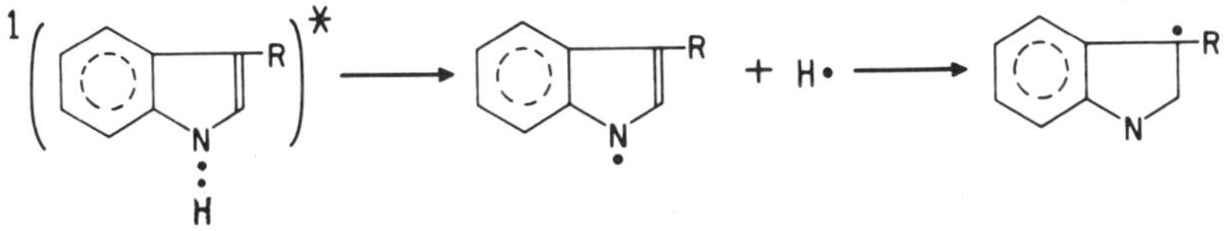

Fig. 7. N—H bond cleavage by photostimulation in 300–365 nm range. Electron abstraction resulting from decay of the first excited singlet state of an indole-type molecule (e.g., tryptophan residue). From Pailthorpe and Nicholls (1971).

show only a small to moderate decrease in Trp content relative to other amino acid alterations; this, together with the Trp replacement by Kyn, would tend to support the importance of energy transfer from Trp. Therefore, two to three deexcitation pathways are being utilized.

The residue distribution observed in brunescent cataract suggests a model in which the radiation induces Trp excited states, which act as the mediator for this damage. Consideration of the specific changes in this process should be the basis for future work.

Histidine. A drop in the His content was one of the first demonstrated changes in the protein of the brunescent lens (Section 4.3.1). The connection between this His loss and Trp excitation is well known. That His interacts with Trp (S_1*) was demonstrated by Sellers and Ghiron (1973). The propensity for His to quench Trp fluorescence was assumed to be due to the action of His as a free radical scavenger. Most recently, Walrant and Santus (1974) have shown that protein His is oxidized upon 300-nm irradiation, with the presence of Trp being required. The depletion of His is therefore a sensitized photooxidation, since His is not the primary absorber. At exactly which stage in the deexcitation of Trp the transfer of energy to His takes place is not clear. The authors suggested that the photosensitizing species was NFK (T_1*) rather than Trp; however, there is no clear evidence for this theory. Since the NFK stems from Trp*, it is difficult to determine the origin of the excitation energy. The same basic difficulty clouds the argument of some authors (van Heyningen, 1973*b*) on whether NFK can photosensitize further Trp oxidation. Since Trp itself is converted to NFK directly by radiation absorption, who is to say that the Trp did not absorb the radiation rather than NFK? If NFK could originate photosensitized damage, it would mean, too, that radiation in the visible region of the spectrum would be damaging to the lens. This is not borne out in reality by the action spectrum.

Tyrosine. No marked decrease in Tyr content was noted in the protein of the brunescent lens. As discussed earlier, this is not unexpected, considering the poor Tyr absorption at 300 nm ($> 2\%$) and the poor quantum yield for damage. Equally important, Tyr has a fluorescent emission which overlaps with Trp absorption, therefore favoring energy transfer. For that small population of Tyr residues that do become excited, a similar array of deexcitation processes are available to it and to Trp. The most favored process, however, is fluorescence quenching and, secondarily, energy transfer (Kronman and Holmes, 1971). Therefore, it is not surprising that no change in Try residues was detected.

Recalling Pirie's (1972*b*) observation of unidentified nonsulfhydryl cross-linkages in the brunescent protein, it is of interest to note that Shimizu (1973) has suggested the possibility of Tyr photoelectron abstraction which proceeds from the Tyr (s_1*) state. However, since most biomolecular reactions proceed via the triplet state, the possibility of Tyr (T_1^*) as the intermediate cannot be excluded. Cross-linkages can result in the form of bityrosine moieties (Fig. 8), which have emission characteristics with λ_{max} of absorption at 283 nm and λ_{max} of fluorescence at 410 nm. However, it is unknown whether this process is of any importance in the brunescent lens.

Cysteine. When one comes to the description of changes in the sulfur-bearing amino acids in the brunescent cataract protein, there is some confusion. Lerman (1972a) found a decrease in free sulfhydryl content with an increase in disulfide linkages. These findings, however, pertain only to the γ-crystallin in the insoluble protein, and these findings may not have been representative of the rest of the albuminoid. Dilley and Pirie (1974) detected no change for methionine, and were unable to detect changes in the amount of Cys. We have determined a marked increase in Cys residues in the insoluble protein fraction. If the finding of decreased sulfhydryl content is valid, it would be inconsistent with what is known about near-UV radiation-mediated effects on Cys via Trp. As less than 1% of the incident radiation is absorbed by Cys at 295 nm, one expects little effect from direct absorption by Cys. On the other hand, it has been shown that disulfide bonds are quite labile to disruption following UV-induced excitation of Trp. Such a reaction of

$$-S-S \xrightarrow{Trp^*} 2-SH$$

amounts to a sensitized photoreduction of cystine. The observation of Kronman and Holmes (1971) that disulfide bridges decrease the quantum yield of Trp fluorescence once again points to an interaction between the first excited singlet state of Trp with cystine by increased nonradiative transitions from this state. It is reasonable to assume that electrons produced by either photoejection or electron abstraction from the Trp (S_1^*) state interact with the disulfide bridge to produce this cleavage. Studies by Dose (1968), Grossweiner and Usui (1970), and Ghiron *et al.* (1971) have confirmed the effect that cystine destruction is favored following Trp excitation by radiation of $\lambda > 280$ nm (a region where cystine essentially does not absorb radiation). Ghiron *et al.* (1971) have pointed out that cystinyl residues of Trp-containing proteins have the highest susceptibility of disruption following the absorption of UV radiation, even when the residues are separated

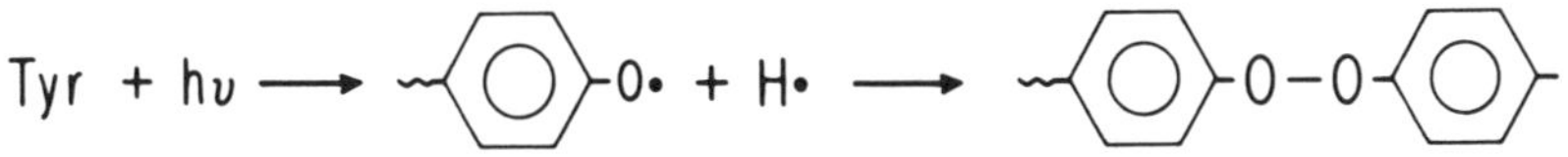

$$\text{Tyr} + h\upsilon \longrightarrow \text{—}\langle\text{O}\rangle\text{—O}\bullet + \text{H}\bullet \longrightarrow \text{—}\langle\text{O}\rangle\text{—O—O—}\langle\text{O}\rangle\text{—}$$

Fig. 8. Cross-linkage formation in the photostimulated, S_1-excited state of a tyrosine residue. From Shimizu (1973).

from each other in the primary sequence. However, their proximity is important as evidenced by the dependence on thiol formation yield on the tertiary structure of the protein. Thiol formation is increased five- to tenfold in proteins containing two Trp residues as compared with proteins containing no Trp residues. Our observation of an increased number of Cys residues is, therefore, in good agreement with the radiation model for lens protein damage acting through excited states of Trp.

Our present description of the effects of UV radiation and the inferences of the processes involved only begin the search for an understanding of the events in the formation of brunescent cataract and the pigmentary aging of the lens. Doubt still exists about how the changes in the lens proceed via a radiation mechanism. In the above sections, we have tried to outline the evidence for such a theory and the progress in these investigations. If brunescent cataract does not proceed via a radiation mechanism, all of the above findings would have to constitute an incredible collection of coincidences.

4.3.3. UV Damage to Lens Epithelium

The only work describing UV radiation damage to the epithelial surface of the lens was written by Zigman and Vaughan (1974), who showed that radiation served to inhibit epithelial cell conversion to fiber cells. In mice, cortical opacities appeared in 50 weeks. Such an effect might be expected for an acute, high-level exposure, and also would be accompanied by damage to the cornea. Since the lens epithelial cells are metabolically active and capable of repair, low-dosage radiation, as in solar radiation, probably will not affect these cells. Pigmentary aging and brunescent cataract formation in the human have a propensity to develop in the nuclear rather than in the cortical regions (Pirie, 1972*a*).

Zigman *et al.* (1973) have shown that lens membrane irradiated *in vivo* demonstrated a decreased ability to take up ^{14}C-labeled amino acids. They proposed that the UV-induced alteration in membrane permeability caused a decrease in protein synthesis. A further step could be to suppose that radiation absorbed by the membrane inhibits respiration and the energy production necessary for active amino acid transport across the membrane.

4.4. Photosensitized Damage to the Lens

Thus far, we have considered only the effects of damage to the ocular tissues following the direct absorption of radiation. Obviously, UV radiation can also bring about such tissue damage via photosensitized reactions, but proven examples of such processes in the eye are few. Mention has already been made of speculation of the role of NFK in photosensitizing damage to lens protein. In the past, the photosensitizing role of the phenothiazines, and chlorpromazine in particular, has been advanced, but this also is unproven. Barron *et al.* (1972) have shown that the pigmentation of human corneas and lenses following use of chlorpromazine results not from a photosensitized reaction but rather from simple deposition of pigmented crystallin material in these tissues, which in time is reversible with withdrawal of the drug.

The only drug thus far definitely known to result in photosensitized damage to the lens is psoralen (furocoumarin), and its family of compounds. Freeman (1966) has shown that it is possible to induce cataract in laboratory animals by injecting 8-methoxypsoralen on a daily basis, with exposure to UV radiation (320–400 nm). When low doses of the drug are used, the cataract is known to regress in time. The mechanism involved in inducing this damage is unknown. However, the UV photosensitizing effect of this drug on the skin has been extensively studied, and it may be assumed that the fundamental mechanisms of cell injury in these two tissues are the same.

Walter *et al.* (1973) have shown that psoralens plus near-UV radiation (320–400 nm) inhibit the S phase (scheduled DNA synthesis) of the epidermal cell cycle; they prevent DNA replication via pyrimidine damage. The unirradiated psoralen tends to form loose complexes with pyrimidine bases of DNA and RNA. Radiation of wavelengths 320–400 nm induces a photoreaction in which the excited psoralen molecule forms a C_4 cycloaddition product involving the 5,6 double bond of the pyrimidine, and either the 4′,5′ or 3,4 double bond of psoralen. On subsequent irradiation, the other end of the psoralen molecule can bind to another pyrimidine on the opposite DNA strand. Such interstrand cross-linkages have been isolated by Pathak *et al.* (1967). These observations suggest that the psoralens exert damaging effects on the epithelial cells of the lens by inhibiting repair processes. Bridges (1971) also has suggested that the psoralens may inhibit excision repair. A further important aspect of psoralen photochemical reaction is the report of Parrish *et al.* (1974) of large-scale therapeutic trials on humans with this drug combined with near-UV radiation for the treatment of psoriasis. Although Harber (1974) noted the danger of ocular damage from accidental eye exposures in the course of such treatments, this new treatment of psoriasis was heralded as a significant advance. It remains to be seen when the first cases of cataract in these patients, secondary to accidental radiation exposure, will occur.

5. RADIATION EFFECTS ON THE RETINA

It would seem that a consideration of the damaging effect on the retina would be central to any discussion of the UV radiation effects on the eye. On the basis of spectral transmission data, such a possibility was totally excluded until recently, and it was even said that "Those UV radiation which might harm the retina do not reach it, and those radiations which reach the retina do it no harm" (Wald, 1952). In support of the above statement, Wald noted in early experiments that no significant difference could be found in the thresholds for either rod or cone vision following near-UV irradiation (290–365 nm). However, the effect of UV exposures on normal human subjects in causing erythropsia, a transient red vision sensation, has been well known, and was most recently described by Kamel and Parker (1973). The incidence and intensity of the erythropsia was found to increase in patients in whom the filter effect of the lens was negated, whether by cataract surgery, dilation of the pupil (mydriasis), or coloboma (congenital fissure) of the iris. In the normal individual, the lens filters out radiation between 293 and 400 nm (Kamel and Parker, 1973) and thus prevents it from reaching the retina. From Table 1, we see that at $\lambda = 350$ nm, only 1.6% of the radiation incident on the eye reaches the retina. For the normal person, therefore, the UV radiation effect on the retina may become manifest only following high-intensity radiation exposure, and the wavelengths of importance are $\sim$350–365 nm. The aphakic eye, on the other hand, is especially at risk of damage at wavelengths ranging from 293 to 400 nm. As pointed out earlier, it was noted by Wald (1952) that the aphakic eye is a thousandfold more sensitive to radiation of $\lambda\sim$365 nm than the normal eye, and it also receives appreciable exposure at 300-nm radiation. For patients with aphakic eyes, damage can become manifest at normal intensities of solar radiation exposure.

The morphological changes observed with near-UV radiation exposure to the retina in various animal systems have been documented by several groups. Zigman and Vaughan (1974) have noted, for mice irradiated with 365-nm radiation, that there was a thinning of the outer segments by 10 weeks postirradiation. By 16 weeks, phagocytic wandering cells were noted digesting the remnants of the rod outer segments. These same processes were observed earlier as a result of visible light irradiation by Marshall *et al.* (1972) in studies on the pigeon retina, and by O'Steen and Karcioglu (1974), who studied the albino rat retina by electron microscopy. Both of these groups agreed that the damage was almost exclusively limited to the photoreceptor outer segments. Zuclich and Connolly (1976) also observed damage to photoreceptor outer segments in rhesus monkeys when irradiating with a variety of laser wavelengths in the region 350–365 nm. With exposures up to 90 s, extensive damage to the photoreceptor layer was noted, while the other parts of the retina were more or less intact. However, their attempts to quantify this effect were unsuccessful. They found great variations between animals with regard to the sensitivity of the retina to damage by UV radiation in this portion of the spectrum. In the pigeon, the cones were preferentially damaged. A loss of parallel membrane arrangement at the distal ends of the rod outer segments was observed, as was a vesicular degeneration of the cone outer segments. At a low-radiation intensity, no damage to the outer nuclear layer or cone pedicles was observed. At these low doses, the radiation damage to the retina was reversible with time. O'Steen and Karcioglu (1974) have shown that, in the rat eye, the repair process involves the removal of debris by two cell populations: (vascular) mononuclear cells which appeared on the scene first, and pigment epithelial cells. The pigment epithelial cells proliferated by mitotic activity and were seen among the degenerated outer segments, separated from Bruch's membrane. The phagosomes of the pigment epithelial cells contained lamelated discs from the photoreceptor outer segments. The authors questioned whether these pigment epithelial cells later migrated back to Bruch's membrane, but there is no evidence that this is the case. Weisse and Stotzer (1974) demonstrated that when the retina of the rat was chronically exposed (3 yr) to high-intensity near-UV radiation, damage proceeded to general atrophy of the first neuron, partial degeneration of the second neuron, and destruction of the retinal structure, as well as vascularization of the pigment epithelium. In all of the studies cited above, the intact eye was exposed to the radiation. Due to lens filtration considerations, therefore, one can assume that the damaging radiation was of wavelengths 350–365 nm. Since the rod visual pigment has an electronic transition with an absorption maximum near 350 nm (β band), it can be assumed that the damage corresponded to radiation-induced bleaching and depletion of the visual pigments. That the sole damage was confined to the photoreceptor outer segments acts as support for this assumption.

Studies of the effects of specific wavelengths of radiation on metabolic processes in the retina are few. Zigman and Bagley (1971) were the first to observe the suppression of protein synthesis in dogfish retinal rods irradiated at 340–380 nm. They also observed a decrease in incorporation of ^{14}C-labeled uridine into RNA. These authors proposed that these changes resulted from radiation damage to a component of the cytochrome oxidase system; this damage interrupted respiration and withheld the energy needed for synthesis and metabolism. The same experiment was repeated by Matuk *et al.* (1974), who obtained strikingly different results. They studied the incorporation of [^{14}C]leucine into rat retinal protein of cell extracts and obtained a much more detailed wavelength dependence. [^{14}C] amino acid incorporation, and therefore protein synthesis, was depressed only following irradiation at 320 nm, probably by direct absorption by proteins or nucleic acids. The respiratory system also could be affected at this wavelength. At 340 nm, no change in ^{14}C-labeled amino acid incorporation was observed, while at 360 nm, amino acid incorporation and protein synthesis were actually observed to increase. This increase would be expected for absorption by the β band of rhodopsin in the photoreceptors. Under high-intensity 360-nm radiation, the debris of the rod outer segments is removed by the phagocytes, while normal synthesis of discs in the outer segments is accelerated.

By virtue of the broader band of radiation incident on the retina of the aphakic eye, more extensive damage is expected. Not only will the photoreceptors be involved, but also the metabolic machinery. As in any photochemical process, the events involved are bound to be more complex than the scheme mentioned above. Recently, Patterson *et al.* (1974) suggested that retinal damage to turkey eyes irradiated with long-wave UV proceeded via lipid peroxidation, as deduced by free radical scavenger studies. The story of the UV radiation effects on the retina has barely begun to unfold.

6. CONCLUSIONS

Ultraviolet radiation effects (200–400 nm) have been documented for the most important tissues of the eye, the cornea, lens, and retina. Because of sequential absorption by the various layers, the wavelength for maximum effect is progressively shifted to the red for the deeper structures, and the metabolic or structural processes affected also change.

The nature of the damage to the cornea is not clear (with maximum damage at $\lambda = 289$ nm), but may involve the cellular respiratory apparatus as well as the genetic apparatus (nucleoprotein).

In the lens, the action spectrum for damage peaks at the Trp absorption maximum of 300 nm. Both the pigment accumulation in an aging lens that retains its transparency and the formation of brunescent cataracts probably proceed via Trp absorption and photooxidation, and/or redistribution of excitation energy. In agreement with this mechanism, Trp photooxidation products have been isolated. A detailed physicochemical model outlining the effects of such absorption in the lens protein damage has been discussed in Section 4.3.2.

UV radiation has been shown to affect the retina in the outer segments exclusively at low to moderate doses. Such damage corresponds to widescale bleaching of photopigments, with absorption centered at 350 nm, the

β-band absorption maximum of rhodopsin. The threat of such damage is greatest in the aphakic eye, which is deprived of lens filtration protection.

UV radiation therefore affects all tissues of the eye. Following low radiation dosage, the cornea and retina, which have rapid regenerative properties, sustain no lasting injury. The lens, with its low metabolic activity, is less capable of repair. The effects of long-term or cumulative exposures on the lens are therefore more serious and are likely to result in damage, either pigmentary changes or a cataract.

The study of UV radiation damage to the eye is a relatively new and exciting field of research, which has only recently begun to blossom with renewed interest, modern methodology, and concerted effort. It is hoped that many of the questions left unanswered in this chapter will soon be resolved.

7. REFERENCES

Bachem, A., 1956, Ophthalmic ultraviolet action spectra, *Am. J. Ophthalmol.* **41**:969–974.

Bando, A., 1973, The relationship between coloration and fluorescence in the human lens, *Acta Soc. Ophthalmol. Jpn.* **77**:873–876.

Barron, C., Rubin, L., and Steelman, R., 1972, Clorpromazine of the eye of the dog. III. Natural daylight versus artificial light, *Exp. Mol. Pathol.* **16**:158–162.

Bridges, B. A., 1971, Genetic damage induced by 254 nm UV light in *Escherichia coli:* 8-Methoxypsoralen as protective agent and repair inhibitor, *Photochem. Photobiol.* **14**:659–662.

Carlson, J. G., and Hollaender, A., 1948, Mitotic effects of UV radiation of the 2250 Å region, with special reference to the spindle and cleavage, *J. Cell. Comp. Physiol.* **31**:149–173.

Cassen, T., and Kearns, D. R., 1969, Phosphorescence and energy transfer in enzymes, *Biochim. Biophys. Acta* **194**:203–212.

Claesson, S., Wettermark, G., and Juklin, L., 1959, Action of ultraviolet light on skin: Effect of the histamine liberator 48/80 and methotrimeprazine, *Nature (London)* **183**:1451–1452.

Cogan, D. G., 1950, Lesions of the eye from radiant energy, *J. Am. Med. Assoc.* **142**:145–151.

Cogan, D. G., and Kinsey, V. E., 1946, Action spectrum of keratitis produced by ultraviolet radiation, *Arch. Ophthalmol.* **35**:670–677.

Dartnall, H. J. A., 1957, *The Visual Pigments,* Methuen, New York.

Dilley, K. J., and Pirie, A., 1974, Changes to proteins of the human lens nucleus in cataract, *Exp. Eye Res.* **19**:59–72.

Dose, K., 1968, The photolysis of free cystine in the presence of aromatic amino acids, *Photochem. Photobiol.* **8**:331–335.

Duke-Elder, S., 1954, *Textbook of Ophthalmology,* Vol. VI: *Injuries,* Chapter LXX, pp. 6443–6467, C. V. Mosby, St. Louis.

Ellinger, F., 1951, Die Histaminhypothese der biologischen Strahlenwirkungen, *Schweiz. Med. Wochenschr.* **81**:61–65.

Feitelson, J., 1971, The formation of hydrated electrons from the excited state of indole derivatives, *Photochem. Photobiol.* **13**:87–96.

Freeman, R. G., 1966, Morphologic changes resulting from photosensitization of the eye with 8-methoxypsoralen—A comparison with conventional ultraviolet injury, *Tex. Rep. Biol. Med.* **24**:588–596.

Freeman, R. G., and Knox, J., 1964, Ultraviolet induced corneal tumors in different species and strains of animals, *J. Invest. Dermatol.* **43**:431–436.

Friedenwald, J. S., Buschke, W., Crowell, J., and Hollaender, A., 1948, The effects of ultraviolet irradiation on the corneal epithelium, *J. Cell. Comp. Physiol.* **32**:161–173.

Genter, E. I., 1971, Resistance of the corneal epithelium to UV radiation, *Tsitologiia* **13**:206–211.

Ghiron, C. A., Volkert, W., and Lahmeyer, H., 1971, Studies on the mechanism of cystine destruction and inactivation of trypsin irradiated with 280 nm light, *Photochem. Photobiol.* **13**:431–436.

Grof, P., and Kovacs, A., 1967, On the mode of action of UV light: Effect of UV rays on most cells *in vivo, Acta Physiol. Acad. Sci. Hung.* **32**:35–44.

Grossweiner, L., and Usui, Y., 1970, The role of the hydrated electron in photoreduction of cystine in the presence of indole, *Photochem. Photobiol.* **11**:53–56.

Hamerski, W., 1969, Investigations on histochemical changes in experimental corneal lesions induced with UV radiation and on prevention of photophthalmia, *Pol. Med. J.* **8**:1469–1476.

Hamerski, W., 1971, Experimental studies of the content of nucleic acids in the cornea subjected to UV rays, *Klin. Oczna* **41**:639–642.

Hamerski, W., and Zajaczkowska, A., 1969, Electrophoretic investigations of proteins of the corneal epithelium in experimental photophthalmia, *Pol. Med. J.* **8**:1464–1468.

Harber, L. C., 1974, Photochemotherapy of psoriasis, *N. Engl. J. Med.* **291**:1251–1252.

Helénè, C., 1973, Energy transfer between nucleic acid bases and tryptophan in aggregates and in oligopeptide–nucleic acid complexes, *Photochem. Photobiol.* **18**:255–262.

Hemmingsen, E. A., and Douglas, E. I., 1970, UV radiation thresholds for corneal injury in Antarctic and temperate-zone animals, *Comp. Biochem. Physiol.* **32**:593–600.

Hirabayashi, K., and Graham, J., 1969, Mediation of radiation erythema, *Int. J. Radiat. Biol.* **16**:85–91.

Hopkins, T., and Lumry, R. W., 1972, Exciplex studies. V. Electron ejection from indole and methyl indole derivatives. *Photochem. Photobiol.* **15**:555–556.

Jagger, J., 1967, *Introduction to Research in Ultraviolet Photobiology,* Prentice-Hall, Englewood Cliffs, N.J.

Kamel, I., and Parker, J., 1973, Protection from UV exposure in aphakic erythropsia, *Can. J. Ophthalmol.* **8**:563–565.

Kashet, E. R., and Brodie, A. F., 1963, Oxidative phosphorylation in fractionated bacterial systems. X. Different roles for the natural quinones of *E. coli* W in oxidative metabolism, *J. Biol. Chem.* **238**:2564–2570.

Kinsey, V. E., 1948, Spectral transmission of the eye to ultraviolet radiations, *Arch. Ophthalmol.* **39**:508–513.

Konev, S. V., and Volotovskii, I. D., 1966, Investigation of the role of singlet and triplet excited states of tryptophan in the photoinactivation of trypsin, *Biophysics* **11**:909–915.

Kronman, M., and Holmes, L., 1971, The fluorescence of native, denatured and reduced-denatured proteins, *Photochem. Photobiol.* **14**:113–134.

Kulczycka, N., 1961, Experimental investivation on the cataractogenic effects of UV rays in young mice. *Acta Biol. Cracow* **4**:59–78.

Kurzel, R. B., Wolbarsht, M. L., and Yamanashi, B. S., 1973*a*, Spectral studies on normal and cataractous intact human lenses, *Exp. Eye Res.* **17**:65–71.

Kurzel, R. B., Wolbarsht, M. L., Yamanashi, B. S., Staton, G. W., and Borkman, R. F., 1973*b*, Tryptophan excited states and cataract in the human lens, *Nature (London)* **241**:132–133.

Langham, M., 1967, *The Cornea,* p. 162, Johns Hopkins University Press, Baltimore.

Lerman, S., 1972a, Lens proteins and fluorescence, *Isr. J. Med. Sci.* **8:**1583–1589.

Lerman, S., 1972b, Lens proteins in aging and cataract formation, in: *Contemporary Ophthalmology—Honoring Sir Stewart Duke-Elder* (J. G. Bellows, ed.), Williams and Wilkins, Baltimore.

Marshall, J., Mellerio, J., and Palmer, D., 1972, Damage to pigeon retinae by moderate illumination from fluorescent lamps, *Exp. Eye Res.* **14:**164–169.

Matuk, Y., Parker, J. A., and Goldlist, G. I., 1974, Wavelength dependent effect of near UV on C^{14}-leucine incorporation into rat retina, *J. Opt. Soc. Am.* **64(**10**):**1373.

McLaren, A. D., and Takahashi, W. N., 1957, Inactivation of infectious nucleic acid from tobacco mosaic virus by ultraviolet light (2537 Å), *Radiat. Res.* **6:**532–542.

McLaren, A. D., Gentile, P., Kirk, D. C., and Levin, N. A., 1953, Photochemistry of proteins. XVII. Inactivation of enzymes with ultraviolet light and photolysis of the peptide band, *J. Polym. Sci.* **10:**333–334.

Noguchi, Y., 1973, Radiation effects of UV rays on the crystalline lens of rabbit eyes, *Act. Soc. Ophthalmol. Jpn.* **77:**34–40.

Oguchi, M., Shimizu, Y., Seki, H., Kawase, S., Sakurai, M., and Uchiyama, Y., 1973, On the fluorescent color of the crystalline lens, a new detection method for GSH, its distribution in the eye, and systemic embryology of the crystalline lens, *Acta Soc. Ophthalmol. Jpn.* **77:**186–191.

O'Steen, W. K., and Karcioglu, Z. A., 1974, Phagocytosis in the light damaged albino rat eye: Light and electron microscopic study, *Am. J. Anat.* **139:**503–518.

Pailthorpe, M., and Nicholls, C., 1971, Indole N—H bond fission during the photolysis of tryptophan, *Photochem. Photobiol.* **14:**135–145.

Parrish, J. A., Fitzpatrick, T., Tannenbaum, L., and Pathak, M., 1974, Photochemotherapy of psoriasis with oral methoxysalen and long wavelength UV light, *N. Engl. J. Med.* **291:**1207–1211.

Pathak, M., Worden, L., and Kaufman, K., 1967, Effects of structural alterations on the photosensitizing potency of furocoumarins (psoralens) and related compounds, *J. Invest. Dermatol.* **48:**103–118.

Patterson, P. S. P., Sweasey, D., Roberts, B. A., and Pattison, M., 1974, The protective effect of promethazine treatment against photoperoxidation of lipid in turkey eyes, *Exp. Eye. Res.* **19:**267–272.

Pirie, A., 1968, Color and solubility of the proteins of human cataracts, *Invest. Ophthalmol.* **7:**634–650.

Pirie, A., 1972a, The effect of sunlight on proteins of the lens, in: *Contemporary Ophthalmology—Honoring Sir Stewart Duke-Elder* (J. G. Bellows, ed.), Williams and Wilkins, Baltimore, Md.

Pirie, A., 1972b, Photooxidation of proteins and comparison of photooxidized proteins with those of the cataractous human lens, *Isr. J. Med. Sci.* **8:**1567–1573.

Pitts, D. G., 1976, *UV Ocular Effects from 300 nm to 400 nm,* NIOSH Contract CDC-99-74-12, Final Report.

Pitts, D. G., and Gibbons, W., 1973, Corneal light scattering measurements of UV radiant exposures, *Am. J. Optom.* **50:**187–194.

Pitts, D. G., and Tredici, T., 1971, The effects of ultraviolet radiation on the eye, *Am. Ind. Hyg. Assoc. J.* **32:**235–246.

Rohrschneider, W., 1936, Linsenschadigung durch ultraviolette Strahlen im Tierversuch, *Arch. Ophthalmol. (Berlin)* **135:**282–292.

Said, F. S., and Weale, R. A., 1959, The variation with age of the spectral transmissivity of the living human crystalline lens, *Gerontologia* **3:**213–231.

Santus, R., Bogin, M., and Aubailly, M., 1972, Influence of energy transfer on the photoionization of tryptophan and tyrosine in basic media, *Photochem. Photobiol.* **15**:61–69.

Sato, K., 1973, Fluorescence in human lenses, *Exp. Eye Res.* **16**:167–172.

Sellers, D., and Ghiron, C. A., 1973, Role of the tryptophan fluorescent state in the UV induced inactivation of β-trypsin, *Photochem. Photobiol.* **18**:393–402.

Setlow, R. B., 1960, Ultraviolet wavelength-dependent effects on proteins and nucleic acids, *Radiation Res. Suppl.* **2**:276–289.

Sherashov, S. G., 1970, Spectral sensitivity of the cornea to ultraviolet radiation, *Biofizika* **15**:543–544.

Shimizu, O., 1973, Excited states in photodimerization of aqueous tyrosine at room temperature, *Photochem. Photobiol.* **18**:125–133.

Sliney, D. H., 1972, The merits of an envelope action spectrum for ultraviolet radiation exposure criteria, *Am. Ind. Hyg. Assoc.* (October).

Steen, H., 1974, Wavelength dependence of the quantum yield of fluorescence and photoionization of indoles, *J. Chem. Phys.* **61**:3997–4002.

Stein, P. J., Henkens, R. W., Yamanashi, B. S., and Wolbarsht, M. L., 1977, Studies on brunescent cataract. II. Fluorescent studies on normal and brunescent lens proteins, *Ophthalmic Res.* (in press).

Tomicic, H., Pieba, M., Romero, C., Soto, A., and Toha, J. C., 1973, Radioprotection (UV and gamma rays) of DNA molecule by indole and indole derivatives, *A. Naturforsch.*[c] **28**:379–385.

Trendelenberg, W., 1943, *Der Desichtsinn,* Springer-Verlag, Berlin.

van Heyningen, R., 1970, Fluorescent glucosides in the human lens, *Natue (London)* **230**:393–394.

van Heyningen, R., 1973*a*, Assay of fluorescent glucosides in the human lens, *Exp. Eye Res.* **15**:121–126.

van Heyningen, R., 1973*b*, Photooxidation of lens proteins by sunlight in the presence of fluorescent derivatives of kynurenine, isolated from the human lens, *Exp. Eye Res.* **17**:137–147.

Wald, G., 1952, Alleged effects of the near ultraviolet on human vision, *J. Opt. Soc. Am.* **42**:171–177.

Walrant, P., and Santus, R., 1974, *N*-Formyl kynurenine, a tryptophan photooxidation product as a photodynamic sensitizer, *Photochem. Photobiol.* **19**:411–417.

Walter, J., Voorhees, J., Kelsey, W., and Duell, E., 1973, Psoralen plus black light inhibits epidermal DNA synthesis, *Arch. Dermatol.* **107**:861–865.

Weisse, I., and Stotzer, H., 1974, Age and light dependent changes in the rat eye, *Virchows Arch. (Pathol. Anat. Physiol.)* **362**:145–156.

Weiter, J. J., and Finch, E. D., 1975, Paramagnetic species in cataractous human lenses, *Nature (London)* **254**:536–537.

Wolbarsht, M. L., 1976, Function of intraocular color filters, *Fed. Proc.* **35**:44–50.

Wolbarsht, M. L., and Yamanashi, B. S., 1977, Intensity sharing, upper and lower limit and genetic dependence of visual pigment absorption spectra, *Biophys. J.* **17**:17A.

Yamamoto, K., 1973, Aging of soluble lens proteins, *Acta Soc. Ophthalmol. Jpn.* **77**:897–906.

Zigler, J. S., Jr., Sidbury, J. B., Jr., Yamanashi, B. S., and Wolbarsht, M. L., 1977, Studies on brunescent cataract. I. Analysis of free and protein-bound amino acids, *Ophthalmic Res.* (in press).

Zigman, S., and Bagley, S., 1971, New UV light effects on dogfish retinal rods, *Exp. Eye Res.* **12**:155–157.

Zigman, S., and Vaughan, T., 1974, Near-UV effects on lenses and retinas of mice, *Invest. Ophthalmol.* **13**:462–465.

Zigman, S., Schultz, J., and Yulo, T., 1973, Possible roles of near UV light in the cataractous process, *Exp. Eye Res.* **15**:201–208.

Zuclich, J. A., and Connolly, J. S., 1976, Ocular hazards of near-UV laser radiation, *J. Opt. Soc. Am.* **66**:79.

4

Lethal and Mutagenic Effects of Near-Ultraviolet Radiation*

Robert B. Webb

Division of Biological and Medical Research, Argonne National Laboratory, Argonne, Illinois 60439

* Research supported by the U.S. Energy Research and Development Administration.

1. INTRODUCTION

Ultraviolet (UV) radiation* of wavelengths longer than 295 nm from sunlight is a ubiquitous part of the natural environment of most organisms. Although beneficial and possibly beneficial aspects of sunlight have long been recognized and studied (Daniels, 1974; Wurtman, 1975), harmful effects, except for sunburn, have received relatively little attention until recently. Photochemical possibilities of natural components of the cell suggest that efficient mechanisms for the partial prevention or repair of resultant damage from exposure to solar radiation must exist for biological entities to survive regular exposure to natural sunlight. Recent work has identified DNA lesions induced by near-UV radiation (Section 2.4). Furthermore, mechanisms of protection and repair of near-UV-induced lesions have been reported (Sections 2.3 and 2.6).

It was demonstrated a half-century ago (Hausser and Vahle, 1927) that sunburn is caused by the short-wavelength (< 320 nm) component of sunlight. Effectiveness was maximum at 300 nm, and declined sharply to low values at 290 and 310 nm. Effectiveness rose again at wavelengths shorter than 290 nm. That landmark result was extended by many similar action spectra for skin erythema in man and a variety of animals (e.g., see Blum, 1959; Freeman *et al.*, 1966; Parrish *et al.*, 1974). More recently, skin cancer,

---

* In this chapter, the following definitions will be followed: far-UV: UV-C, 200–290 nm, mid-UV: UV-B, 290–320 nm, near-UV: UV-A, 320–400 nm; visible light: 400–750 nm. Other definitions are used by some workers (Jagger, 1973).

which accounts for one-half of the cancer cases in the United States (Scotto *et al.,* 1974) and changes in appearance of the skin concomitant with aging have been associated with the same UV component (< 320 nm) of sunlight (Blum, 1959; Daniels, 1964; Urbach *et al.,* 1974). As action spectra for skin cancer and skin aging are known only in a preliminary way (Epstein, 1970; Setlow, 1974), it has been assumed that they follow the action spectrum for erythema (Section 6.1). In contrast to the previous assumption, Setlow (1974) proposed that the appropriate spectrum for skin cancer is one that coincides with the action spectrum for the production of changes in DNA. In support, Setlow and Hart (1975) gave initial evidence that cancer (thyroid carcinoma) in fish can be induced by pyrimidine dimers, relying on the specificity of enzymatic photoreactivation (Setlow and Setlow, 1972).

Most organisms appear to be well adapted to survive in their natural solar environment (Caldwell, 1971). However, even a moderate change in the ozone layer of the upper atmosphere would increase the amount of radiation in the 290–320 nm range that reaches the earth's surface, and might overwhelm the protection and repair systems of many forms of life. With the present spectral distribution of solar radiation, many kinds of organisms are easily killed by exposure to unfiltered sunlight (Blum, 1941; Buchbinder *et al.,* 1941; Harm, 1969). The most lethal component of sunlight for most organisms is the wavelength range below 310 nm (Buchbinder *et al.,* 1941; Ashwood-Smith *et al.,* 1967). *Escherichia coli* K12 AB2480 *uvrA recA*, a strain of bacteria lacking the capability for the dark repair of pyrimidine dimers, is inactivated 2 orders of magnitude by a 1-min exposure to unfiltered sunlight (Harm, 1969). A large fraction of this damage can be photoreactivated under appropriate conditions, indicating that pyrimidine dimers are the major lethal lesions produced in this strain (Harm, 1969). However, wild-type strains of *E. coli* K12 (Harm, 1969) and *Saccharomyces cerevisiae* (Resnick, 1970) show little evidence of the involvement of pyrimidine dimers in inactivation by sunlight. Harm (1969) and Resnick (1970) concluded that wild-type cells repair the usual germicidal damage by sunlight so efficiently that inactivation arises mostly from nonrepairable damage, perhaps in some component other than DNA (see Sections 2.4 and 2.9). When wavelengths in sunlight shorter than 320 nm were removed by a filter, the killing rate was substantially reduced both for cells and for transforming DNA (Rupert and Harm, 1966; Ashwood-Smith *et al.,* 1967; Jagger, 1975). However, the amount of decrease in the killing rate for wild-type cells was less than would be predicted if only the absorption by DNA was a factor in sunlight inactivation.

Action spectra for inactivation of a wide variety of cells show a close correspondence to the absorption spectrum of DNA, implicating DNA as the chromophore for the effects observed (Giese, 1968). However, few of the action spectra for cellular damage included wavelengths longer than 300 nm. Hollaender (1943) demonstrated that *E. coli* can be readily inactivated with high fluences of radiation at wavelengths between 350 and 500 nm (primarily 365 nm). Luckiesh (1946) published a preliminary action spectrum for lethality in *E. coli* extending from 200 to 700 nm. Hollaender (1943) and Luckiesh (1946) proposed that inactivation at wavelengths longer than 320 nm is caused by nonspecific physiological processes.

Hollaender and Emmons (1946) demonstrated the mutagenicity of unfiltered sunlight on conidia of the fungus *Aspergillus terreus*. However, they found that removal of wavelengths shorter than 320 nm by filtration eliminated the mutagenic effects of sunlight. Other early investigations of the mutagenicity of wavelengths longer than 320 nm were either negative or inconclusive. See reviews by Duggar (1936), Blum (1941), and Zelle and Hollaender (1955) for discussions of early work on lethal and genetic effects of long wavelength UV radiation. More recent discussions of mid- and near-UV radiation are in reviews by Harrison (1967), Eisenstark (1971), Jagger (1972), Webb (1972), Smith (1974*a,b*), Krinsky (1976), and Giese (1971, 1976).

This chapter primarily will treat work on lethal and mutagenic effects of near-UV radiation on microorganisms without added sensitizing agents in which fluences and wavelength ranges were reported. The numerous publications describing biological effects by unmeasured and undefined radiant emissions from a wide variety of light sources will not be included. Major emphasis will be given to work with monochromatic or narrow-band radiation. However, evidence is given in Section 5 that effects obtained with narrow-band radiation may not be the final answer: strong synergistic effects may occur between radiations of different wavelengths. Thus studies using broad-spectrum sources in which the emission spectrum is accurately defined may yield information unavailable from studies using monochromatic radiation only.

For discussions of photodynamic action emphasizing effects with exogenous sensitizers, see reviews by Blum (1941), Spikes (1968), Spikes and Livingston (1969), and Knowles (1975). Reviews of photosensitization with psoralens, which do not require oxygen, include those of Musajo and Rodighiero (1972) and Cole and Sinden (1975).

2. LETHAL EFFECTS

2.1. Action Spectra for Lethality

Hollaender (1943) and Luckiesh (1946) established that high fluences of near-UV and visible light can kill bacteria in the absence of added sensitizers. The sensitivity of the repair-proficient bacterium *E. coli* B/r to monochromatic or narrow-band radiation in the far-UV, mid-UV, near-UV, and visible ranges is depicted in Fig. 1. The curve shapes, which on this log–log plot differ from the more familiar curves on semilog plots, are similar at 254 nm and 313 nm, but show somewhat larger shoulders at 365, 460, and 550 nm. The data for inactivation at 254 nm and 365 nm shown in Fig. 1 are consistent with the results obtained by Hollaender (1943) for a different strain of *E. coli*. However, the decline in sensitivity from 365 to 550 nm is much greater than that reported for a different strain of *E. coli* by Luckiesh (1946).

Survival curve shapes of a variety of bacterial strains show large variations at different wavelengths (Hollaender, 1943; Webb and Lorenz, 1970; Webb and Brown, 1976; Webb *et al.*, 1976; Mackay *et al.*, 1976). Therefore, a single sensitivity constant is not an adequate expression for the comparison of lethal effects of a wide range of wavelengths. Various approaches have been used to deal with this problem. In some cases, a more complete description of the survival curves was attempted, using one of the following expressions:

$$S = ne^{-kF} \tag{1}$$

$$S = 1 - (1 - e^{-kF})^n \tag{2}$$

where S is the surviving fraction; k is the inactivation constant, the slope of the log transform survival curve in reciprocal fluence (F) units; and n is the shoulder constant, numerically equal to the y-axis intercept of the extrapolation of the exponential part of the survival curve. Most UV survival curves are not fitted well in the low-dose region by Eq. (2); points

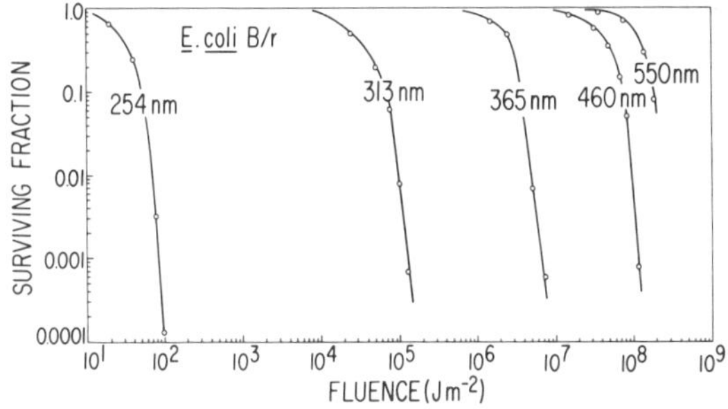

Fig. 1. Inactivation of stationary-phase *E. coli* B/r (in buffer suspension) at wavelengths of 254 nm (Hg line), 313 nm (Hg line), 365 nm (Hg line), 460 nm (band width at half-maximum, 20 nm), and 550 nm (band width at half-maximum, 40 nm). Irradiation was at 25°C under aerobic conditions. Fluence rates were as follows: 254 nm, 1.5 W m⁻²; 313 nm, 75 W m⁻²; 365 nm, 1250 W m⁻²; 460 nm, 3500 W m⁻²; 550 nm, 6500 W m⁻². Modified from Webb and Brown (1976, and unpublished observations).

generally fall below the theoretical line produced by this expression (Webb and Lorenz, 1970). Therefore, to use Eq. (2), very low weight must be given to the low fluence points in order to obtain an accurate description of the final slope.

As the inactivation constant k, which is based on the final slope [Eq. (2)], does not include information about any shoulder present, other approaches have been employed (Webb and Brown, 1976). The shoulder region of the survival curve, in many cases, represents the repair potential of the strain: the shoulder results from a decrease in the capacity of the cellular processes to repair damage induced at greater fluences of radiation. Thus in such cases the shoulder constant n from Eq. (2) does not describe a target number (see reports by Haynes, 1966, 1975, for a detailed discussion of the effect of repair processes on survival curves). Another approach to the expression of radiation sensitivity includes the empirical use of the reciprocal of fluences that yield survival values of 0.50, 0.37, or 0.10 (Gates, 1930; Hollaender, 1943; Setlow, 1964; Peak *et al.*, 1973*a,b,c*; Webb and Brown, 1976). Comparisons in this chapter are based on final slope (k), $1/F_{37}$, or $1/F_{10}$ values, depending on the expression deemed most appropriate, and in consideration of the method used by the author to describe his data.

An action spectrum for lethality in *E. coli* B/r Hcr, an excision-defective strain, over the wavelength range of 240–550 nm, is shown in Fig. 2. The inactivation constant is the final slope [k, Eq. (2)] of the log-transformed survival curves. These survival curves show a small shoulder [extrapolation number, $n = 2$, Eq. (2)] at wavelengths between 240 and 313 nm (Webb and Lorenz, 1970). At wavelengths longer than 325 nm, the shoulders are much larger ($n = 50$–150). The plots do not reflect the significantly larger shoulders on survival curves obtained at the longer wavelengths. In contrast to the action spectrum by Luckiesh (1946), the action spectrum for inactivation of *E. coli* B/r Hcr based on final slopes shows significant shoulders at 340, 410, and 500 nm. An action spectrum of *E. coli* B/r Hcr based on $1/F_{37}$ values revealed similar shoulders at the same wavelengths (Webb and Brown, 1976).

Sensitivity was independent of fluence rate at 365 nm over the range of 150–1500 W m^{-2} for the strains tested (*E. coli* B/r, *E. coli* B/r Hcr, and *E. coli* K12 AB2480) (Brown and Webb, 1972; Webb and Brown, 1976).

The action spectrum for lethality of stationary-phase cells of *E. coli* B/r Hcr closely follows the absorbance of DNA over the range of 240–313 nm (Fig. 2). Thus the ratio of sensitivity based on quantum units at 260 nm vs. 313 nm of 2.9×10^3 for this strain is very close to the ratio of absorbance of DNA at these two wavelengths. However, the absorbance of DNA at wavelengths longer than 320 nm cannot be measured because of light scatter. Although an absorbance ratio between radiation at 265 and 365 nm of $1 \times$

10^6 has been estimated by Setlow (1974) based on biological measurements and chemical assays of DNA photoproducts by Brown and Webb (1972) and Tyrrell (1973), the absorbance of thymidine at 330 nm is 1×10^{-8} of its absorbance at 260 nm (R. B. Webb, unpublished data), which makes it unlikely that direct absorption by DNA can account for the biological effects observed. The sensitivity ratios based on quantum units in *E. coli* B/r Hcr are 1.9×10^8 for 260 nm vs. 550 nm, and 1.3×10^9 for 260 nm vs. 650 nm.

Bacterial strains deficient in recombination repair show action spectra for lethality that depart significantly from the absorbance of DNA in the 280–300 nm range (Mackay *et al.*, 1976). For example, both stationary-phase and exponential-phase cells of *Salmonella typhimurium* KSU2480 (*recA*) are 4 times more sensitive at 290 nm than would be expected from DNA absorption (Fig. 3). The maximum departure from DNA absorbance

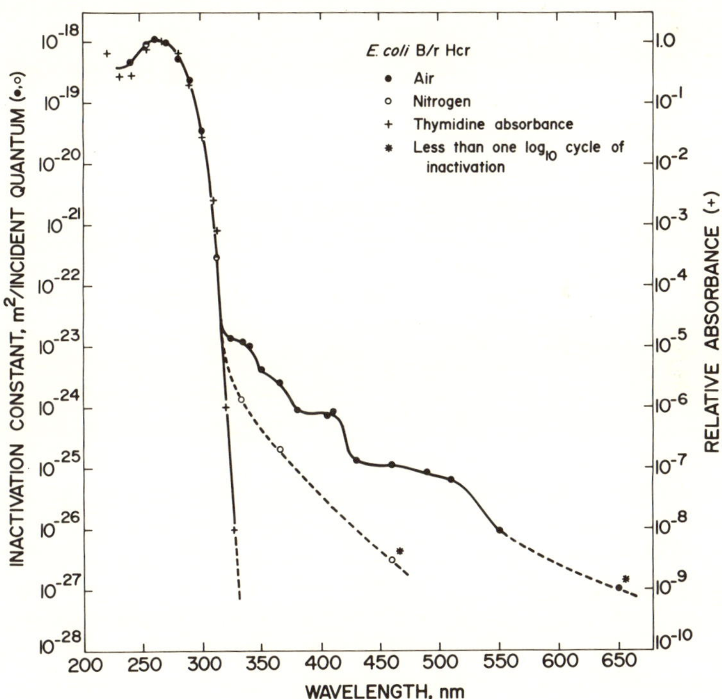

Fig. 2. Ultraviolet and visible action spectra for the inactivation of stationary-phase cells of *E. coli* B/r Hcr. The inactivation constant is the final slope *k* from Eq. (2). Cell suspensions (in buffer) were irradiated at 25°C under aerobic or anaerobic conditions. Points identified with an asterisk were estimated from incomplete survival curves by assigning a value of *n* shown by the nearest complete survival curve. Modified from Webb and Brown (1976).

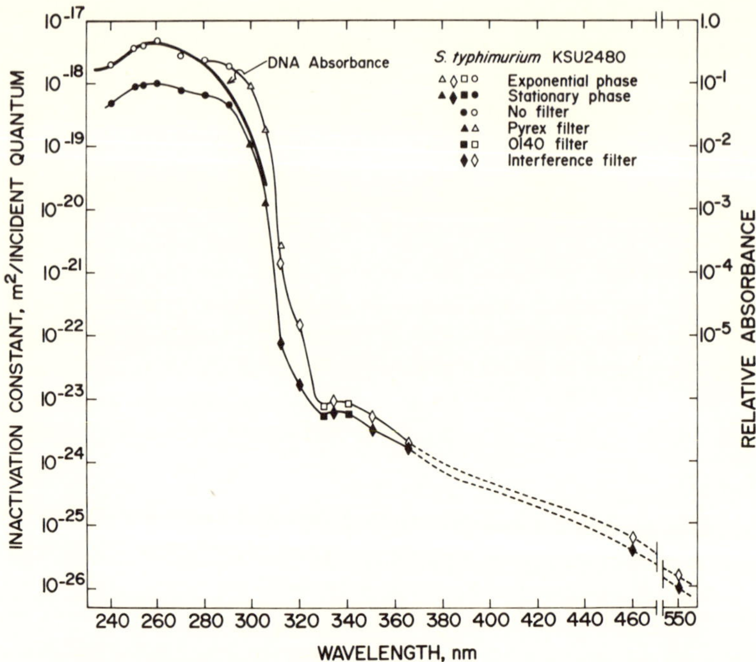

Fig. 3. Ultraviolet and visible action spectra from inactivation of stationary- and exponential-phase cells of *Salmonella typhimurium* KSU2480 (*recA*). The inactivation constant for wavelengths shorter than 315 nm is *k* from Eq. (1) fitted to the initial exponential component of the complex survival curves. For wavelengths longer than 315 nm, *k* is from Eq. (2), expressing the final slope of the shouldered survival responses. Cell suspensions were irradiated at 25°C under aerobic conditions. Modified from Mackay *et al.* (1976).

occurs between 290 and 300 nm, which suggests that the unexpected sensitivity in this range may not involve DNA or protein absorption. Earlier studies of the radiation-resistant bacterium *Micrococcus radiodurans* showed that the far-UV action spectrum corresponds more closely to protein absorption than to DNA absorption (Setlow and Boling, 1965). It was suggested that because of the highly efficient repair of photochemical damage to DNA in this bacterium, the cause of death might be damage to some protein component of the repair systems. Damage to repair systems would increase the lethal effect of lesions induced in the DNA. Action spectra of many other systems, including mammalian cell cultures, also correspond more closely to protein than to nucleic acid absorption. Nevertheless, for most cells, as well as viruses, other evidence strongly supports DNA as the major chromophore and target for lethal effects of far-UV radiation (see Smith and Hanawalt, 1969; Setlow and Setlow, 1972).

The sensitivity of bacterial strains deficient in recombination repair to narrow-band UV wavelengths longer than 320 nm is only two times greater than the sensitivity of wild-type strains (Mackay *et al.*, 1976). The relatively high sensitivity to broad-spectrum near-UV sources of recombination-deficient strains in exponential phase is discussed in Sections 2.3.1 and 2.5.

2.2. Oxygen Dependence for Lethality

Lethal effects of near-UV radiation are reported to be strongly enhanced by oxygen for a number of different microorganisms. A carotenoid-deficient colorless mutant of *Sarcina lutea* was readily inactivated by sunlight in the presence of oxygen, but no inactivation occurred in its absence (Mathews and Sistrom, 1959*b*). Both carotenoid-proficient (yellow) and carotenoid-deficient (white) strains of *S. lutea* were more sensitive to narrow-band near-UV (365 nm) radiation in the presence of oxygen than in its absence (Denniston *et al.*, 1972).

Eisenstark and co-workers (Eisenstark 1970, 1971, 1973; Ferron *et al.*, 1972) have investigated the near-UV sensitivity of recombination-deficient (*rec*) strains of *Bacillus subtilis, E. coli* and *S. typhimurium* bacteria. These strains in exponential phase were found to be sensitive to near-UV radiation from fluorescent blacklight (BLB) lamps in the presence of oxygen. The cells were much less sensitive under anoxic conditions. Stationary-phase cells showed little near-UV sensitivity at the fluences employed (Eisenstark and Ruff, 1970).

Several strains of *E. coli* also have shown a strong oxygen dependence for inactivation by near-UV radiation (365 nm) (Webb and Lorenz, 1970; Tyrrell, 1976*a*). The oxygen-dependent inactivation of *E. coli* B/r Hcr at wavelengths longer than 313 nm is shown in Fig. 2.

Inactivation at 254 nm was independent of the presence of oxygen (Zetterberg, 1964; Webb and Lorenz, 1970). Furthermore, oxygen dependence at 313 nm was below detection. Thus oxygen dependence is a clear distinction between biological effects from far-UV and near-UV radiation. The oxygen dependence reported for a variety of cells suggests that lethal effects at wavelengths longer than 320 nm may involve photodynamic processes mediated by endogenous sensitizers (see Section 2.10).

2.3. Repair of Lethal Damage

2.3.1. Sensitivity of Repair-Deficient Strains

Early findings on effects of wavelengths longer than 320 nm on bacteria suggested that physiological rather than genetic damage was the basis of the lethal action. Hollaender (1943) observed that radiation at wavelengths between 350 and 490 nm (the predominant biologically effective wavelengths were near 365 nm) at fluences that caused little inactivation produced long division delays in *E. coli* and increased the sensitivity of the irradiated cells to physiological saline. Later, Harrison (1967) reported that two strains of *E. coli* that differ in repair capability, B/r (repair proficient) and B_{s-1} (*uvrB lexA*), in exponential phase showed equal sensitivity to broad-band near-UV radiation (330–380 nm). Peak (1970) also observed that these two strains in exponential phase had equal sensitivities to 365-nm radiation. However, significant differences have been found in the sensitivities of other bacterial strains that differ in repair capability. Eisenstark (1970) reported that recombination-deficient strains of *S. typhimurium* and *E. coli* were readily inactivated by a broad-spectrum near-UV source (fluorescent BLB), while wild-type and excision-deficient cells did not show significant inactivation under the conditions employed. In stationary phase, *E. coli* B/r Hcr was 2.5-fold more sensitive than *E. coli* B/r to 365-nm radiation; however, the difference was not as great as that observed at wavelengths shorter than 313 nm (Webb and Lorenz, 1970; Webb and Brown, 1976).

Recent work clearly shows that DNA repair plays a major role in near-UV lethality in *E. coli* (Webb and Brown, 1976; Webb *et al.*, 1976). The lethal response of an isogenic series of *E. coli* K12 strains differing in repair capability to irradiation at 365 nm is shown in Fig. 4. Although the range of sensitivity at 254 nm (Howard-Flanders and Boyce, 1966; Rupp and Howard-Flanders, 1968) is greater than the range of sensitivity shown at 365 nm, the differences shown indicate that repair capability is an important factor in the lethal effects of 365-nm radiation. The differences in curve shape of the five strains present the difficulties in attempting to compare relative sensitivities. The final slope is an incomplete expression of the sensitivity of the strain, especially when the shoulder is likely to represent much of the repair potential of the strain (Haynes, 1975). To allow in part for the shoulder, therefore, sensitivity is sometimes expressed as $1/F_{37}$, where F_{37} is the fluence that results in a surviving fraction of 0.37. As the F_{37} values occur in the shoulder region, much of the repair potential of the strain will be included. Table 1 presents F_{37} and $1/F_{37}$ (sensitivity) values of the five strains of *E. coli* shown in Fig. 4 for both 254- and 365-nm radiations. Comparisons of the different strains at 365 nm reveal that the ratio of $1/F_{37}$ values of the most sensitive *E. coli* K12 strain, AB2480 (deficient in recombination and excision repair), and the least sensitive strain, AB1157 (repair proficient), is 67. The ratios of $1/F_{37}$ values at 365 nm of the double repair-deficient mutant (AB2480) and the single repair-deficient mutants tested are 23 for AB2463 (recombination deficient) and 35 for AB1886

TABLE 1. Sensitivity of Five Strains of *Escherichia coli* in Stationary Phase to 254-nm and 365-nm Radiations[a]

Wavelength	Parameters	Strain				
		K 12 AB2480 *recA uvrA*	K12 AB463 *recA*	K12 AB1886 *uvrA*	K12 P3478 *polA*	K12 AB1157 *rec$^+$ uvr$^+$ pol$^+$*
365 nm	F_{37} (J m^{-2})	2.68×10^4	6.71×10^5	9.5×10^5	5.8×10^5	1.8×10^6
	$1/F_{37}$ (m^{2}J^{-1})	3.73×10^{-5}	1.49×10^{-6}	1.05×10^{-6}	1.7×10^{-6}	5.5×10^{-7}
	k^b	3.73×10^{-5}	3.7×10^{-6}	9.4×10^{-6}	9.1×10^{-6}	2.2×10^{-6}
	$\dfrac{1/F_{37}\ (2480)}{1/F_{37}\ (\text{strain})}$	1	23	35	22	67
254 nm	F_{37} (J m^{-2})	0.029	0.83	1.29	3.8	36.9
	$1/F_{37}$ (m^{2}J^{-1})	34.5	1.20	0.78	0.26	0.027
	k^b	46.1	0.44	1.17	0.68	0.095
	$\dfrac{1/F_{37}\ (2480)}{1/F_{37}\ (\text{strain})}$	1	29	44	133	1280
Ratio 254/365	$\dfrac{1/F_{37}\ (254)}{1/F_{37}\ (365)}$	9.2×10^5	7.5×10^5	7.8×10^5	1.5×10^5	4.9×10^4
	k^b_{254}/k^b_{365}	1.2×10^6	3.2×10^5	1.8×10^5	7.5×10^4	4.3×10^4

[a] Values are derived from Fig. 4, data of Webb and Brown (1976), and R. B. Webb (unpublished data).
[b] The constant k is from equation (2).

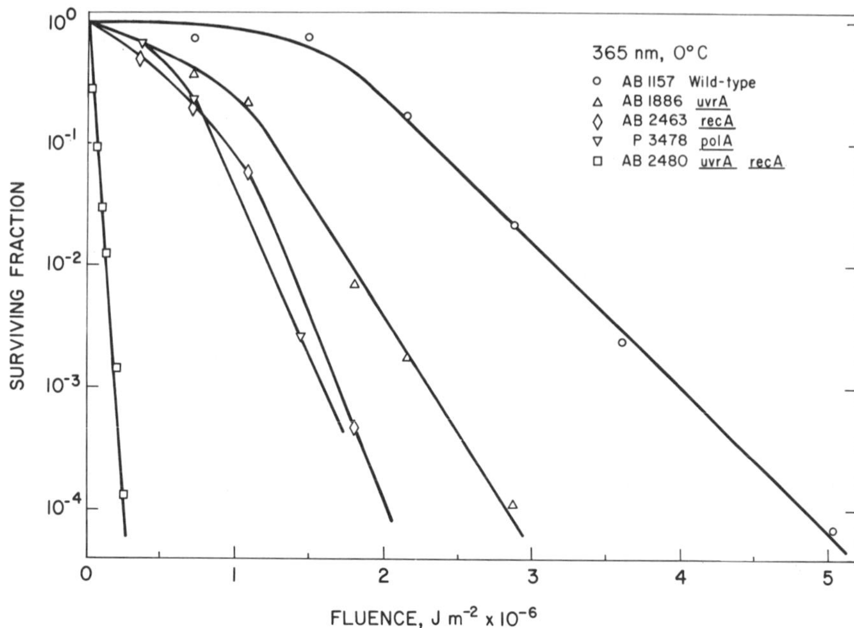

Fig. 4. Inactivation by 365-nm radiation of strains of *E. coli* K12 differing in repair capability. Stationary-phase cells (suspended in buffer) were irradiated at 0°C (to reduce concomitant photoreactivation) under aerobic conditions. Fluence rates were 1000–1250 W m^{-2}. Radiation source was a 2.5-kW Xe/Hg arc lamp (Hanovia) coupled to a predispersion prism (Schoeffel) and a 500-mm monochromator (Bausch and Lomb). Biologically effective short-wavelength stray light was eliminated with a Corning 7-51 band pass filter. Modified from Webb *et al.* (1976).

(excision deficient). The $1/F_{37}$ ratios of the double repair-deficient mutant and the single repair-deficient mutants at 365 nm are very close to the $1/F_{37}$ ratios at 254 nm for the same strains (Table 1). However, sensitivities at higher fluences based on final slopes [k, Eq. (2)] are considerably less at 365 nm than at 254 nm, which suggests that at higher 365-nm fluences both recombination repair and excision repair may be less effective. The difference in sensitivity of strains AB2480 and AB1157 [ratio of $1/F_{37}$ (AB2480) to $1/F_{37}$ (AB1157)] is much greater at 254 nm (1280) than at 365 nm (67). This result is consistent with reduced repair effectiveness at high 365-nm fluences.

Comparison of final slopes, k from Eq. (2), of the strains represented in Fig. 4 at 254 and 365 nm reveals that the ratios of inactivation constants ($k_{254} : k_{365}$) of the single repair mutants are 3.2×10^5 for AB2463 and 1.8×10^5 for AB1886, a decrease from the ratio based on $1/F_{37}$ values of these strains. The sensitivity ratio of the wild-type strain AB1157 at 254 and 365 nm is significantly less than observed with the single and double repair-deficient mutants. The ratio $k_{254}:k_{365}$ is 4.3 $\times$ 10^4 and the ratio $1/F_{37}$ (254):1/F_{37} (365) is 4.9 $\times$ 10^4 (Table 1).

2.3.2. Sensitivity of Exponential-Phase and Stationary-Phase Cells

Repair-proficient strains of *E. coli* show greater sensitivity to far-UV radiation in exponential growth than in stationary phase (Durham and Wyss, 1956; Ginsberg and Jagger, 1965; Hanawalt, 1966; Morton and Haynes, 1969; Tyrrell *et al.*, 1972*a*). The difference is the decreased size of the shoulder on survival curves of exponential-phase cells; the final slopes are similar for the two growth phases. Harrison (1967) and Peak (1970) reported that cells of repair-proficient strains of *E. coli* also show a greater sensitivity to near-UV radiation in exponential phase than in stationary phase. As depicted in Fig. 5, the relative increase in sensitivity during exponential growth is much greater for 365-nm than for 254-nm radiation (Peak, 1970).

At 254 nm, the increase in sensitivity in exponential growth is associated with excision-repair capability (Hanawalt, 1966; Morton and Haynes, 1969; Tyrrell *et al.*, 1972*a*). Strains capable of excision repair, including the repair-proficient strain *E. coli* K12 AB1157 and the recombination repair-deficient strain *E. coli* K12 AB2463, showed a substantial

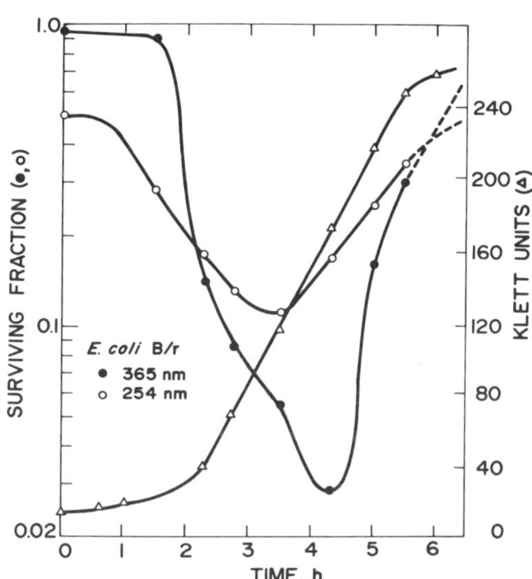

Fig. 5. Inactivation of *E. coli* B/r at 254 and 365 nm (plus traces of 313- and 334-nm radiation) at a constant fluence through the growth cycle. Cells were removed, washed, and resuspended in buffer, and irradiated in a minimum time. Modified from Peak (1970).

increase in 254-nm sensitivity during exponential growth (Tyrrell *et al.*, 1972*a*). However, strains deficient in excision repair, *E. coli* K12 AB1886 and *E. coli* K12 AB2480, did not show an increase in sensitivity during exponential growth (Tyrrell *et al.*, 1972*a*). On the contrary, strain AB1886, deficient in excision repair but proficient in recombination repair, revealed a significant *decrease* in sensitivity during growth. The double repair-deficient mutant AB2480 was equally sensitive throughout the growth cycle.

M. J. Peak and R. B. Webb (unpublished data) tested the 365-nm sensitivity of four strains of *E. coli* K12 at different stages of the growth cycle. The pattern of sensitivity for 365-nm radiation was similar to that reported by Tyrrell *et al.* (1972*a*) for 254-nm radiation in that strains capable of excision repair, AB1157 and AB2463 (*recA*), showed much greater near-UV sensitivity in exponential phase; while strains deficient in excision repair, AB1886 (*uvrA*) and AB2480 (*recA uvrA*), were approximately equal in sensitivity in exponential and stationary phase.

Eisenstark and co-workers (Eisenstark, 1970, 1973; Mackay *et al.*, 1976) reported that *rec* strains of *E. coli, S. typhimurium,* and *B. subtilis* are readily inactivated by a broad-spectrum near-UV source (fluorescent BLB bulbs) in exponential growth (cells were irradiated in nutrient medium), but are not inactivated at even greater fluences in stationary phase. Repair-proficient and excision-repair-deficient strains showed no significant inactivation in exponential or stationary phase under the conditions and fluences employed. The difference in sensitivity of exponential- and stationary-phase strains is considerably greater for inactivation by broad-spectrum near-UV radiation than for either monochromatic 254- or 365-nm radiation (Eisenstark, 1970; Mackay *et al.*, 1976). See Section 2.5 for a discussion of mechanisms for the near-UV sensitivity of *recA* strains of bacteria.

The greater sensitivity of strains capable of excision repair in exponential phase to near-UV radiation accounts for the failure of some workers to observe a greater near-UV sensitivity of repair-deficient strains than repair-proficient strains (Harrison, 1967; Peak, 1970; Peak *et al.*, 1973*b*; Tyrrell and Webb, 1973). In exponential phase, *uvr* and related repair-proficient strains showed almost the same sensitivity to 365-nm radiation (Tyrrell and Webb, 1973; Peak *et al.*, 1973*b*). However, even in exponential phase, the double repair-deficient mutant *E. coli* K12 AB2480 (*uvrA recA*) was much more sensitive to near-UV radiation than the isogenic single repair-deficient and repair-proficient strains. In exponential phase (based on $1/F_{37}$ values), strain AB1886 (*uvrA*) was no more sensitive, strain AB2463 (*recA*) was fivefold more sensitive, and strain AB2480 (*uvrA recA*) was fiftyfold more sensitive than the isogenic repair-proficient strain AB1157 (M. J. Peak and R. B. Webb, unpublished data).

Hanawalt (1966) suggested that the cause of the 254-nm sensitivity of repair-proficient strains in exponential growth was the onset of DNA replication after the occurrence of excision, but before the completion of the polymerization and rejoining steps. Attempted replication of the resultant single-strand regions of DNA would result in lethal configurations. A similar mechanism may be responsible for the greater exponential-phase sensitivity to near-UV radiation. Acriflavine in the plating medium after 365-nm irradiation has a greater sensitizing effect on uvr^+ strains than after 254-nm irradiation (see Section 2.3.3). The onset of excision after 365-nm irradiation may be delayed for several hours in cells suspended in buffer (R. M. Tyrrell, personal communication). Whether DNA synthesis is delayed more than excision has not been determined. Damage to metabolic aspects of DNA replication also has been suggested as the basis of the near-UV sensitivity of $recA$ strains in exponential phase (Eisenstark, 1973; Ferron et al., 1972). It is unlikely, however, that damage to metabolic processes independent of DNA lesions is responsible for near-UV lethality in exponential- or stationary-phase bacterial cells (see Sections 2.4 and 2.9).

2.3.3. Effects of Repair Inhibitors

It is well known that bacterial strains that are proficient in the dark repair of DNA lesions show greater sensitivity to far-UV and X radiation when certain compounds are present in the postirradiation growth medium. Such postirradiation sensitizing agents include acridine dyes (Alper, 1963; Alper and Hodgkins, 1969; Alper et al., 1972; Harm, 1967; Day and Deering, 1968; Peak, 1970), caffeine (Harm, 1967; Peak, 1970), and 8-methoxypsoralen (Igali et al., 1970). Host-cell reactivation of bacteriophage inactivated at 254 nm is inhibited by certain basic dyes (Patrick and Rupert, 1967). The excision of pyrimidine dimers is inhibited by acridine orange (Setlow, 1964). Quinacrine strongly inhibits the rejoining of X-ray-induced DNA single-strand breaks mediated by the $recA$ gene product (Fuks and Smith, 1971). However, only a weak effect of quinacrine was obtained on the DNA polymerase 1 mediated rejoining of DNA single-strand breaks. Various kinds of evidence suggest that acriflavine interferes with some step or steps in both excision and recombination repair (Witkin, 1963; Harm, 1967; Smith, 1971; Alper et al., 1972). Chloramphenicol inhibits the rejoining of single-strand breaks by the excision repair process (Youngs et al., 1974), but does not block the excision of pyrimidine dimers (Swenson and Setlow, 1970). Evidence has been presented that chloramphenicol inhibits $recB$-mediated repair of DNA damage from higher fluences (> 20 J m^{-2}) of far-UV radiation (Ganesan and Smith, 1972; Van der Schueren *et al.*, 1973). The metabolic inhibitor 2,4-dinitrophenol (DNP) sensitizes *E. coli* K12 cells to both far-UV and X radiation when added to the postirradiation growth medium. Sensitization by DNP to X radiation was found to occur by inhibiting the growth-medium-dependent repair of DNA single-strand breaks (Van der Schueren and Smith, 1974). DNP, which sensitizes both wild-type and *polA* strains of *E. coli* K12 to far-UV radiation, was found to interfere with the rejoining of DNA single-strand breaks produced by the excision of dimers (Van der Schueren and Smith, 1974).

There are fewer reports on the effects of repair inhibitors with near-UV radiation. Peak (1970) found that *E. coli* B/r is more sensitive to 365-nm radiation (from a Philips HP 125-W mercury lamp) when plated in the presence of acriflavine or caffeine. In comparison with the 254-nm response, caffeine sensitization was smaller and acriflavine sensitization was larger at 365 nm.

The presence of acriflavine in the growth medium following 365-nm irradiation greatly increased the sensitivity of *uvrA*, *recA*, and repair-proficient strains of *E. coli* (Webb and Brown, 1976). The increase in sensitivity was associated with the loss of the large shoulders characteristic of 365-nm survival curves in the absence of a repair inhibitor. In accord with the earlier report of Peak (1970), acriflavine in the postirradiation growth medium produced a greater effect after irradiation at 365 nm than at 254 nm.

Sensitization of *E. coli* by quinacrine to 365-nm radiation paralleled the effects of acriflavine except for the *recA* strain K12 AB2463, which was sensitized only a small amount (Webb and Brown, 1976). This relationship is similar to the quinacrine inhibition of the repair of X-ray-induced single-strand breaks in DNA reported by Fuks and Smith (1971). Quinacrine strongly inhibited repair associated with the presence of the *recA* gene product at a concentration (76 μg/ml) that had little effect on repair mediated by DNA polymerase I. Neither acriflavine nor quinacrine caused significant sensitization to 365-nm radiation in the double repair-deficient mutant K12 AB2480 (Webb and Brown, 1976).

The greater sensitizing effect of acriflavine or quinacrine at 365 nm compared with the response at 254 nm may be caused by a delay in the onset of excision repair after 365-nm irradiation. This delay may parallel growth and division delay known to be induced by 365-nm radiation (reviewed by Jagger, 1972). The nutrient-independent branch of excision repair of 254-nm-induced lesions occurs rapidly at 25°C in nonnutrient buffer (Youngs *et al.*, 1974). Therefore, a significant amount of excision repair may occur before the cells are plated. If such excision repair is delayed after irradiation at 365 nm, its onset might not occur until after the cells are plated, and the presence of a repair inhibitor in the medium should have an increased effect.

The greater sensitivity of repair-deficient bacterial strains to 365-nm radiation, together with the large sensitizing effect of DNA repair inhibitors in strains capable of dark repair, provides evidence that lesions induced in DNA play the major role in the lethal action of 365-nm radiation. However, other kinds of damage—membrane effects, enzyme inactivation, or growth or division delay—may interact with near-UV-induced DNA lesions.

2.4. Near-UV-Induced Lesions

2.4.1. Pyrimidine Dimers

Pyrimidine dimers are induced in bacterial DNA by 365-nm radiation at a rate of 5.5×10^{-5} thymine-containing dimers per *E. coli* genome per J m^{-2} (Tyrrell, 1973). The ratio of dimer induction at 254 nm to that at 365 nm based on energy fluence was found to be 7.1×10^5, a value close to the ratio (9×10^5) for inactivation for the radiation-sensitive double repair-deficient mutant *E. coli* K12 AB2480 (*uvrA recA*) (Brown and Webb, 1972). The induction of dimers at 254 and 365 nm is depicted for *E. coli* B/r Hcr in Fig. 6. The production of thymine-containing pyrimidine dimers at dif-

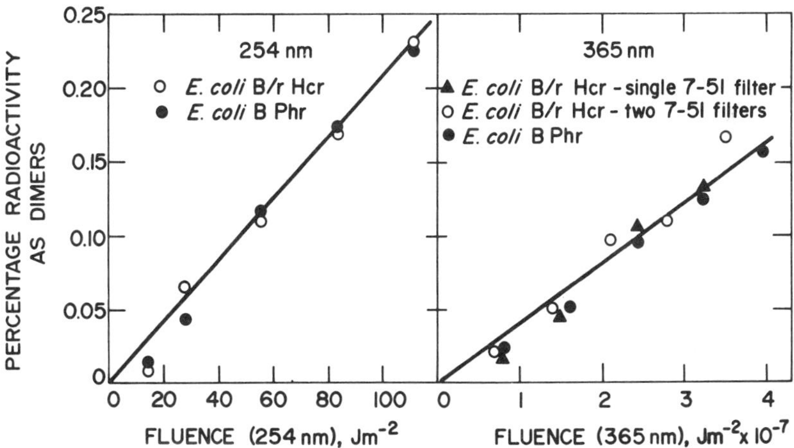

Fig. 6. Induction of thymine-containing pyrimidine dimers in *E. coli* B/r Hcr and *E. coli* B Phr by 254- and 365-nm radiation. Exponential-phase cells (washed and resuspended in buffer) were irradiated under aerobic conditions at 25°C (254 nm) or 0°C (365 nm). Fluence rates were 9 W m^{-2} at 254 nm and 1000 W m^{-2} at 365 nm. Dimer yields were assayed by two-dimensional paper chromatography. Redrawn from Tyrrell (1973).

ferent UV wavelengths is summarized in Table 2. As 365 nm is within the efficient part of the action spectrum for photoreactivation (PR), the 365-nm irradiations were delivered at 0–1°C to reduce concomitant PR (Brown and Webb, 1972). Dimer induction is also shown in Fig. 6 for *E. coli* B *phr*, a strain that lacks the PR enzyme. The induction rate was the same in strains B/r Hcr and B *phr*, indicating that concomitant PR was not a significant factor in these measurements. To confirm the nature of the 365-nm photoproduct, the "dimer" region of the chromatogram was eluted and subjected to 235-nm radiation. Virtually all of the photoproduct was converted to thymine, as would be predicted from the work of Wulff and Fraenkel (1961). Further evidence that the thymine-containing photoproduct induced at 365 nm was the same as the cyclobutane-type pyrimidine dimer induced at 254 nm was provided by the removal of the 365-nm-produced dimers in *E. coli* DNA by yeast PR enzyme and 380–440 nm illumination *in vitro* (Tyrrell *et al.*, 1973).

The possibility of a significant contribution to 365-nm dimer production by the more efficient short-wavelength component of stray light from the monochromator was eliminated by comparing dimer induction with one and two stray light filters (Corning 7-51). Any stray light of less than 320 nm trasmitted by the first filter would be greatly reduced by the second filter. As shown in Fig. 6, dimer induction was the same whether one or two filters were used. Therefore, wavelengths shorter than 320 nm did not contribute significantly to the dimer yields at 365 nm. Different ratios of the two classes of dimers containing thymine [thymine–thymine (T<>T) and uracil–thymine (U<>T)] were induced at 254 and 365 nm. The ratio of T<>T and U<>T dimers at 254 nm was 1.1–1.2 over a tenfold fluence range, in agreement with the value of 1.2 measured by Setlow and Carrier (1966) for *in vivo* irradiation of DNA. However, at 365 nm, the ratio of T<>T to U<>T was 5. (Tyrrell, 1973). Pyrimidine dimer induction at 365 nm is similar to the acetophenone-sensitized induction at 313 nm in that the dimers formed in both cases are mostly of the T<>T type (Table 2) (Tyrrell, 1973; Rahn, 1973).

Only a few measurements of cytosine–cytosine (C<>C) dimers have been made. Setlow (1966) obtained values for C<>C dimers of 7% and 6% of the total pyrimidine dimers at 265 and 280 nm, respectively, in aqueous solutions of DNA at pH 7. An estimate for the distribution of dimers in *E. coli* irradiated at 265 nm *in vivo* is C<>C:10%; C<>T:40%; T<>T:50% (Setlow and Carrier, 1966). The ratio for T<>T to C<>T of 5:1 at 365 nm suggests that the contribution of C<>C dimers would be expected to be very small.

No thymine-containing dimers were detectable in *E. coli* DNA irradiated *in vivo* at 405 nm after a fluence of 10^8 J m^{-2} (R. M. Tyrrell,

TABLE 2. Production of Pyrimidine Dimers and Single-Strand Breaks (or Alkali-Labile Bonds) in Double-Stranded DNA in Cells and in Aqueous Solution by UV and Visible Radiation of Different Wavelengths

Wavelength (nm)	Dimers[a]					SS breaks[b]				Dimers per SS break
	In vitro yield[c]	*In vivo* yield[c]	Relative yield	$T\langle\;\rangle T$ $U\langle\;\rangle T$	References[d]	*In vitro* yield[c]	*In vivo* yield[c]	Relative yield	References[d]	
254	40		1		(1)	6.3×10^{-2e}		1	(2)	
		39	1	1.1	(3)		4.9×10^{-2}	1	(4)	800
265	46		1.2	1.7	(5)					
		46	1.2	1.2	(6)					
280	16		0.40	2.6	(5)					
313	2.8×10^{-2}		7.0×10^{-4}	1.2	(5)	9.8×10^{-4}		1.6×10^{-2}	(9)	25
							1.3×10^{-3}	2.7×10^{-2}	(7)	21[f]
365	1.0×10^{-4}		2.5×10^{-6}		(8)	6.0×10^{-5e}		9.5×10^{-4}	(4)	1.7
		5.5×10^{-5}	1.4×10^{-6}	5	(3)		3.0×10^{-5}	6.1×10^{-4}	(4)	1.8
							7.5×10^{-5g}	1.5×10^{-3}	(7)	0.73
405		$<1 \times 10^{-6}$			(8)		1.1×10^{-5}	2.2×10^{-4}	(7)	<0.09

[a] Thymine-containing pyrimidine dimers ($T\langle\;\rangle T + U\langle\;\rangle T$).

[b] Single-strand breaks or alkali-labile bonds.

[c] DNA lesions per 2.5×10^9 daltons (assumed value for *E. coli* genome) per J m^{-2}.

[d] References: (1) Setlow (1968), (2) Brent (1972), (3) Tyrrell (1973), (4) Tyrrell *et al.* (1974), (5) Setlow and Carrier (1966), (6) Setlow (1966), (7) R. D. Ley (personal communication), (8) R. M. Tyrrell (personal communication), (9) Zierenberg *et al.* (1971).

[e] Based on extracted phage DNA.

[f] Based on dimer yield *in vitro*.

[g] SS-break yield in *E. coli* P3478 *polA1*.

personal communication). An estimate of the sensitivity of the chromatography assay places the upper limit at approximately 1×10^{-6} dimers per 2.5×10^9 daltons per J m^{-2} (Table 2). The absence of photoreactivability after 405-nm inactivation (0°C) of *E. coli* K12 AB2480 under conditions in which a large PR sector was obtained at 365 nm supports the result from the chemical assay (Brown and Webb, 1972; M. S. Brown, unpublished data).

Recently, it has been reported that thymine glycols, monomeric ring-saturated products of the 5,6-dihydroxydihydrothymine type, are formed in DNA of human cells both *in vitro* and *in vivo* by far-UV irradiation (Hariharan and Cerutti, 1974, 1977*a*). Estimates for total ring saturation in monomeric thymine photoproducts can be obtained from products of the 5,6-dihydroxydihydrothymine type by the alkali–acid degradation assay of Hariharan and Cerutti (1974). Thymine glycols were formed *in vitro* at 4 times the rate they were formed *in vivo*. In HeLa S-3 cells, one thymine glycol was formed for 8.4 pyrimidine dimers at 240 nm, one for 20 dimers at 260 nm, one for 17 dimers at 280 nm, and one for 1.4 dimers at 313 nm (6-nm band width at half-maximum) (P. A. Cerutti, personal communication). Thymine glycols also were produced at 334 and 365 nm; however, the yield was considerably less than for pyrimidine dimers at these wavelengths. Thymine glycol photoproducts, which are produced by both γ and UV radiation, can be excised by extracts of xeroderma pigmentosum fibroblasts (Hariharan and Cerutti, 1976). When thymine glycols were produced in the single-stranded phage ϕX174 by osmium tetroxide, it was determined that 2.3 thymine glycols were required per lethal event (Hariharan *et al.*, 1977). The possibly important role of thymine glycols in the inactivation of bacteria at wavelengths longer than 300 nm has not been determined (P. A. Cerutti, personal communication).

2.4.2. Single-Strand Breaks (or Alkali-Labile Bonds)

In addition to pyrimidine dimers, 365-nm radiation induces single-strand breaks (or alkali-labile bonds) in bacterial DNA at a rate slightly greater than one-half that for dimers (Tyrrell *et al.*, 1974). Single-strand breaks were induced linearly with 365-nm fluence over the range tested (Fig. 7) at a rate of $1.18 \pm 0.07 \times 10^{-6}$ single-strand breaks per 10^8 daltons per J m^{-2}. Assuming 2.5×10^9 as the molecular weight of a single *E. coli* genome, the rate of break induction was 3.0×10^{-5} single-strand breaks per genome per J m^{-2}. Degradation of DNA was much too small to account for the single-strand breaks observed.

Single-strand breaks were also produced by 365-nm radiation in intact-phage DNA and extracted-phage DNA at rates of $6.5 \pm 2.1 \times 10^{-7}$ and 2.4

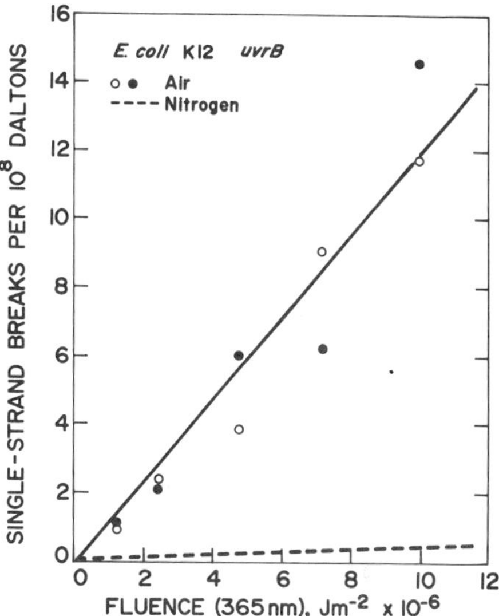

Fig. 7. Induction of single-strand breaks (or alkali-labile bonds) in DNA of exponential-phase *E. coli* SR22B (*uvrB5*) by 365-nm radiation (0°C) under aerobic (O,●) and anaerobic (---) conditions. Open and closed circles represent separate experiments. Modified from Tyrrell *et al.* (1974).

$\pm$ 0.3 $\times$ 10^{-6} single-strand breaks per 10^8 daltons per J m^{-2}, respectively (Tyrrell *et al.*, 1974). The presence of 0.1 M 2-aminoethylisothiouronium bromide-HBr (AET) during 365-nm irradiation of both intact-phage and extracted-phage DNA reduced the rate of single-strand-break induction by factors of 3 and 9, respectively. The reduced DNA single-strand-break induction in the presence of AET, a free-radical scavenger, suggests that free radicals may be involved in the induction of DNA breaks by 365-nm radiation. No detectable DNA single-strand breaks were formed in DNA of bacteria irradiated at 365 nm under stringent anaerobic (nitrogen) conditions (Tyrrell *et al.*, 1974; R. D. Ley and R. B. Webb, unpublished data).

The repair of single-strand breaks induced in bacterial DNA by aerobic 365-nm radiation at 0°C was observed after holding the irradiated cells in phosphate buffer at 30°C (R. D. Ley, personal communication). Repair was almost complete in a wild-type strain, *E. coli* W3110, after a 10-min holding in buffer. Comparable results were obtained in a related *uvrA* strain. However, no repair of DNA breaks was observed in buffer in *E. coli* P3478 *polA1,* a strain deficient in DNA polymerase I. Furthermore, the 365-nm single-strand-break yield was twofold greater in the *polA* strain (Table 2), suggesting that some repair of the single-strand breaks may occur at 0°C during irradiation in strains capable of such repair. The repair in buffer of 365 nm-induced single-strand breaks appears to parallel the DNA polymerase I dependent repair of single-strand breaks induced by X radiation (type II repair) described by Town *et al.* (1973). The observation by R.

D. Ley (personal communication) that DNA single-strand breaks induced in *E. coli* cells at 365 nm were readily repaired in buffer in wild-type and *uvrA* strains but not in a *polA* strain is consistent with the lesions being true breaks rather than alkali-labile bonds. However, other endonucleases occur that could account for the repair of alkali-labile lesions in DNA.

The rate of single-strand-break (or alkali-labile-bond) induction increased strikingly relative to that of pyrimidine dimers from 254 to 405 nm (Table 2). At 254 nm, the ratio of dimers to that of breaks was 800; at 313 nm, that ratio declined to 21; and at 405 nm, the dimer-to-break ratio was less than 0.1 (R. D. Ley, personal communication).

2.4.3. Damage to Repair Systems

In addition to the induction of DNA lesions, near-UV radiation can inhibit the full functioning of DNA repair systems (Tyrrell, 1973, 1974, 1976a; Tyrrell and Webb, 1973; Tyrrell *et al.*, 1973; Webb *et al.*, 1977b). The capacity of the PR enzyme to monomerize pyrimidine dimers was inactivated by near-UV radiation both *in vivo* and *in vitro* at fluences that inactivate repair-proficient strains of *E. coli* (Tyrrell *et al.*, 1973). The *in vitro* 365-nm inactivation of the yeast PR enzyme, shown in Fig. 8, is oxygen dependent (Tyrrell, 1976a). The 365-nm inactivation rate of the *E. coli* PR enzyme *in vivo* was 3.2×10^{-7} m² J⁻¹, which was not significantly different from the inactivation rate of the yeast PR enzyme *in vitro* (3.7×10^{-7} m² J⁻¹) (Tyrrell *et al.*, 1973). The similarity of the inactivation rate of the PR enzyme under *in vitro* and *in vivo* conditions suggests that the enzyme complex may be, or may contain, the chromophore. The nature of the damage is not known.

In parallel with the inhibition of PR enzyme activity determined by the photoreactivation of dimers by chemical assay (Tyrrell *et al.*, 1973), the capacity for the photoreactivation of 254-nm lethality in exponential-phase cells of *E. coli* was inhibited by prior exposure to 365-nm radiation (Brown, 1977). No PR after 254-nm inactivation could be demonstrated after a 365-nm fluence of 1.7×10^6 J m⁻² in exponential-phase cells. However, stationary-phase cells retained PR capacity to a greater 365-nm fluence of 3×10^6 J m⁻², the greatest used in the experiment. The difference between exponential- and stationary-phase cells is likely due to the presence of a greater number of PR enzyme molecules in stationary-phase cells (Tyrrell *et al.*, 1972b).

The capacity to excise pyrimidine dimers is impaired by fluences of 365-nm radiation in the biological range (Fig. 9) (Tyrrell and Webb, 1973). After a 365-nm fluence of 1.5×10^6 J m⁻², only 15% of the capacity to excise pyrimidine dimers remained. The rate of inactivation of the same repair-proficient strain increased sharply at fluences greater than 1.5×10^6 J m^{-2}. These results suggested that the sharp increase in 365-nm inactivation at higher 365-nm fluences may be caused by damage to the excision-repair system, thereby increasing the lethal effects of DNA lesions induced by 365-nm radiation. If a decrease in capacity to excise dimers is not reversed before the onset of DNA replication, cells irradiated at 365 nm should show an increased sensitivity to subsequent irradiation at 254 nm. Results from such an experiment are shown in Fig. 10. Exponential-phase cells of *E. coli* B/r were strongly sensitized to 254-nm radiation by a prior exposure to 1×10^6 J m^{-2} of 365-nm radiation. Comparable experiments with other repair-proficient strains K12 AB1157 and W3110, *uvrA* strains B/r Hcr and K12 AB1886, and the *recA* strain K12 AB2463 strongly suggest that both excision-repair and recombination-repair processes are significantly impaired by fluences of 365-nm radiation greater than 1×10^6 J

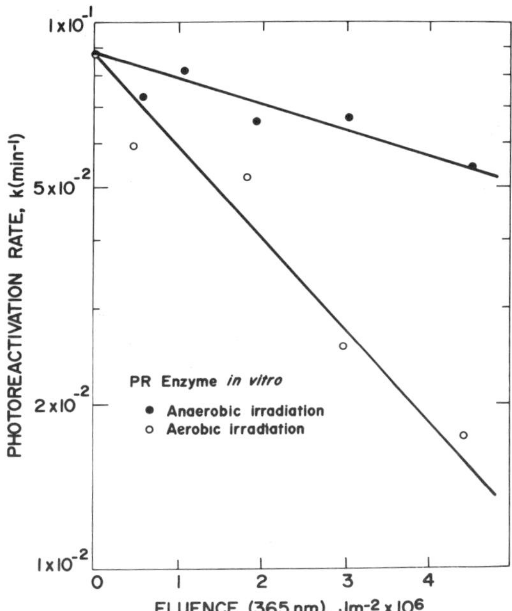

Fig. 8. Inactivation of yeast photoreactivating (PR) enzyme *in vitro* at 365 nm in the presence and absence of oxygen. Photoreactivation rate was assayed on ³H-labeled *E. coli* DNA irradiated at 365 nm to induce pyrimidine dimers. The irradiated PR enzyme was mixed with dimer containing DNA. After 30 min, the mixture was illustrated with PR light (370–420 nm) isolated from a 500 W quartz iodine lamp for 15 min at a fluence rate of 150 W m^{-2} and assayed for thymine-containing dimers (Tyrrell, 1973). Redrawn from Tyrrell (1976*a*).

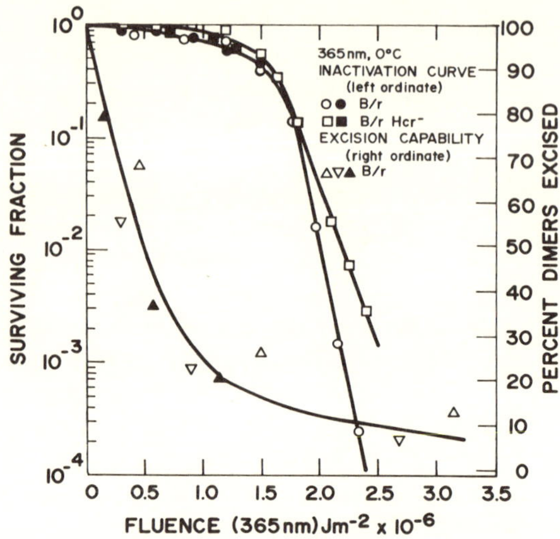

Fig. 9. Inactivation of *E. coli* B/r and *E. coli* B/r Hcr in exponential phase (washed and resuspended in buffer) by 365-nm radiation (fluence rate 1000 W m⁻²). Decline in the efficiency of excision repair is shown on the same 365-nm fluence scale (right ordinate). A dimer substrate for excision was provided by irradiating *E. coli* B/r cells with 50 (▲), 72 (△), and 75 (▽) J m⁻² of 254-nm radiation. After holding 80 min in supplemented medium at 37°C, less than 10% of the dimers remained in the control samples (100% on the right-hand ordinate). Redrawn from Tyrrell and Webb (1973).

m⁻² (Webb *et al.,* 1977b). However, the sensitizing effect of a prior exposure to 365-nm radiation on the 254-nm sensitivity was much less with K12 AB1157 than with B/r (Webb *et al.,* 1977b; R. M. Tyrrell, personal communication).

Near-UV-induced photoprotection (see Jagger, 1972) and possible damage to repair systems were observed in both exponential- and stationary-phase cells of the *recA* strain *E. coli* K12 AB2463 (Webb *et al.,* 1977b). The effect of prior exposure to different fluences of 365-nm radiation on the survival from a constant fluence of 254-nm radiation is shown in Fig. 11. For 365-nm fluences up to 7.5 × 10⁵ J m⁻², both exponential- and stationary-phase cells became less sensitive to 254-nm radiation. This protoprotection was greater for exponential-than for stationary-phase cells. At 365-nm fluences greater than 7.5 × 10⁵ J m⁻², the cells increased in sensitivity to 254-nm radiation. This increase in sensitivity is interpreted as further evidence for damage to the excision repair system by near-UV radiation. Irradiation of stationary-phase cells under anoxia at 365 nm did not alter the protective effect (photoprotection) at fluences below 8 × 10⁵ J m⁻²; However, the increase in 254-nm sensitivity found after greater 365-nm fluences in this *recA* strain was strongly affected by anoxia. Similar biological evidence with other strains indicates that inhibition by 365-nm radiation of the capacity to repair 254-nm damage by both the recombination and excision repair systems also is strongly oxygen dependent (Webb and Brown, 1977*b*).

Tyrrell (1974) demonstrated that near-UV (365-nm) radiation can inhibit the repair of DNA single-strand breaks induced by X radiation in cells of *E. coli*. The components of repair were assayed following the rational of Smith and co-workers (Town *et al.*, 1973; Martignoni and Smith, 1972, 1973). Reduction in DNA polymerase I mediated repair of breaks (buffer at 25°C) recovered after holding for 60 min before exposure to X radiation. Damage to *recA*-mediated repair (full medium at 37°C) did not recover after holding for 60 min. Survival data corresponded to the alkaline sucrose-gradient results (Tyrrell, 1974, 1976*a*). In parallel with inactivation of the PR enzyme *in vitro*, the 365-nm-induced sensitivity to electron and X irradiation was strongly oxygen dependent (Tyrrell, 1976*a*).

The synergistic interaction of far-UV and X radiation is well established (Haynes, 1964; Baptist and Haynes, 1972). This synergism was presented as evidence for the overlap of sites of action of far-UV and X radiation (Haynes, 1964). The possibility that far-UV radiation damages repair enzymes was considered by Haynes (1966). More recently, Martignoni and Smith (1972) demonstrated an interference, by an exposure to 254-nm radiation, with the *rec* (full medium) repair of X-ray-induced DNA single-strand breaks. This synergism was attributed to the

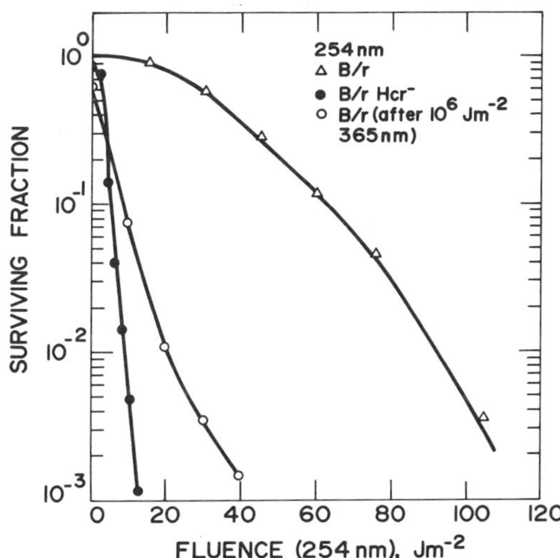

Fig. 10. Inactivation of exponential-phase (washed and resuspended in buffer) *E. coli* B/r and *E. coli* B/r Hcr after exposure to 254-nm radiation. The 254-nm sensitivity of strain B/r is also shown immediately following an exposure to 1×10^6 J m⁻² of 365-nm radiation. Redrawn from Tyrrell and Webb (1973).

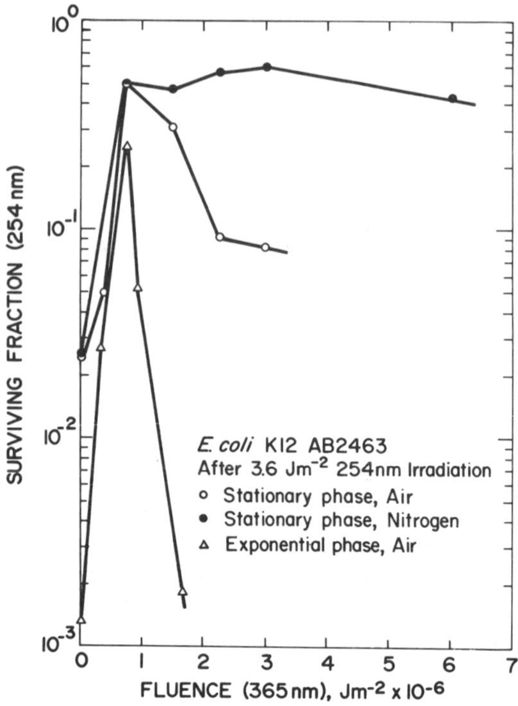

Fig. 11. Survival of exponential- and stationary-phase cells (washed and resuspended in buffer) of *E. coli* K12 AB2463 after exposure to a constant fluence (3.6 J m^{-2}) of 254-nm radiation as a function of the fluence of a prior exposure to 365-nm radiation at 0°C. Stationary-phase cells are shown for both aerobic and anaerobic 365-nm irradiation. Exponential-phase cells are shown for aerobic conditions only. The open symbols are for air, the closed ones for nitrogen. From Webb *et al.* (1977*c*).

interference by far-UV lesions with *rec* repair of X-ray-induced DNA single-strand breaks. Synergism between far-UV and X irradiation occurred regardless of order of irradiation for cells suspended in buffer (Haynes, 1964). In contrast, significant synergism between near-UV and far-UV radiation occurred only if the near-UV exposure was given first (Webb *et al.*, 1977*b*).

2.4.4. Damage to Membranes

Hollaender (1943) reported that cells of *E. coli* exposed to 350–405 nm radiation (the biologically effective wavelength was primarily 365 nm) are sensitive to the toxic effect of physiological saline at 37°C. This induced toxicity to saline was detectable at near-UV fluences that caused little inactivation in the absence of saline. However, with monochromatic 365-nm radiation, saline toxicity occurred only at fluences that inactivate past the shoulder region (R. M. Tyrrell, personal communication). This saline toxicity, which was detectable immediately after irradiation, was interpreted by Hollaender as evidence for cell membrane damage. Hollaender (1943) also observed that far-UV radiation (265 nm) induced saline sensitivity in cells of *E. coli*. In this case, however, saline sensitivity did not appear immediately after irradiation, but developed after holding the irradiated cells in buffer for a period of time.

Delayed cell membrane damage by far-UV radiation (254 nm) was also reported by Swenson and Schenley (1974). Triton X-100 added to cells of *E. coli* B/r immediately after irradiation at 254 nm resulted in a pronounced loss of turbidity after 40 min. This loss of turbidity was associated with a cessation of respiration after far-UV irradiation. Triton X-100 had no effect on the viability of respiring cells, either unirradiated or irradiated. It is unlikely that far-UV radiation caused direct damage to any part of the cell membrane, as respiration did not stop, and there was no corresponding loss of turbidity in *recA* and *lexA* cells (Swenson and Schenley, 1974; Swenson, 1976).

Nonspecific membrane damage by near-UV and visible radiation has been reported for bacteria and yeast at fluences somewhat greater than those required to kill wild-type cells under the usual culture conditions. Large fluences of far-UV, near-UV, and visible radiation resulted in loss of potassium retentivity in cells of *S. cerevisiae* (Bruce, 1958). Similarly, cells of *E. coli,* in which galactosidase transport was rendered inoperative became "leaky" after exposure to 365-nm radiation (contaminated with 1–2% of 313- and 334-nm wavelengths) at a fluence of 1.2×10^6 J m^{-2}, a fluence somewhat greater than required to kill 99.9% of the cells under the conditions of the experiment (Koch *et al.,* 1976). A fluence of 6×10^5 J m^{-2} of near-UV radiation had no effect on the generalized permeability of *E. coli.*

Carotenoidless mutants have been reported to be much more sensitive to near-UV or visible radiation than wild-type pigmented cells (see Section 2.9) (Mathews and Sistrom, 1959*b*). Because the carotenoid pigments are localized in the cell membrane of *S. lutea,* and certain enzymes associated with the membrane were preferentially destroyed in carotenoidless mutants, cell membrane damage was proposed as the basis of cell killing by "visible light" (Mathews and Krinsky, 1965).

2.4.5. Damage to Transport and Metabolic Systems

Inhibition of transport by near-UV and visible radiation has been reported in bacteria and yeast by Barran *et al.* (1974), Sprott *et al.* (1975), and Kubitschek and co-workers (Koch *et al.,* 1976; Doyle and Kubitschek, 1976). Transport inhibition in all reported cases occurred at fluences that caused little inactivation in repair-proficient strains. Kubitschek *et al.*

(1965) observed close parallels in transport inhibition of the carbon source and induction of growth delay in *E. coli*, which led them to propose that this transport inhibition may be the cause of near-UV-induced growth delay. In addition, Kubitschek and Doyle (1977) observed that a peak in the action spectrum for transport inhibition of succinate in *E. coli* occurred at 340 nm, which is also the peak for the action spectrum for growth delay. Other workers (Ramabhadran and Jagger, 1976) have presented evidence that 4-thiouracil, contained in certain species of tRNA, may be the chromophore and the target for growth delay induced by near-UV radiation. It was observed that RNA synthesis was much more sensitive to near-UV irradiation than protein or DNA synthesis (Ramabhadran, 1975). The near-UV action spectrum for RNA-synthesis inhibition was similar to the action spectrum for growth delay (Jagger, 1972) and to the absorption spectrum of *E. coli* valyl tRNA, which contains 4-thiouracil. Although other components of the cell show absorption peaks in the 330–340 nm region, for example, vitamin K2 and NADH, these are not considered likely candidates for the chromophore for growth delay (Ramabhadran and Jagger, 1975; Ramabhadran *et al.,* 1976). (See discussion of near-UV-induced growth delay by Swenson, 1976.)

Action spectra for the inhibition of two separate leucine transport systems revealed two significant peaks: one near the typical maximum for protein inactivation, 290 nm, and the other at approximately 365 nm, which may reflect a nonprotein chromophore with an absorbance peak near 365 nm (Robb *et al.,* 1976). Both transport systems showed the same action spectrum. Evidence is presented that for both systems, the permease itself was inhibited at near-UV fluences that did not cause detectable killing of the cells.

Transient inhibition of respiration also occurs at fluences of near-UV radiation that cause little cell killing (Kashket and Brodie, 1962; Jagger, 1964, 1972). Inhibition of respiration was proposed as the basis of growth delay (see Jagger, 1972); however, under certain conditions, near-UV radiation inhibits growth without affecting respiration (Jagger, cited by Swenson, 1976), and in other circumstances respiration inhibition does not parallel growth delay (Doyle and Kubitschek, 1976; Kubitschek and Doyle, 1977).

Coetzee and Pollard (1975) observed a high level of sensitivity of *E. coli* tryptophanase to near-UV radiation. The near-UV action spectrum peak was 365 nm both for irradiation of intact cells and for the purified enzyme *in vitro*. The enzyme, either purified or from induced cells lysed in a minimal medium, was more sensitive than when irradiated in intact cells, which suggested that tryptophanase is protected by components of the cell. Tryptophanase was inactivated with monochromatic 365-nm radiation at three times the rate with broad-spectrum near-UV radiation (303–405 nm).

This difference in effect between monochromatic and broad-spectrum radiation is in contrast to findings for cell inactivation and for the production of photoproducts of tryptophan that are toxic for *recA* strains. Repair-proficient, *uvrA*, and *recA* strains of *E. coli* are five to 30 times more sensitive to broad-spectrum near-UV radiation from a fluorescent BLB source than to monochromatic 365-nm radiation (MacKay *et al.*, 1976; Yoakum *et al.*, 1977; R. B. Webb and G. H. Yoakum, unpublished observations). The action spectrum for tryptophanase inactivation was distinctly different from the absorption spectrum of either the enzyme or the required cofactor, pyridoxal phosphate (Coetzee and Pollard, 1975). The selectively of enzyme inactivation was demonstrated by the high resistance of tryptophan synthetase to near-UV radiation both *in vivo* and *in vitro*.

The induced formation of galactosidase (S. J. Webb and Bhorjee, 1967) and tryptophanase (Swenson and Setlow, 1970) was also inhibited by relatively low fluences of near-UV radiation. The action spectrum for the inhibition of tryptophanase has a peak at 334 nm (Swenson and Setlow, 1970), in common with the action spectra for growth delay, transport inhibition, and bacterial lethality.

2.4.6. Inactivation of Phage Production Capacity

The ability of bacteria to support the multiplication and liberation (capacity) of bacteriophage can be impaired by both far-UV (Anderson, 1948) and near-UV radiations (Hill, 1956; S. J. Webb and Bhorjee, 1967; Ferron *et al.*, 1972; Day and Muel, 1974). The capacity to produce phage is much more resistant than colony-forming ability to far-UV radiation (Anderson, 1948); while phage capacity is more sensitive than colony-forming ability to near-UV radiation (Hill, 1956; Ferron *et al.*, 1972). Hill (1956) found that *E. coli* B cells lost the capacity to produce phage T1 after near-UV irradiation (320–460 nm) at fluences that produced little cell killing. This ability to produce phage was regained after incubation of the irradiated cells for 90 min in nutrient broth at 37°C. There was no increase in cell number during this incubation in these experiments. Holding the cells in buffer did not result in a recovery of phage reproductive capacity. The recovery of the capacity to produce phage paralleled the recovery of near-UV inhibition of the ability to repair X-ray-induced DNA single-strand breaks (Tyrrell, 1974). Adsorption of phage T1 or T7 was not inhibited in near-UV-irradiated cells of *E. coli* B at fluences that inhibited phage production (Hill, 1956; Day and Muel, 1974). Furthermore, Hill (1956) reported that near-UV-irradiated bacteria were killed by adsorbed phage, although no phage progeny were released.

Induction of phage λ in a lysogenic strain of *E. coli* K12 by far-UV (254-nm) radiation also was reported to be inhibited by prior irradiation with sublethal fluences of unfiltered "white" light (Dulbecco and Weigle, 1952). However, a near-UV component from the visible-light source likely produced the observed effects. Coetzee and Pollard (1974) obtained a similar inhibition of phage λ induction by near-UV radiation.

An action spectrum for inactivation of the capacity of *E. coli* B for supporting growth of T7 phage shows a peak at 270 nm, a minimum at 255 nm, and a shoulder at 340 nm (Day and Muel, 1974). The shoulder at 340 nm is similar to the 334-nm peak observed in the action spectra for photo-protection (Jagger and Stafford, 1962), growth delay (Jagger *et al.*, 1964), inhibition of tryptophanase induction (Swenson and Setlow, 1970), inhibition of succinate transport (Kubitschek and Doyle, 1977), and bacterial killing (Webb and Brown, 1976). Whether these apparently different processes share a common chromophore has not been determined. However, several components of cells of *E. coli* have absorption peaks in the 330–340 nm range (see Section 2.10). The components of the cell's reproductive system that are involved in phage reproduction, which are damaged by near-UV irradiation, have not been identified.

2.5. Effects of L-Tryptophan Photoproducts (TP)

Photoproducts especially toxic for *rec* and *lex* (*exr*) mutants of *E. coli, S. typhimurium,* and *B. subtilis* are produced by the aerobic irradiation of L-tryptophan at wavelengths between 260 and 370 nm, with an action spectrum peak at 290 nm (Yoakum and Eisenstark, 1972; Eisenstark, 1973; Yoakum *et al.*, 1977); effectiveness fell below detection at wavelengths shorter than 254 nm. The failure of the action spectrum for TP production to follow the absorption spectrum of L-tryptophan suggested a complex photochemical process involving one or more intermediate products that served as chromophores (Yoakum *et al.*, 1977). This interpretation is consistent with their observation that broad-spectrum near-UV radiation (310–405 nm) from a fluorescent (BLB) source* was much more effective for TP production than any single wavelength over the wavelength range tested.

Photoproducts of tryptophan were shown to inhibit replication gap closure in the DNA of *E. coli* (Yoakum *et al.*, 1974), and to inhibit the full-

---

* Not all fluorescent lamps labeled BLB have the same spectral emission. Recently, the phosphor in a widely used fluorescent blacklight lamp was changed, which substantially altered the spectral emission; the lamp designation including the BLB label was not changed (Forbes *et al.*, 1976).

medium repair (requiring the *recA* gene product) of DNA single-strand breaks induced by X radiation (Yoakum *et al.*, 1975). Inhibition of replication gap closure was proposed as the basis of the toxic effects in *recA* mutants. In parallel with the inhibition of DNA single-strand break repair, postirradiation treatment with TP sensitized cells of *E. coli* W3110 (repair proficient) to X radiation.

When present during irradiation, TP can sensitize repair-deficient and repair-proficient strains of *E. coli* to 365-nm radiation (Yoakum, 1975). Moreover, the photoproducts greatly enhanced the induction of single-strand breaks (or alkali-labile bonds) by near-UV (365-nm) radiation. Figure 12 demonstrates the single-strand break yield assayed by an alkaline sucrose-gradient technique (McGrath and Williams, 1966) in *E. coli* W3110 irradiated at 365 nm in the presence and absence of TP. When TP was present during 365-nm irradiation, the single-strand break yield increased by a factor of 11.5.

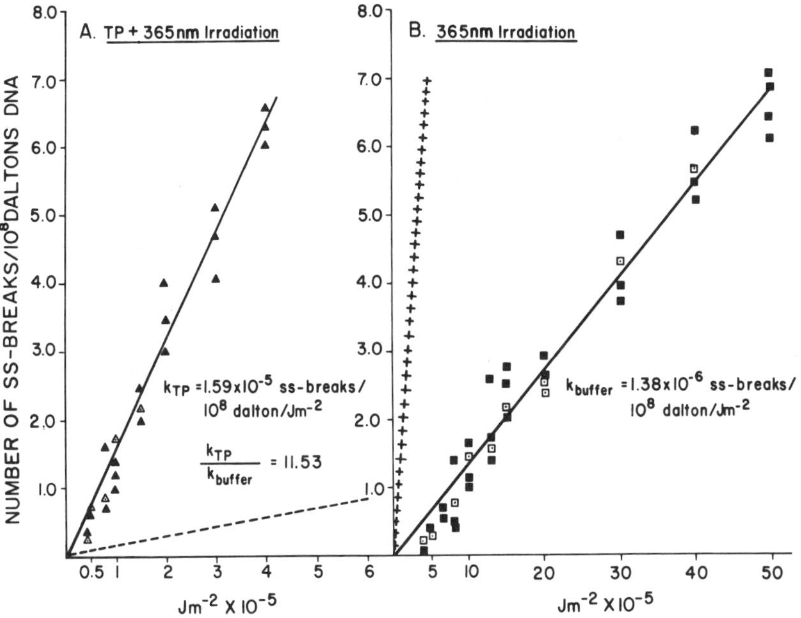

Fig. 12. Yield of single-strand (SS) breaks in the DNA of cells of *E. coli* W3110 irradiated at 365 nm at 0°C in the presence and absence of near-UV irradiated tryptophan containing toxic photoproducts (TP). Single-strand-break induction in the presence of TP is shown in panel A (▲,△) and in the absence of TP is shown in panel B (■,□). The dashed line in panel A represents the low-fluence SS breaks produced in the absence of TP (from panel B). The line of pluses in panel B represents SS breaks produced in the presence of TP (from panel A). From Yoakum (1975).

Tryptophan photoproducts enhanced the induction of mutations (resistance to phage T5) by broad-spectrum near-UV radiation in chemostat cultures of *E. coli* B/r Hcr (Yoakum *et al.,* 1974). At the concentration used, TP did not significantly increase the dark-mutation rate.

A possible role of photoproducts from endogenous tryptophan in the UV inactivation of *recA* strains was suggested by parallels between the induction of TP and lethality in *recA* cells (Mackay *et al.,* 1976; Yoakum *et al.,* 1977). The shoulder on the action spectra for lethality of both exponential- and stationary-phase *recA* cells in the 280–300 nm range (see Fig. 3) corresponds closely to the 290-nm peak of the action spectrum for the induction of TP (Yoakum *et al.,* 1977). Furthermore, broad-spectrum (310–405 nm) radiation from a fluorescent near-UV source (BLB) was much more efficient in both the inactivation of *recA* cells and the induction of TP (Yoakum *et al.,* 1977). As toxic photoproducts of tryptophan were produced at very low levels at monochromatic wavelengths longer than 320 nm, they are not likely to play a significant role in near-UV lethality in either *recA* or *rec⁺* strains.

Recent efforts to purify and identify the active component(s) of near-UV-irradiated L-tryptophan have demonstrated that the photoproduct(s) are highly labile and tend to decompose upon purification (Zigman and Hare, 1976; L. Glatzer, personal communication). However, Eisenstark and co-workers (McCormick *et al.,* 1975) have identified hydrogen peroxide as a toxic component of near-UV-irradiated tryptophan. This finding was confirmed and extended by Glatzer (1977). Whether hydrogen peroxide is the primary toxic photoproduct, or a breakdown product of a more complex labile molecule, has not been determined. However, there are significant differences in the action of near-UV-irradiated tryptophan (TP) and hydrogen peroxide. First, although both TP (Yoakum, 1975) and hydrogen peroxide (Ananthaswamy and Eisenstark, 1976) strongly sensitized the 365-nm induction of single-strand breaks (or alkali-labile bonds) in *E. coli* DNA, hydrogen peroxide (Ananthaswamy and Eisenstark, 1977) but not TP (Yoakum, 1975) induced breaks in the unirradiated controls. Second, TP had comparably low toxicity for *uvrA* and repair-proficient strains, while hydrogen peroxide was much more toxic for *uvrA* than for repair-proficient strains (G. H. Yoakum, personal communication). Zigman and Hare (1976) also separated near-UV photoproducts on Sephadex G10 columns and found potent antimitotic and growth-inhibitory activity in the 250–500 dalton range. Of two major species present, the one with absorption peaks at 225, 255, and 315 nm had the greatest biological activity.

It is not yet possible to resolve differences reported by different workers, since different test systems were employed. It is possible that meaningful amounts of hydrogen peroxide are formed only after extensive near-UV irradiation of tryptophan solutions (G. H. Yoakum, personal communication); photoproduct preparations used in the reported biological experiments were produced with much shorter irradiations (Yoakum and Eisenstark, 1972; Yoakum *et al.*, 1974, 1975; Yoakum, 1975). Further work will be required to identify the biologically active components of near-UV-irradiated tryptophan.

2.6. Protection Against Near-UV-Induced Lethality

Stanier and co-workers (Sistrom *et al.*, 1956) observed that carotenoid-deficient mutants of the photosynthetic bacterium *Rhodopseudomonas spheroides* were rapidly killed when exposed simultaneously to light and air. Killing did not occur either in the dark in air or after exposure to light under anoxia. It was concluded that carotenoid pigments normally present in photosynthetic organisms protect against damaging photooxidations sensitized by chlorophyll (Stanier and Cohen-Bazire, 1957; Stanier and Straight, 1967). Dworkin (1958) suggested that photokilling of the carotenoid-deficient mutant of *R. spheroides* resulted from disruption of the plasma membrane.

Mathews and Sistrom (1959*a,b*) tested the hypothesis that carotenoid pigments may protect nonphotosynthetic bacteria from the lethal effects of an endogenous photosensitizer. They compared the sensitivity of yellow (carotenoid-containing) wild-type and white (carotenoid-deficient) mutant strains of *S. lutea* to UV and visible radiations. The white mutant was readily killed by direct, unfiltered sunlight in air, whereas the yellow strain was not (Mathews and Sistrom, 1959*b*). Neither strain was killed under anoxic conditions at the fluences employed. A similar result was obtained using broad-spectrum radiation from a 650- or 1000-W tungsten–quartz iodine lamp (Fig. 13) (Mathews, 1964*a*; Mathews-Roth and Krinsky, 1970). The yellow and white strains showed identical survival curves when irradiated with far-UV (254 nm) or X radiation (Mathews and Sistrom, 1960). See review on protection by carotenoids by Krinsky (1976).

Although Mathews and Sistrom (1959*a,b*) and Mathews (1964*a*) observed strong carotenoid protection with "visible light" sources, these sources (sun and tungsten-halogen lamps) emit substantial amounts of near-UV radiation. Recent measurements comparing the sensitivity of yellow and white strains of *S. lutea* in stationary phase to monochromatic radiation revealed that carotenoid protection primarily occurred with radiation in the near-UV range (Denniston *et al.*, 1972; Webb *et al.*, 1977*a*). An action spectrum from protection by carotenoids is shown in Fig. 14. A broad peak for protection appeared between 300 and 370 nm; a narrow peak appeared at 410 nm; no protection occurred below 300 nm or above 420 nm. Between 420 and 510 nm there was a small but reproducible decrease in sensitivity of the white vs. the yellow strain.

The small amount of protection against monochromatic radiation shown in Fig. 14 is in sharp contrast to the fourfold or greater protection (ratio of F_{10} values) observed by Mathews and Sistrom (1959b), Mathews (1964a), and Mathews-Roth and Krinsky (1970) using broad-spectrum sources (Fig. 13). When the two strains used in the experiments shown in Fig. 14 were irradiated with a broad-spectrum near-UV source (fluorescent BLB), a protection ratio of 3.5 was obtained (Webb *et al.*, 1977a). The greater protection observed using broad-spectrum sources suggests that carotenoid pigments protect more efficiently against the interaction of two or more wavelengths than against any single wavelength within that range. See Section 5.2 for a discussion of synergism.

Protection by carotenoid pigments was also observed in *Mycobacterium* sp. by Wright and Rilling (1963). Pigmented forms of this bacterium were obtained by cultivation under low-level illumination, while pigment-free cells resulted from cultivation in total darkness. Cells lacking carotenoid pigments were readily killed by broad-spectrum radiation (360–590 nm) when air was present. No killing occurred under anoxia. Pigmented cells showed little lethality at the fluences used.

The slime bacterium *Myxococcus xanthus* also produces a carotenoid pigment after growth under continuous weak illumination. Dark-grown cells

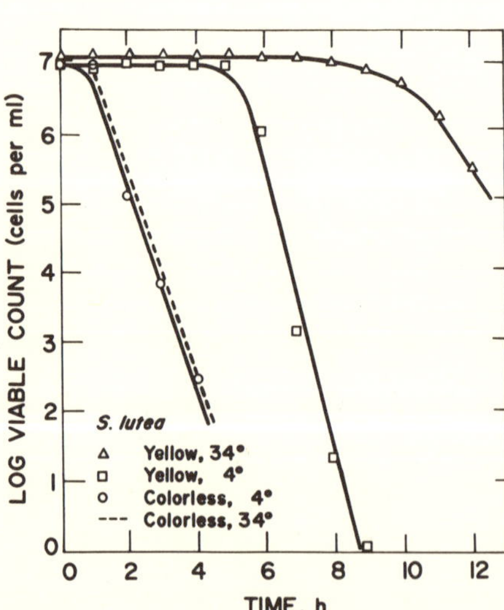

Fig. 13. Inactivation of yellow (carotenoid-containing) and colorless (carotenoid-deficient) strains of *S. lutea* in exponential phase (washed and resuspended in buffer) after irradiation with an unfiltered 1000-W Sylvania "Sun Gun." The fluence rate is unknown. Modified from Mathews-Roth and Krinsky (1970).

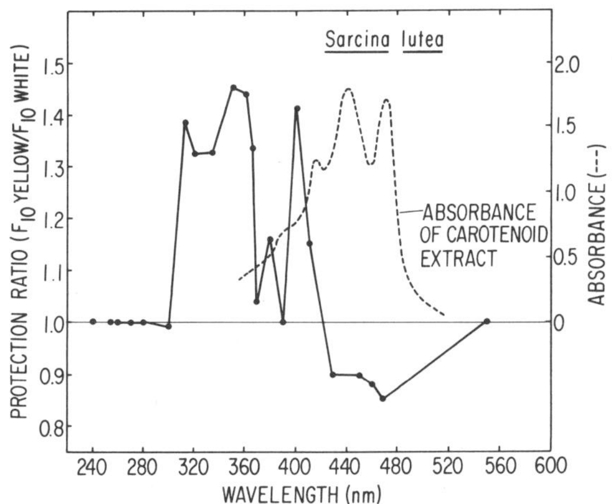

Fig. 14. Action spectrum for carotenoid protection in *S. lutea*. Protection ratio is the ratio of fluences that produce 10% survival in the yellow vs. the white strains. Stationary-phase cells of the two strains (washed and resuspended in buffer) were irradiated together to minimize dosimetric errors. Each point is the mean of two or three experiments. The radiation source was a 2.5-kW xenon arc lamp with the emission dispersed by a 500-mm Bausch and Lomb monochromator. The band width at half-maximum was 10 nm below 410 nm and 18 nm above 410 nm. From Webb *et al.* (1977*a*).

do not develop pigmentation (Burchard and Dworkin, 1966). In the absence of carotenoid pigmentation, cells over the growth cycle showed sensitivities (i.e., lysis) to near-UV-visible radiation that correlated with the appearance of a porphyrin. No lysis occurred in pigmented cells. An action spectrum for carotenoid-free cells of *M. xanthus* showed a sharp peak at 410 nm and a smaller peak at 510 nm, which is consistent with a porphyrin as the photosensitizer (Burchard *et al.*, 1966).

Some cells containing carotenoid pigments are not significantly protected against broad-spectrum near-UV (fluorescent BLB) nor visible (fluorescent cool white) radiation (Blanc *et al.*, 1976). Suspensions of *Neurospora crassa* conidia containing carotenoids were readily inactivated by the near-UV radiation. However, the white mutant conidia were only slightly more sensitive; the ratio of protection based on F_{10} values was only 1.1. Very little inactivation was obtained with visible radiation in the absence of an exogenous sensitizer by the fluorescent source used.

The protective properties of carotenoid pigments against chlorophyll photobleaching were found to require nine or more conjugated double bonds, eight double bonds being ineffective (Claes, 1960, 1961). A similar minimum number of conjugated double bounds also were required for protection against lethal photooxidations in *R. spheroides* (Crounse *et al.*,

1963) and *S. lutea* (Mathews-Roth and Krinsky, 1970; Mathews-Roth *et al.*, 1974a). A parallel observation was made by Foote *et al.* (1970), including data of Claes (1960, 1961), in which maximum quenching of singlet oxygen and of protection against photobleaching of chlorophyll *a* required at least nine conjugated double bonds. Seven conjugated double bonds or less showed little quenching or protective effect. Based on such evidence, Foote and Denny (1968) and Foote *et al.* (1970) proposed that carotenoid pigments protect biological systems by quenching singlet oxygen, with the formation of the triplet excited state of the carotenoids. It was further proposed that the carotenoid energy is released to the liquid medium. In addition, carotenoids can directly quench the excited triplet state of both exogenous and endogenous sensitizers (Foote, 1976). Furthermore, Krinsky (1976) proposed that singlet oxygen can also react chemically with carotenoids. However, the quenching of singlet oxygen was proposed as the major pathway of protection by Foote (1976) and Krinsky (1976).

The presence of compounds that provide protection against damage to DNA, repair systems, membranes, and metabolic systems by UV and visible radiation would have obvious survival value to biological systems exposed to natural sunlight. The widespread occurrence of carotenoid pigments suggests such a protective function (Stanier and Cohen-Bazire, 1957; Stanier and Straight, 1967). Also, the protective effect of histidine against inactivation of transforming DNA by near-UV radiation (see Section 3.3) suggests that such protective compounds may be ubiquitous.

2.7. Lethality in Partially Dehydrated Cells

Partially dehydrated cells of a variety of bacteria are more sensitive to far-UV radiation than cells in liquid suspension (Wells and Wells, 1936; Koller, 1939; Kaplan and Kaplan, 1956). See reports by Wells (1955) and S. J. Webb (1972) for more detailed discussions of the early literature. Kaplan and Kaplan (1956) observed that cells of *Serratia marcescens* showed a much greater sensitivity to far-UV (254-nm) radiation when equilibrated at 33% relative humidity than at higher humidities. At low relative humidities, the survival curves were simple exponentials instead of the strongly shouldered type characteristic of this bacterium in aqueous suspension. S. J. Webb and Tai (1970) observed similar relationships in *E. coli* and found that for cells at 30% relative humidity, the excision-deficient strain *E. coli* WP2 Hcr showed the same response to 254-nm radiation as the repair-proficient strain *E. coli* WP2.

The effect of relative humidity on the lethal response was found to be even greater for near-UV radiation (broad-spectrum BLB source, 320–405 nm) than for 254-nm radiation (S. J. Webb, 1963; S. J. Webb and Tai, 1968, 1969, 1970). The sensitivity reported for *E. coli* B/r was a thousand-fold greater to broad-spectrum near-UV radiation at 30% relative humidity (S. J. Webb and Tai, 1970) than others have observed for the same strain with a similar near-UV source in aqueous suspension (R. B. Webb, unpublished data). In contrast to the absence of oxygen enhancement at 254 nm, cells of *E. coli* at relative humidities less than 50% were more sensitive to near-UV radiation in the presence of oxygen than in its absence (S. J. Webb and Tai, 1970). The finding that the near-UV inactivation of *E. coli* at higher humidities was oxygen independent (S. J. Webb, and Tai, 1970) contrasts sharply with the strong oxygen enhancement of the near-UV response of the same strains in aqueous suspension observed by others (R. B. Webb and Lorenz, 1970; Tyrrell, 1976*a*; R. B. Webb, unpublished data). S. J. Webb and Tai (1970) concluded that the lesions responsible for the death of cells in aqueous suspension are different from those associated with cells at relative humidities below 50%. The mutational response of partially dehydrated cells is discussed in Section 4.3.3.

2.8. Lethality in Mammalian Cells

Recent work by Danpure and Tyrrell (1975, 1976) demonstrated that two mammalian cell lines, Chinese hamster (V-79) and HeLa, when irradiated in inorganic buffer at 365 nm, are approximately ten times more sensitive than exponential-phase cells of *E. coli*. The radiation-induced loss in colony forming ability in both mammalian cell lines was strongly oxygen dependent, suggesting that lethality might occur through a photodynamic mechanism (Fig. 15). Such a mechanism has been proposed for the 365-nm inactivation of *E. coli* (Webb and Lorenz, 1970). Under aerobic conditions, both V-79 and HeLa cells were 10^4 times more sensitive at 254 nm than at 365 nm (Danpure and Tyrrell, 1975; 1976), a difference somewhat less than that observed for *E. coli* under similar conditions (see Table 1).

Chinese hamster cells (V-79) were sensitized to 254-nm radiation by a prior exposure of the cells to a fluence of 365-nm radiation that inactivated less than 50% of the population. However, HeLa cells were not sensitized to 254-nm radiation by prior irradiation at 365 nm (Danpure and Tyrrell, 1976). The reason for this difference is not known.

Photoproducts toxic to three mammalian cell lines (human $D98/AH_2$, mouse 3T6, and Chinese hamster 753-B-3M) were produced by irradiating Dulbecco's modified Eagle's medium (MEM) with a broad-spectrum near-UV source (fluorescent BLB) (Wang *et al.,* 1974; Stoien and Wang, 1974). These toxic photoproducts were produced at relatively low fluences of near-

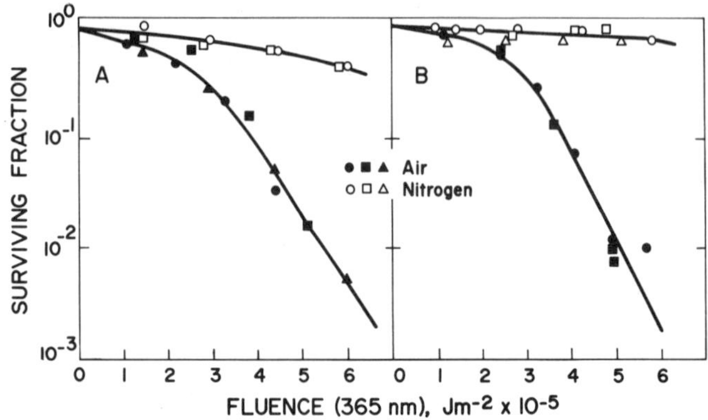

Fig. 15. Fluence–survival responses of HeLa (A) and V-79 (B) mammalian cells in buffer suspension at 365 nm in the presence and absence of oxygen. Redrawn from Danpure and Tyrrell (1976).

UV radiation $(2 \times 10^4$ J m$^{-2})$. Human cells were also killed by irradiating them with a broad-spectrum near-UV source while suspended in complete nutrient medium. However, direct irradiation of the cells suspended in non-nutrient buffer at near-UV fluences to 4×10^4 J m^{-2} did not produce detectable inactivation (Wang *et al.*, 1974). Active components in the MEM medium were identified as riboflavin and tryptophan or tyrosine (Stoien and Wang, 1974); omission of riboflavin prevented the toxicity. In contrast, recombinationless mutants of bacteria exposed to near-UV-irradiated tryptophan alone showed a toxic response (see Section 2.5).

2.9. Lethal Consequences of Near-UV-Induced Lesions

2.9.1. Role of Pyrimidine Dimers

Repair capability determines the role of pyrimidine dimers in near-UV-induced lethality (Brown and Webb, 1972; Webb *et al.*, 1976). The extent of photoreactivation of 365-nm inactivation in four strains of *E. coli* K12 that differ in repair capability is shown in Fig. 16. An enzymatic photoreactivation is specific for cyclobutane-type pyrimidine dimers (Setlow, 1967), the large photorepair sector of 0.83 obtained with the double repair-deficient strain AB2480 (*uvrA recA*) demonstrates a major lethal role for dimers in this strain.

Photoreactivation can be expressed as

$$PR_s = k_0 - k_{PR} \tag{3}$$

where PR_s is the photoreactivation sector, k_0 the logarithm slope of the survival curve in the absence of photoreactivation, and k_{PR} the slope of the survival curve in the presence of maximum photoreactivation. The slopes of the inactivation curves are obtained from Eq. (2). The photoreactivation sector after 365-nm inactivation in AB2480 was comparable to that after 254-nm inactivation. A large PR sector after 365-nm inactivation was also observed with another double-repair-deficient strain *E. coli* B_{s-1} (*uvrB lexA fil*) (Tyrrell *et al.*, 1973). Characteristics of PR after 365- and 254-nm inactivation were found to be similar in saturation fluence rate, temperature dependence, and maximum effectiveness of a single flash of light (Brown and Webb, 1972, 1977).

It is clear that dimers can account for most of the lethality in the double-repair-deficient strain AB2480 at both 254 and 365 nm. Pyrimidine dimers are induced at 254 nm in *E. coli* DNA at the rate of 39 dimers per genome per J m^{-2} (Tyrrell, 1973). If only the sector that can be photoreactivated is considered, there are 28 lethal events per J m^{-2} ($PR_s \times 1/F_{37} = 0.81 \times 35 = 28$). Thus 1.4 dimers per genome correspond to one lethal event ($39 \div 28 = 1.4$) (Table 2). If the final slope [k, Eq. (2)] of the 254-nm

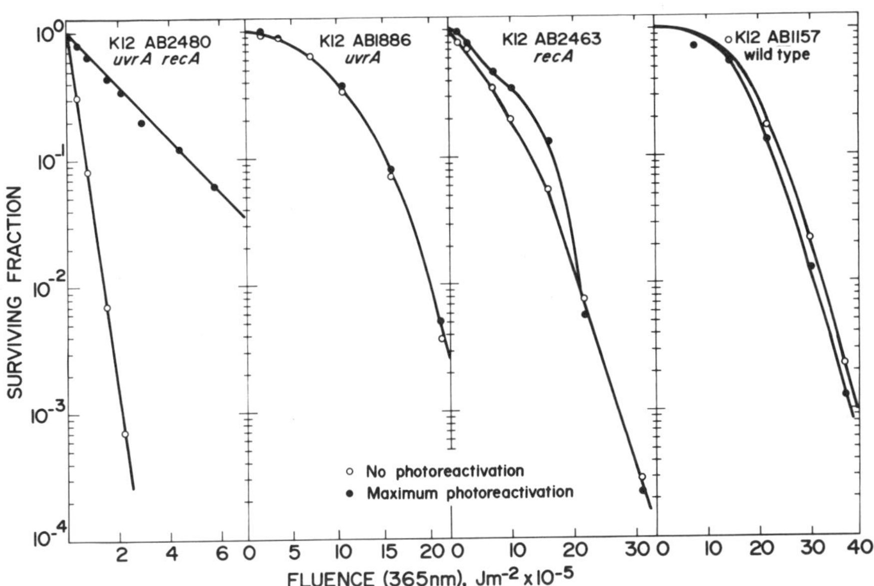

Fig. 16. Fluence–survival responses at 365 nm of four strains of *E. coli* K12 in stationary phase (suspended in buffer) differing in repair capability with and without photoreactivation treatment. The 365-nm irradiation was at 0°C at fluence rates of 1000–1200 W m^{-2}. Photoreactivation illumination (380–440 nm) was at 25°C at fluence rates of 20 W m^{-2} for AB2480 and 75 W m^{-2} for the other strains. Redrawn from Webb *et al.* (1976).

inactivation curves is used instead of $1/F_{37}$ values, the calculation yields 1.0 dimer per lethal event. The small shoulder (Y-axis extrapolation value 1.8) seen routinely at 254 nm with this strain (Brown and Webb, 1972; Tyrrell and Davies, 1974) has not been observed at 365 nm (Brown and Webb, 1972; Tyrrell, 1976a; Webb and Brown, 1976). The small shoulder at 254 nm is consistent with the requirement that one dimer must be induced in each DNA strand to inactivate the cell.

At 365 nm, photoreactivable damage is induced at 3.1×10^{-5} lethal events per J m^{-2} (Table 1). As dimers are induced at 365 nm at the rate of 5.5×10^{-5} dimers per genome per J m^{-2} (Tyrrell, 1973), one photoreactivable lethal event corresponds to 1.8 dimers per genome ($5.5 \times 10^{-5} \div 3.1 \times 10^{-5} = 1.8$). In contrast to the shoulder at 254, the absence of a shoulder at 365 nm suggests that a single lesion has the potential to inactivate the cell. Based on this interpretation, a dimer induced at 365 nm has a 0.56 probability of being a lethal event in this strain. These calculations suggest that in this strain with no known capability to repair pyrimidine dimers, there is a significant difference in the biological effects of dimers induced at 365 and 254 nm. Tyrrell (1973) reported that the ratio of T<>T to T<>C induced at 365 nm was 5:1, whereas the ratio at 254 nm was 1.1:1. However, there is no basis to propose that this difference could account for the differences in biological response observed. In contrast to the large PR sector after 365-nm inactivation obtained in strains K12 AB2480 and B$_{s-1}$, only a small amount of photorepair was obtained with the recombination-deficient strain AB2463 (*recA*). The *rec*$^{+}$ strains, AB1886 (*uvrA*) and AB1157 (repair proficient), showed no photorepair after 365-nm inactivation. Similarly, no photorepair was obtained with two other *rec*$^{+}$ strains of *E. coli* B/r and B/r Hcr. Radiation at 365 nm within the biological range can damage the PR enzyme (Tyrrell *et al.*, 1973), and can sensitize cells to subsequent PR illumination (Webb and Brown, 1977a). (See Section 5.2 for a discussion of synergism between 365- and 405-nm radiations.) However, both mechanisms have been excluded as an explanation for the absence of photorepair under the conditions of the experiments shown in Fig. 16 (Webb *et al.*, 1976). Figure 17 demonstrates the presence of functional PR enzymes after a fluence of 2×10^{6} J m^{-2} of 365-nm radiation. PR illumination (405 nm) was delivered under anoxia to prevent lethality by 405-nm radiation (Section 5.2). Although no photorepair could be demonstrated after 365-nm irradiation alone, strong photorepair was obtained after sequential exposure to 365-nm and 254-nm radiation.

The absence of photorepair under conditions in which functional PR enzymes could be demonstrated led to the conclusion that pyrimidine dimers, even under conditions that minimize concomitant PR, are not significant lethal lesions at 365 nm in repair-proficient and excision-deficient

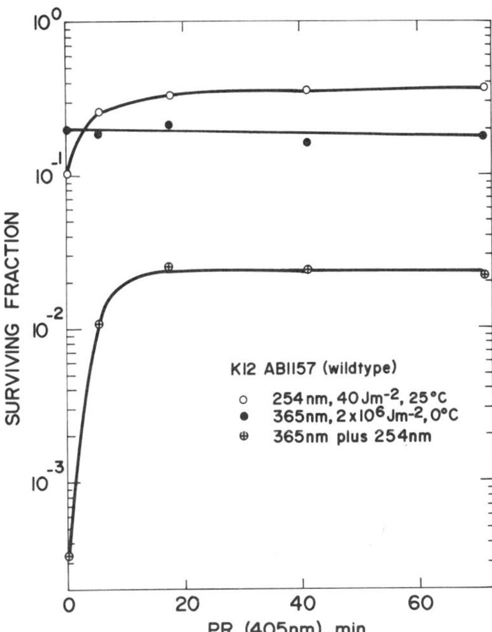

Fig. 17. Photoreactivation after inactivation of cells of strain K12 AB1157 (wild-type) in stationary phase (suspended in buffer) at 254 nm (O), 365 nm (●) and 365 nm followed by 254 nm (⊕). Photoreactivation illumination was at 405 nm with the cells held at 25°C under stringent anoxia (N_2). Cell suspensions were maintained at 0°C for the 365-nm irradiations and at 25°C for 254- and 405-nm irradiations. Fluence rates were 1000 W m^{-2} for 365-nm radiation, 10 W m^{-2} for 254-nm radiation, and 75 W m^{-2} for 405-nm radiation. Redrawn from Webb *et al.* (1976).

strains (Webb *et al.*, 1976). At lower 365-nm fluence rates and higher irradiation temperatures, concomitant PR can effectively remove the induced dimers during irradiation (Brown and Webb, 1972; Tyrrell *et al.*, 1973).

The number of pyrimidine dimers per "lethal event" (F_{37}) at 365 nm in *uvrA rec*[+] strains, 52 (Table 3), together with the absence of photorepair (Fig. 16) suggests that dimers are repaired efficiently by the dark repair systems in *E. coli* until fluences are reached that inhibit dark repair. The number of dimers at F_{37} in *uvrA recA, uvrA rec*[+], and *uvr*[+] *recA* strains is the same (or slightly greater) for aerobic 365-nm radiation as for 254-nm radiation (Table 3). However, when stationary-phase cells were irradiated anoxically at 365 nm, the number of dimers per F_{37} increased dramatically to 153 for the *uvr*[+] *recA* strain and to 285 for the *uvrA rec*[+] strain; values 5.7 and 4.8 times greater, respectively, than the number of dimers per F_{37} at 254 nm. These results suggest that there is a much more efficient repair of dimers induced at 365 nm that at 254 nm. The small number of dimers at F_{37} in repair-proficient strains at 365 nm (aerobic and anoxic) relative to 254 nm can be accounted for by greater damage to repair systems at the higher fluences required to kill these strains.

TABLE 3. Sensitivity ($1/F_{37}$), Thymine-Containing Pyrimidine Dimers, and Single-Strand Breaks, at 365 nm (0°C) and 254 nm (25°C), of Strains of *E. coli* K12 and *E. coli* B/r That Differ in Repair Capability[a]

Wavelength (nm)	Parameter	O_2/N_2	PR[b]	Strain					
				AB2480 *recA uvrA*	AB2463 *recA*	AB1886 *uvrA*	B/r Hcr *uvrA*	AB1157 *rec⁺ uvr⁺*	B/r *rec⁺ uvr⁺*
365	$1/F_{37}$	O_2	−	3.73×10^{-5}	1.49×10^{-6}	1.05×10^{-6}	8.3×10^{-7}	5.5×10^{-7}	3.03×10^{-7}
		O_2	+	6.3×10^{-6}	1.07×10^{-6}	1.02×10^{-6}	9.0×10^{-7}	5.7×10^{-7}	3.13×10^{-7}
		N_2	−	3.42×10^{-5}	3.6×10^{-7}	1.92×10^{-7}	9.1×10^{-7}	8.7×10^{-8}	8.5×10^{-8}
		N_2	+	—	—	—	8.1×10^{-9}	—	—
	Dimers per	O_2	−	1.47	37	52	66	100	182
	genome per F_{37}	N_2	−	1.61	153	285	545	634	649
	SS breaks	O_2	−	0.8	19	30	36	55	99
	per genome	O_2	+	4.8	26	29	33	54	97
	per F_{37}	N_2	−	0	0	0	0	0	0
		N_2	+	0	0	0	0	0	0
254	$1/F_{37}$	O_2	−	34.5	1.20	0.78	0.52	0.027	0.021
		O_2	+	6.8	0.31	0.24	0.15	0.018	0.014
		N_2	−	34.5	—	—	0.52	—	0.021
		N_2	+	6.8	—	—	0.15	—	—
	Dimers per genome per F_{37}	O_2	−	1.13	33	50	75	1440	1860
	SS breaks	O_2	−	0.0014	0.042	0.064	0.096	1.8	2.3
	per genome per F_{37}	O_2	+	0.0074	0.16	0.21	0.33	2.8	3.6

[a] The fluence rate was 1000–1200 W m⁻² at 365 nm and 0.05–0.6 W m⁻² at 254 nm. Data from Webb and Lorenz (1970), Tyrrell (1973), Tyrrell *et al.* (1974), Webb and Brown (1976), Webb *et al.* (1976), and R. B. Webb (unpublished observations).

[b] Values with maximal photoreactivation (PR) are indicated with a plus; those without photoreactivation are indicated with a minus.

2.9.2. Role of DNA Single-Strand Breaks

There is a strong oxygen dependence for inactivation at wavelengths longer than 313 nm in all repair-proficient and single repair-deficient mutants (*uvrA rec⁺* and *urv⁺ recA*) of *E. coli* tested (Webb and Lorenz, 1970; Tyrrell, 1976*a*; Webb *et al.*, 1977*c*). Oxygen-enhancement ratios (OER) at 365 nm for stationary-phase cells varied from 4.5 in K12 AB2463 to 8.5 in B/r Hcr (Fig. 2). The OER appears to be somewhat smaller (1.5–2.5) for exponential-phase cells (Tyrrell, 1976*a*). Significant oxygen dependence was not observed for the 365-nm inactivation of the double repair-deficient mutant K12 AB2480 (*uvrA recA*) in the absence of photo-reactivation (Tyrrell, 1976*a*; Webb *et al.*, 1977*c*).

If oxygen-dependent damage to repair systems instead of oxygen-dependent lesions accounts for the oxygen enhancement of lethality (Section 2.4.3), inactivation of *rec⁺* strains at 365 nm under aerobic conditions past the shoulder region should eliminate the oxygen dependence of further inactivation. If the oxygen-dependent induction of lesions accounts for the oxygen enhancement of lethality, inactivation of *rec⁺* strains past the shoulder region should remain oxygen dependent. Results from such experiments are shown in Fig. 18, in which stationary-phase cells of *E. coli* B/r Hcr were irradiated at 365 nm (0°C) in air with fluences of 2.3×10^6 J m^{-2} or 3.3×10^6 J m^{-2}, after which irradiation was continued under stringent nitrogen anoxia. The OER after aerobic irradiation with 2.3×10^6 J m^{-2} was about 8, a value comparable to the OER obtained without prior irradiation. However, the anoxic sensitivity $(1/F_{37})$ was 2.5 times greater after a prior aerobic exposure of 2.3×10^6 J m^{-2}. The strong oxygen enhancement of inactivation after aerobic exposures of 2.3 and 3.3×10^6 J m^{-2} at 365 nm demonstrates that the lethal lesions are induced by an oxygen-dependent mechanism. Furthermore, the increased sensitivity under anoxic conditions is consistent with an involvement of oxygen-dependent damage to repair systems in the lethal action of 365-nm radiation.

Biological measurements based on that part of the damage that can be photoreactivated after 365-nm inactivation in the presence and absence of oxygen indicate that the dimer yield in strain K12 AB2480 is at least as great in the absence of oxygen as in its presence (Webb *et al.*, 1977*c*). Preliminary chemical assays (R. M. Tyrrell, personal communication) also indicate that the 365-nm induction of dimers is not enhanced by oxygen. Therefore, pyrimidine dimers are not candidates for the oxygen-dependent lesions induced by 365-nm radiation.

The complete oxygen dependence of the induction of DNA single-strand breaks (or alkali-labile bonds) by 365-nm radiation (Section 2.4.2) makes them strong candidates for the major lethal lesions at this wavelength.

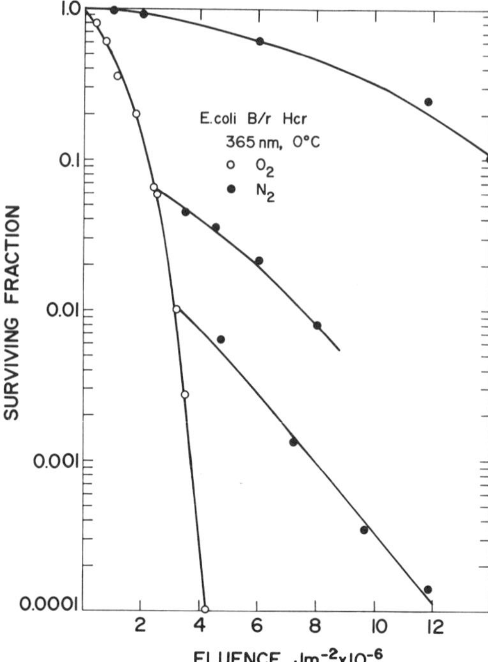

Fig. 18. Inactivation of stationary-phase cells of *E. coli* B/r Hcr (suspended in buffer) at 0°C by 365-nm radiation in the presence and absence of oxygen. In the two lower anaerobic curves, irradiation was initially aerobic for fluences of 2.3×10^6 and 3.3×10^6 J m^{-2}, after which irradiation was continued under anoxia. From Webb *et al.* (1977c).

The double repair-deficient mutant K12 AB2480 appears to be able to repair or bypass a substantial fraction of the DNA single-strand breaks (or alkali-labile bonds) induced at 365 nm. Single-strand breaks that are induced at 61% of the rate of dimers (Tyrrell *et al.*, 1974; Yoakum, 1975) can account for no more than 17% of the lethal damage in this strain, as photoreactivable damage accounts for 83% of the lethal effects. Nonphotoreactivable damage is induced at the rate of 6.3×10^{-6} lethal events per J m^{-2} (k_{PR}). Single-strand breaks are induced at the rate of 3.0×10^{-5} per genome per J m^{-2}. Therefore, if the DNA single-strand-break yield, measured at 365-nm fluences greater than 10^6 J m^{-2}, remains linear at the lower fluences that inactivate strain K12 AB2480, 4.8 single-strand breaks are induced per nonphotoreactivable lethal event [$(3.0 \times 10^{-5}) \div (6.3 \times 10^{-6}) = 4.8$]. If the higher rate of induction of DNA single-strand breaks observed in *E. coli* P3478 *polA1* (Table 2) is taken instead of the lower published value in *E. coli* S22 *uvrB*, 12 single-strand breaks are induced per nonphotoreactivable lethal event [$(7.5 \times 10^{-5}) \div (6.3 \times 10^{-6}) = 12$] in strain K12 AB2480.

This relationship suggests that although this strain is unable to repair pyrimidine dimers by the dark repair processes, it is able to repair or bypass a substantial fraction of the lesions induced under aerobic conditions at 365 nm. It is known that strain K12 AB2480 can repair a substantial fraction of DNA single-strand breaks induced by X radiation (Town *et al.,* 1973), and is only a little more sensitive than the *recA* single repair-deficient mutant. The nature of the DNA single-strand breaks or alkali-labile lesions induced at 365 nm has not been determined.

The possible role of thymine glycols in the near-UV inactivation of bacteria has not been determined (Hariharan and Cerutti, 1976*a,b*). However, the large PR sector (0.80–0.85) obtained after 313-nm inactivation (Section 2.9.1) suggests only a minor role for nondimer lesions at this wavelength. The strong oxygen dependence for inactivation seems to eliminate thymine glycols from a major lethal role at 365 nm, as preliminary measurements indicate that the UV-induction of thymine glycols is oxygen independent (P. A. Cerutti, personal communication).

R. D. Ley (personal communication) has observed that a substantial reduction in the number of 365-nm-induced DNA breaks occurred in *E. coli* W3110 (wild type) when the cells were held at 30°C in buffer for 30 min after irradiation at low fluences at 0°C. This evidence for the repair of DNA single-strand breaks supports the biological evidence that cells capable of dark repair can repair or tolerate a moderate number of the aerobic lesions induced at 365 nm.

The sensitization of *E. coli* cells to X rays by a prior exposure to 365-nm radiation, associated with inhibition of repair of X-ray induced single-strand breaks (Tyrrell, 1974), completely recovered in a wild-type strain (*E. coli* W3110) when the 365-nm irradiated cells were held in growth medium for 90 min before X irradiation. Holding of the cells in growth medium after the combined 365-nm and X-radiation treatment with no intermediate incubation did not produce significant recovery. Tyrrell (1976*a*) proposed that, even though repair capability recovers after low to moderate fluences of 365 nm radiation, this recovery may not be effective in the repair of oxygen-dependent lesions, as they appear to be rapidly altered such that they are no longer repairable. Therefore, the temporary inhibition of repair would cause these oxygen-dependent photoproducts to become lethal.

It is proposed that rec^+ strains of *E. coli* are inactivated by near-UV radiation by two classes of lesions. One class, which is induced by an oxygen-dependent mechanism and through interaction with oxygen-dependent repair disruption, accounts for most of the inactivation under aerobic conditions. The second class of lesion does not require oxygen for its induction and accounts for much of the 365-nm lethality under anoxic conditions. The role of repair disruption in inactivation under anoxic conditions has not been investigated. It is proposed that DNA single-strand breaks (or alkali-labile bonds) are the oxygen-dependent lethal lesions.

However, the production of some other oxygen-dependent DNA lesion has not been eliminated. It is proposed that pyrimidine dimers account for much of the oxygen-independent inactivation of the *E. coli* strains tested (Tyrrell, 1976*a*; Webb *et al.*, 1976, 1977*c*). Tyrrell (1976*a*) also proposed that the lethal action of pyrimidine dimers does not depend on how rapidly their repair takes place.

2.9.3. Role of Membrane, Transport, and Metabolic Damage

Damage to transport and metabolic systems occurs at near-UV fluences that cause no detectable killing of repair-proficient strains (Barran *et al.*, 1974; Jagger, 1972; Koch *et al.*, 1976; Doyle and Kubitschek, 1976; Robb *et al.*, 1977). This damage is transient and largely recovers during the growth and division delay that is induced over a similar fluence range. It is unlikely that transient inhibition of transport, respiration, or metabolic enzymes contributes significantly to near-UV lethality. This interpretation does not exclude a possible involvement of respiration failure or other metabolic disturbances under some experimental conditions that may occur as a consequence of derepression of operons (Swenson and Schenley, 1970; Swenson, 1976). Respiration failure may be a factor in the near-UV sensitivity of a haploid strain of *S. cerevisiae* reported by Fong *et al.* (1975). Cells irradiated at 365 nm in exponential growth on a defined glucose medium were approximately two times more sensitive when plated on a nonfermentable medium containing glycerol or lactate than when plated on a fermentable medium containing glucose. Holding of the irradiated cells for 100 min in a glucose medium before plating on a medium containing glycerol or lactate resulted in the same survival observed for cells plated on glucose medium. Apparently, the irradiated yeast cells cannot recover full respiratory function after a shift down to a nonfermentable medium. Also, the 365-nm sensitivity reported for a haploid strain of *S. cerevisiae*, even when plated on a glucose medium, was 30 times greater than others have observed using the identical light source (Doyle and Kubitschek, 1976; H. E. Kubitschek, personal communication). This great sensitivity might be the result either of a photosensitizer in the irradiation medium, or the production of a phototoxin during irradiation.

Although membrane damage may be the cause of cell death in *E. coli* subjected to holding in saline solution after near-UV irradiation (Hollaender, 1943), there is no convincing evidence that such damage contributes to near-UV lethality under the usual growth conditions. It is clear that lesions in DNA account for lethality in near-UV-irradiated repair-deficient strains of *E. coli* Fig. 4). The relative sensitivity of repair-proficient strains of *E. coli* to near-UV radiation can be accounted for by the damage to repair systems observed (Tyrrell *et al.*, 1973; Tyrrell and Webb, 1973; Webb *et al.*, 1977*b*).

Membrane damage has been proposed as the cause of cell death by near-UV or visible radiation in carotenoidless forms of cells that normally contain carotenoid pigments (Mathews and Sistrom, 1960; Krinsky, 1976). In some cases, carotenoidless forms of *R. spheroides* and *M. xanthus* were observed to undergo lysis upon irradiation (Sistrom *et al.*, 1956; Burchard and Dworkin, 1966). Cells containing carotenoid pigments did not undergo lysis under the same conditions. Nevertheless, a genetic basis of near-UV-induced lethality in cells containing carotenoid pigments has not been excluded.

2.10. Chromophores for Near-UV-Induced Lethal Effects

2.10.1. Chromophores for DNA Damage and Repair Inhibition

As pyrimidine dimers are induced by near-UV radiation, DNA must be considered as a possible chromophore for lethal effects. Direct absorption by DNA very likely is the basis of much of the lethal and mutagenic action (see Section 4.1) by monochromatic radiation at wavelengths up to 320 nm. Stationary-phase cells of several strains of *E. coli* show the same or slightly greater PR ratio at 313 nm as at 254 nm (M. S. Brown and R. B. Webb, unpublished observations), implicating pyrimidine dimers as the major lethal lesion at 313 nm. Although the absorbance of DNA at near-UV wavelengths longer than 330 nm has not been measured directly, it is unlikely that DNA alone absorbs strongly enough to account for the pyrimidine dimers and DNA single-strand breaks at 365 nm.

Five basic photochemical mechanisms have been proposed by K. C. Smith (cited by Pollard, 1974) for consideration in accounting for DNA effects at wavelengths longer than 320 nm: (1) direct excitation, (2) photosensitization, (3) photocombination (e.g., psoralen–pyrimidine adducts), (4) photodynamic action (photooxidation), and (5) products of hydroxy radicals. Data are insufficient to allow assigning a role to any of the five mechanisms in the production of pyrimidine dimers by near-UV radiation. Direct excitation, although unlikely because of the low absorbance of the DNA bases at wavelengths longer than 320 nm (Fig. 2), cannot be ruled out. The possibility has been suggested that pyrimidine dimers could be formed by the direct excitation to the triplet state of thymine after absorption in the ground state (K. C. Smith, as cited by Pollard, 1974). However, photosensitization remains an attractive possibility for the production of dimers by near-UV radiation. As dimers are induced more efficiently in *E. coli* DNA *in vitro* than *in vivo,* any sensitizer must be bound tightly to the DNA (see Table 2). The predominant production of T< >T dimers (Table 2), together with preliminary evidence that they are formed more efficiently in the absence of oxygen than in its presence, supports the possibility that a triplet photosensitizer is involved in the production of pyrimidine dimers by near-UV radiation.

Oxygen dependence implicates photodynamic action in the near-UV production of DNA single-strand breaks, or alkali-labile bonds (Fig. 7). However, the chromophore for such photodynamic action has not been identified. As DNA breaks also are formed at higher rates in DNA *in vitro* than *in vivo* (Table 2), any such chromophore must be tightly bound to the DNA.

Inactivation of the PR enzyme at 365 nm also is oxygen dependent (Fig. 8). The chromophore for enzymatic photoreactivation may be the chromophore for inactivation of the enzyme. Since the yeast enzyme is inactivated by 365-nm radiation at approximately the same rate *in vitro* and *in vivo,* the chromophore seems to be part of, or bound to, the PR enzyme.

The oxygen dependence of the 365-nm inhibition of repair processes that mend single-strand breaks in bacterial DNA induced by X radiation or electrons again suggests a photodynamic process (Tyrrell, 1976*a*). The 365-nm inhibition of the dark repair of near-UV- or far-UV-induced damage also is strongly oxygen dependent (Webb and Brown, 1977*b*). The role of oxygen in the induction of DNA breaks and damage to repair enzymes at wavelengths other than 365 nm has not been investigated.

The existing data do not allow the identification of the possible chromophores for the production of pyrimidine dimers, DNA breaks, and repair-system inhibition from among the many candidates that absorb in the near-UV part of the spectrum. However, the action spectra for lethality (Figs. 2 and 3) indicate that chromophores associated with lethality in *E. coli* B/r Hcr must absorb between 320 and 500 nm. Significant shoulders or peaks occur at 335, 410, and 500 nm. Oxygen dependence suggests that the chromophores for lethality at wavelengths greater than 320 nm are not a part of DNA; however, DNA is an important target molecule for lethal action (see discussions in Sections 2.3, 2.4, and 2.5). There are many candidates for the chromophores for near-UV-induced oxygen-dependent inactivation of the bacterial cell. These include the flavins [riboflavin, lumiflavin, flavin adenine dinucleotide (FAD), flavin mononucleotide (FMN), etc.], pteridines, naphthoquinones, ubiquinones, porphyrins, nicotinamide adenine dinucleotide (reduced form) (NADH), nicotinamide adenine dinucleotide phosphate (reduced form) (NADPH), 4-thiouracil, pyridoxal phosphate, and tryptophan. The possible roles of these compounds in lethal and nonlethal effects of near-UV radiation have been discussed in reports or reviews by Spikes and Livingston (1969), Jagger (1972), Eisenstark (1973), Knowles (1975), Krinsky (1976), Giese (1976), and Ramabhadran and Jagger (1976). Flavins, quinones, and porphyrins are known to sensitize a variety of *in vitro* systems. For example, FMN sensitizes malate dehydrogenase to blue light in the presence of oxygen (Codd, 1972). This effect was strongly reduced by tryptophan, which quenches triplet–triplet transfer in FMN.

A porphyrin was implicated in the oxygen-dependent sensitivity to visible light (from an incandescent source) developed by a respiration-deficient mutant of *S. cerevisiae* Sc-7 in exponential phase (Elkind and Sutton, 1957*a,b*). Stationary-phase cells were not significantly inactivated by extended irradiation with the same light source. Holding of exponential-phase cells in buffer for 1–3 h resulted in the development of light sensitivity in parallel with the accumulation in the cells of a compound with a major absorption peak at 420 nm and minor peaks at 480 and 540 nm. The absorption spectrum was considered to be consistent with a hemoprotein. An action spectrum for inactivation was in good agreement with the 420-nm absorption peak shown by the light-sensitive cells. The presence of the excess porphyrinlike compound did not significantly alter the sensitivity of the cells to 254-nm or X radiation.

2.10.2. Photosensitizing Effects of Natural Pigments

Some organisms that are not normally exposed to direct sunlight have natural pigments that sensitize them to certain wavelengths in sunlight. Such photosensitization may lead to damage or inactivation of the organism. The natural photosensitizers reported generally fall into four main classes: chlorophylls, porphyrins, hypericins, and furocoumarins (see review by Giese, 1971).

The best-studied example of natural photosensitivity is the protozoan *Blepharisma,* in which a photosensitizing agent, zoopurpurin, a red hypericinlike pigment, is produced when the cells grow at very low light levels, but is not produced when the cells are exposed to higher intensities of light (Giese, 1953, 1971). Cells of *Blepharisma* grown in the dark accumulate the pigment and are readily killed when exposed, in the presence of oxygen, to visible light at high fluence rates. No inactivation occurred in the absence of oxygen (Giese, 1946). The oxygen-dependent inactivation closely followed the absorption spectrum of the endogenous pigment. It has been suggested that the red pigment of *Blepharisma* may function to protect the cell against the short wavelength component of sunlight, since the pigment absorbs strongly below 350 nm.

Many instances of sunlight sensitivity have been reported in which the chromophore was not identified. For example, certain cave-dwelling crustaceans did not survive when continuously exposed to sunlight at $\frac{1}{20}$ of the normal noonday intensity (Maguire, 1960). The basis of this sensitivity was not reported. However, as the animals showed the same high sensitivity to an incandescent lamp with the short and long wavelengths filtered out, the lethal effect is unlikely to be due to the short wavelength component of sunlight.

Not all natural pigments protect or sensitize the organism. The bacterium *S. marcescens* showed the same sensitivity to a broad-spectrum incandescent source in the presence and absence of the red pigment, prodigiosin. However, the extracted pigment was capable of photosensitizing a colorless strain of *S. lutea,* but did not sensitize a colorless strain of *S. marcescens* (Mathews-Roth, 1967). These results suggest that different types of cells differ in their response to both endogenous and exogenous sensitizing agents.

3. INACTIVATION OF DNA AND BACTERIOPHAGE

3.1. Action Spectra for Inactivation of Transforming DNA

Transforming DNA can be readily inactivated by monochromatic near-UV radiation (Szybalski and Opara-Kubinska, 1965; Peak *et al.,* 1973*a,b,c;* Cabrera-Juarez and Swenson, 1975). The survival response of transforming DNA from both *B. subtilis* and *Haemophilus influenzae* to near-UV radiation differs strikingly from the response to far-UV radiation (Peak *et al.,* 1973*a;* Cabrera-Juarez and Swenson, 1975). Typical inactivation curves of *B. subtilis* DNA assayed on repair-proficient and repair-deficient (*uvr⁻*) recipient strains for 254- and 365-nm radiation are shown in Fig. 19. At 254 nm and below 320 nm, survival curves show decreasing slopes at higher fluences. These curves usually are linear on the square-root plot of Rupert and Goodgal (1960). This response was analyzed by Bresler *et al.* (1964, 1967) and Rupert (1968) who interpreted the relationship as indicating that the entire DNA segment, not limited to the marker cistron, is the target for marker inactivation. On this basis, the survival curve shape is produced by the varying lengths of DNA segments present, the longer pieces being the more sensitive.

At wavelengths longer than 320 nm, survival curves of transforming DNA no longer show declining slopes at higher fluences. The survival response shown in Fig. 19 for the tryptophan marker of *B. subtilis* DNA for 365-nm radiation, in which the relationship is approximately exponential with a small shoulder, is typical of near-UV radiation for both *B. subtilis*

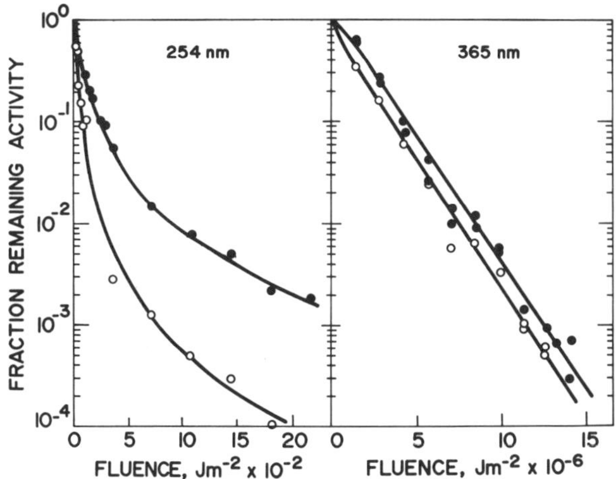

Fig. 19. Inactivation of *B. subtilis* transforming DNA (tryptophan marker) by 254- and 365-nm radiation tested on recipient cells of repair-proficient (*uvr⁺*) (●) and excision-deficient (*uvr⁻*) (○) phenotypes. Irradiation was at 25°C under aerobic conditions. Redrawn from Peak *et al.* (1973*b*).

and *H. influenzae* transforming DNA (Peak and Peak, 1973; Peak *et al.*, 1973*a,b,c;* Cabrera-Juarez and Swenson, 1975). If a linear relationship on the square-root plot of Rupert and Goodgal (1960) suggests that lesions outside the marker can influence its inactivation, exponential relationships are consistent with the requirement that near-UV-induced lesions occur within the marker to bring about its inactivation. Whatever interpretation is given, it appears that the large variation of size that occurs in transforming DNA preparations (Marmur *et al.,* 1961) has relatively little effect on the fluence-survival curves at these longer wavelengths.

There was only a small effect of the *uvr* genotype on the 365-nm inactivation of the tryptophan marker of transforming DNA (Fig. 19), in contrast to the fourfold effect for 254-nm inactivation (Peak *et al.,* 1973*c*).

Action spectra for the inactivation of tryptophan and leucine markers of *B. subtilis* DNA are shown in Fig. 20. Although no peak is evident in the near-UV range, there is a significant shoulder in this region, which may represent a distinct near-UV inactivation mechanism. The tryptophan marker was twice as sensitive as the leucine marker at 254 nm; however, the leucine marker was more sensitive at wavelengths of 334 nm and longer. These results suggest that the lesions responsible for marker inactivation are different in the far-UV and near-UV regions. Dimer yields in extracted *E. coli* DNA at 254 and 365 nm revealed a ratio of 3.8×10^5 (Table 2), while the ratio of sensitivities of *B. subtilis* DNA at 254 and 365 nm was only 7×10^3 for the leucine marker and 1.5×10^4 for the tryptophan marker. Thus

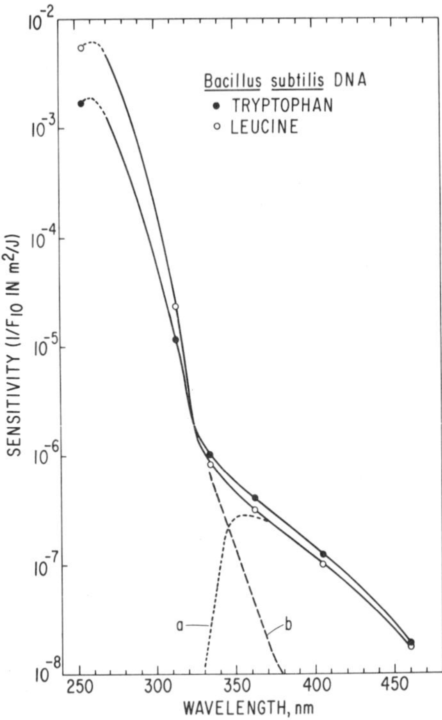

Fig. 20. Action spectra for the inactivation of the tryptophan and leucine markers of *B. subtilis* transforming DNA. Curve "b" was extrapolated by eye from the response of the tryptophan marker at shorter wavelengths. The extrapolation also took into account relative pyrimidine dimer yield at 365 and 254 nm measured for *E. coli* DNA (see Table 2) (Tyrrell, 1973). Curve "a" was derived from "b" and the tryptophan curve at longer wavelengths by difference. F_{10}, fluence to inactivate 90% of the transforming activity. O, Leucine; ●, tryptophan. Modified from Peak *et al.* (1973*a*).

B. subtilis DNA is 25–55 times more sensitive to 365-nm radiation than can be accounted for by dimer induction alone. This interpretation is supported by the absence of photoreactivation using the yeast PR enzyme after 365-nm irradiation. A large PR sector was obtained after 254-nm inactivation (Peak *et al.*, 1973*c*).

The action spectrum for inactivation of *H. influenzae* transforming DNA (Cabrera-Juarez and Swenson, 1975; Cabrera-Juarez *et al.*, 1976) is similar to that of *B. subtilis* DNA depicted in Fig. 20. However, the shoulder in the 334–365 nm region of the action spectrum was more pronounced for *H. influenzae* DNA than for *B. subtilis* DNA. Also, the ratio of the sensitivity of *H. influenzae* DNA in the 334–365 nm region to the 254-nm sensitivity was found by Cabrera-Juarez *et al.* (1976) to be 9 × 10⁴ using the same monochromator and dosimetry as that used by Peak *et al.* (1973*a,b,c*) for *B. subtilis* DNA. The basis for this tenfold difference in relative sensitivity at 254 and 365 nm of the two DNAs probably is related to the different modes of DNA uptake and integration in the two bacteria (see Section 3.5).

Photoreactivation data as a function of inactivation wavelength using the yeast PR enzyme were also obtained for *H. influenzae* DNA by Cabrera-Juarez *et al.* (1976). A large PR sector was obtained at 254, 290, and 313 nm; only a small PR sector was obtained at 334 nm; and no PR was evident at 365 nm and longer wavelengths. These results are similar to those obtained with *B. subtilis* DNA (Peak *et al.*, 1973c), suggesting a negligible role of pyrimidine dimers in the inactivation of both DNAs at 365 nm and longer wavelengths.

3.2. Oxygen Dependence of Transforming DNA Inactivation

Both *B. subtilis* and *H. influenzae* DNA showed oxygen-dependent near-UV inactivation after primary purification (Cabrera-Juarez, 1964; Peak *et al.*, 1973b). After additional purification of *H. influenzae* DNA, this oxygen dependence was lost. However, the oxygen dependence of near-UV inactivation of *B. subtilis* DNA was not lost after repeated stringent purification (Peak *et al.*, 1973b; M. J. Peak, personal communication). If the oxygen dependence of near-UV inactivation is the result of a photodynamic process mediated by a chromophore bound to the DNA, this sensitizer may be more tightly bound to *B. subtilis* DNA than to *H. influenzae* DNA.

The demonstration of oxygen dependence for near-UV-induced lethality of *E. coli* requires stringent anoxic conditions. Conditions sufficient to show the full oxygen dependence for the X-ray inactivation of *E. coli* are insufficient to show an oxygen effect for the near-UV inactivation of cells or transforming DNA (R. B. Webb, unpublished observations). Thus the possibility must be considered that the apparent oxygen independence reported for the near-UV inactivation of purified *H. influenzae* DNA (Cabrera-Juarez, 1964) may be caused by traces of oxygen that were present under the anoxic conditions employed.

3.3. Protection of Transforming DNA

Histidine at concentrations greater than 0.05 M was found to protect against the inhibition of the transforming DNA of *B. subtilis* by near-UV radiation (Peak and Peak, 1973; *et al.*, 1973b). An action spectrum for histidine protection (Peak *et al.*, 1973b) indicated that no protection was afforded at wavelengths shorter than 300 nm. A small amount of histidine protection occurred at 313 nm, increasing at 334 and 365 nm. Protection was maximum for both tryptophan and leucine markers between 365 and 405 nm, declining at longer wavelengths. Recently Cabrera-Juarez and co-workers (1976) repeated the histidine protection experiments of Peak *et al.* (1973b) using *H. influenzae* DNA. The results with *H. influenzae* DNA generally followed that with *B. subtilis* except for the unexpected sensitization of *H. influenzae* by histidine at 334 nm. This striking result, which is in sharp contrast to the large protective effect obtained with *B. subtilis* DNA at 334 nm (Peak *et al.*, 1973*b*), is not explained. The spectra for histidine protection (except for the *H. influenzae* result at 334 nm) correspond generally to the shoulder on the near-UV spectra for inactivation of DNA markers. The spacing of the points for the action spectra of transforming DNA is not sufficiently close to ascertain the presence of minor shoulders or peaks at 335 and 405 nm that are seen for lethality in *E. coli* (Fig. 2) and *S. typhimurium* (Fig. 3). Other amino acids did not afford protection against near-UV inactivation of *B. subtilis* transforming DNA (Peak and Peak, 1973; M. J. Peak and J. G. Peak, personal communication).

Protection against near-UV inhibition of transforming activity was also obtained with 0.1 M AET (Peak and Peak, 1975). The action spectrum for AET protection against marker inactivation of *B. subtilis* DNA corresponds to the action spectrum peak for histidine protection (Peak *et al.*, 1973*b*). There was a small effect at wavelengths shorter than 300 nm, with maximal protection occurring at approximately 350 nm. Protection declined sharply at wavelengths longer than 365 nm.

3.4. Action Spectra for Bacteriophage Lethality

Early work (Wahl, 1946; Wahl and Latarjet, 1947) indicated that several bacteriophages can be inactivated by near-UV and visible light. The action spectrum had a maximum in the near-UV and violet regions and declined to low values at longer wavelengths. However, as phage in these preparations were suspended in complete bacterial growth medium, inactivation could have been sensitized by components of the medium (Luria, 1955).

Recently, M. J. Peak and J. G. Peak (personal communication) have examined the near-UV and blue-visible response of purified bacteriophage T7 suspended in inorganic salts buffer. Survival curves of bacteriophage T7 are simple exponentials at wavelengths shorter than 320 nm. At wavelengths longer than 320 nm, the survival curves are exponential with a small shoulder (extrapolation number less than 2) (Fig. 21). Action spectra for phage T7 inactivation are shown in Fig. 22 as assayed on *E. coli* strains B (repair proficient) and B_{s-1} (*uvrB lexA*). A large host-cell reactivation sector was obtained at 254 and 313 nm. Host-cell reactivation was reduced at 334 nm and was very small or absent at longer wavelengths. However, R. M. Tyrrell (personal communication) observed significant host-cell reacti-

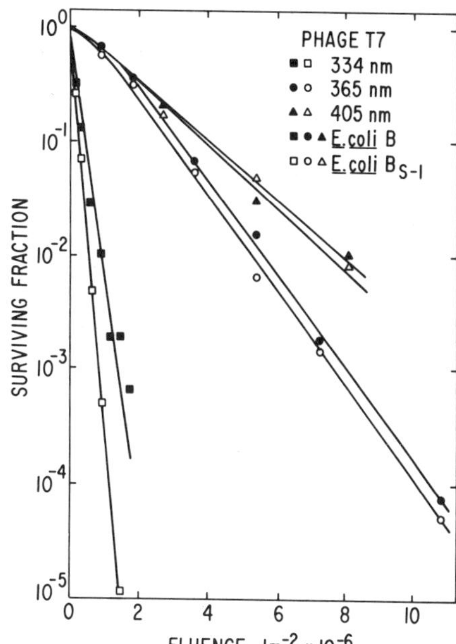

Fig. 21. Inactivation of phage T7 in M9 buffer suspension at 334, 365, and 405 nm assayed on cells of *E. coli* B (repair proficient) and *E. coli* B$_{s-1}$ (*uvrB lexA*). The irradiation procedure was the same as that described by Peak *et al.* (1973*a*). From M. J. Peak and J. G. Peak (personal communication).

vation at 365 nm in phage T1. There was a shoulder in the T7 action spectrum at wavelengths longer than 313 nm, a result similar to the transforming DNA action spectra (see Fig. 20).

The ratio for sensitivity of T7 at 254 and 365 nm was found to be 5 × 10⁴ (Fig. 22), a value significantly greater than the ratio of 7 × 10³ to 1.5 × 10⁴ for the inhibition of marker activity of *B. subtilis* DNA, but significantly less than the ratio of 3.8 × 10⁵ for pyrimidine dimer yield in DNA *in vitro* at these two wavelengths (Table 2). If dimers are induced in phage DNA by 365-nm radiation at the same rate as in extracted *E. coli* DNA, the sensitivity of phage T7 at 365 nm appears to be 7 times too great to be accounted for by dimer induction alone.

3.5. Mechanisms of Inactivation of Transforming DNA and Bacteriophage

The relative sensitivity of transforming DNA to near-UV wavelengths together with the absence of photoreactivation with the PR enzyme militates against pyrimidine dimers as the major inactivating lesion (Peak *et al.*, 1973*a,b,c*). Irradiation at 365 nm had no detectable effect on the T_m value or the shape of the melting and annealing curves of transforming DNA, a result suggesting that cross-linking is not a major 365-nm lesion

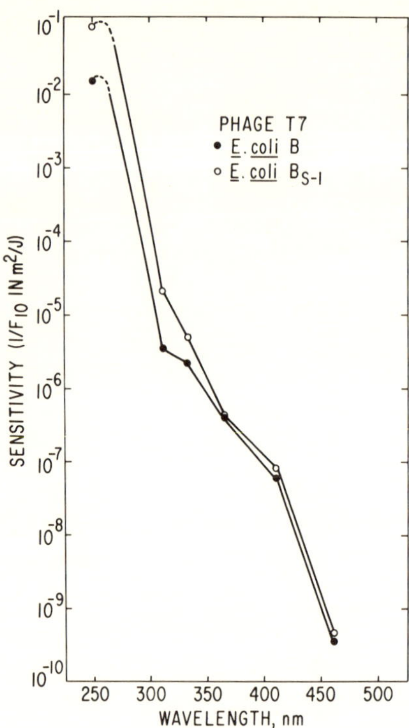

Fig. 22. Action spectra for the inactivation of phage T7 suspended in M9 buffer. Bacteriophage survival was assayed on *E. coli* B (repair proficient) and *E. coli* B$_{s-1}$ (*uvrB lexA*). From M. J. Peak and J. G. Peak (personal communication).

(Peak *et al.*, 1973*c*). The small effect of the *uvr* genotype for transforming DNA or the *uvrB lexA* genotype for phage T7 is consistent with a nondimer lesion, suggesting that the inactivating near-UV-induced lesions either are not repairable by the excision repair system, or the *uvr* gene product, an endonuclease, is not required for the repair of the lesions.

Contrary to expectations, the incorporation of 5-bromouracil in *B. subtilis* DNA significantly *decreased* its sensitivity to broad-spectrum near-UV radiation, an effect opposite to that obtained at 254 nm (Szybalski and Opara-Kubinska, 1965). The mode of inactivation by 254-nm and 330–405 nm radiation is clearly different. Furthermore, as the presence of 5-bromouracil in DNA should increase its absorbance in the near-UV range (see Boyce and Setlow, 1963), the decreased sensitivity observed is consistent with the presence of chromophores other than a component of DNA.

Oxygen enhancement, histidine protection, and AET protection suggest that inactivation of transforming DNA by near-UV radiation occurs by an indirect, photosensitized process (Peak *et al.*, 1973*a,b,c*; Peak and Peak, 1973, 1975). The chromophore may be a non-DNA contaminant of cellular origin. However, such a contaminant could be present only in traces and must be tightly bound to the DNA, as any such compound associated with the DNA copurifies with it (Peak *et al.,* 1973*c*).

Near-UV inactivation of transforming DNA paralleled X-ray responses more closely than far-UV responses. In addition to oxygen enhancement, histidine and AET protect *B. subtilis* DNA against both X-ray and near-UV inactivation (Peak and Peak, 1974). Furthermore, the *uvr* genotype has only a small effect on the inactivation of either near-UV- or X-irradiated transforming DNA. However, cysteine, which is an efficient protective agent against X-ray inactivation, had no significant effect on the near-UV response of *B. subtilis* DNA (Peak *et al.,* 1973*c*).

The production of single-strand breaks (or alkali-labile bonds) by 365-nm radiation in extracted phage T4 DNA was approximately four times more efficient than *E. coli* DNA *in vivo* (R. D. Ley, personal communication). Thus these oxygen-dependent lesions (single-strand breaks or alkali-labile bonds) are induced at a sufficient yield to be considered candidates for the near-UV inactivation of transforming DNA.

The tenfold greater sensitivity to near-UV radiation of *B. subtilis* DNA relative to *H. influenzae* DNA may be related to the form in which the DNA is taken into the cell (M. J. Peak, personal communication). In *B. subtilis,* the donor DNA exists in the single-stranded state within the host cell before assocation with and incorporation into the host-cell DNA. The other strand undergoes degradation at some point after attachment of the double-stranded donor DNA to the membrane of the host cell (Venema *et al.,* 1965; Piechowska and Fox, 1971; Dubnau and Cirigliano, 1972; Davidoff-Abelson and Dubnau, 1973). In contrast, DNA from *H. influenzae* remains a duplex to the point of incorporation into the DNA of the host cell (Notani and Goodgal, 1966; Stuy, 1965, 1974). Incorporation of donor DNA results in a heteroduplex of donor and host-cell DNA that could arise only from the incorporation of a single strand of the donor DNA into the host DNA in both *B. subtilis* (Bodmer and Ganesan, 1964) and *H. influenzae* (Notani and Goodgal, 1966). In *B. subtilis,* the presence of a single-strand break within a specific marker of the transforming DNA may inactivate that marker if the cistron becomes single stranded before the repair systems can operate. In contrast, with *H. influenzae,* the donor DNA duplex is maintained until the point of incorporation, and the repair of DNA single-strand breaks and other photoproducts should occur with high efficiency. Thus single-strand breaks may be the major inactivating lesion induced by near-UV radiation in DNA from *B. subtilis.* The single-strand break may not be an important inactivating lesion in DNA from *H. influenzae.* The nature of the inactivating lesion in *H. influenzae* DNA, except that it is not photoreactivable at wavelengths of 365 nm and longer, is unknown.

Studies of the X-ray and chemical inactivation of transforming DNA support the proposed role of single-strand breaks in the near-UV inactivation of transforming DNA. Inactivation by X radiation of *B. subtilis* transforming DNA can be accounted for by single-strand breaks (Hutchinson and Hales, 1970). *B. subtilis* DNA also was readily inactivated by agents such as hydrogen peroxide that produce primarily single-strand breaks in DNA (Freese *et al.*, 1967). In contrast, inactivation of *H. influence* DNA by X radiation was attributed to the production of double-strand breaks (Randolph and Setlow, 1971). The irradiated DNA was taken up by the cell, but integration in the host DNA was inhibited.

Present data do not permit the identification of DNA lesions and their role in the near-UV inactivation of bacteriophage. Single-strand breaks were produced in the DNA of intact phage T4 at the rate of 6.5×10^{-7} per 10^8 daltons per J m^{-2}, a rate that was one-half of that in the DNA of intact cells of *E. coli* (1.2×10^{-6} single-strand breaks per 10^8 daltons per J m^{-2}) (Tyrrell *et al.*, 1974). The absence of host-cell reactivation at 365 nm and the failure of the ratio of sensitivities at 254 and 365 nm (5×10^4) to correspond to the ratio of the relative yield of pyrimidine dimers at the same wavelengths (3.8×10^5) suggest that pyrimidine dimers are not significant lethal lesions for phage T7 at 365 nm.

4. MUTAGENIC EFFECTS

4.1. Early Work

Unfiltered sunlight was found to be mutagenic for *Asperigillus* conidia by Hollaender and Emmons (1946). Filtration of wavelengths shorter than 320 nm eliminated mutation induction, indicating that the mutagenic wavelengths of sunlight are between 295 and 320 nm. More recently, Ashwood-Smith *et al.* (1967) made similar observations on liquid and frozen suspensions of *E. coli*. Mutation induction was observed with unfiltered sunlight, but vanished when wavelengths shorter than 320 nm were removed. Early efforts to induce mutations at wavelengths longer than 320 nm were either negative or of doubtful validity. These reports have been reviewed by Zelle and Hollaender (1955).

4.2. Mutagenesis at Low Fluence Rates

4.2.1. Mutagenesis in Continuous Cultures

While investigating photodynamic mutagenesis in continuous cultures, Kubitschek (1967) and Webb and Malina (1967) observed that near-UV and

visible light, in the absence of an exogenous sensitizer, are strongly mutagenic. Continuous-culture techniques make possible the study of mutation
in the absence of detectable killing of cells, while most other techniques for
the study of mutagenesis require a substantial amount of cell killing
(Kubitschek, 1970). Figure 23 shows mutagenesis to resistance to bacteriophage T5 in chemostat cultures of *E. coli* B/r/1 *trp* by several wavelength ranges. In this strain, even if wavelengths shorter than 350 nm are
excluded, light from cool white fluorescent lamps at typical laboratory
intensities will approximately double the spontaneous dark mutation rate
(Webb and Malina, 1967).

Wavelengths in both the near-UV and visible ranges are mutagenic in
this system. The mean expression delay is about four generations for this
mutation. The mutant increase ceases four generations after the light is
turned off, suggesting that a photoinduced mutagen with a long lifetime is
not the cause of the near-UV and visible-light mutagenicity observed. The
linear increase of mutant frequency observed is evidence for the nonselec-
tion of mutants. Selection against the mutant population would produce a

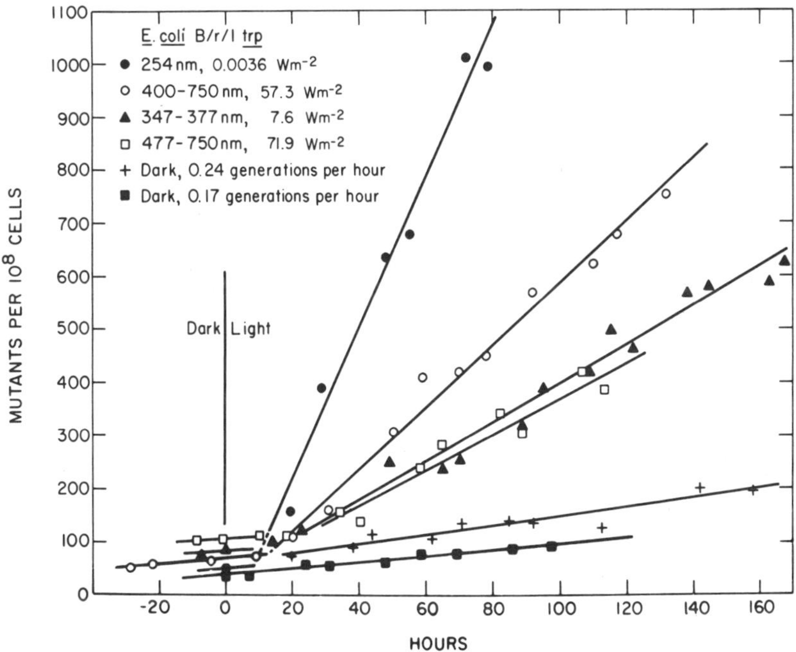

Fig. 23. Induction of mutants resistant to the phage T5 in glucose-limited continuous cultures
(chemostat) of *E. coli* B/r/1 *trp* by continuous irradiation in the far-UV, and mid-UV, near-
UV, and visible radiation ranges. Growth rates were 0.18–0.22 division per hour. Modified
from Webb and Malina (1970).

curve that bends to a plateau, whereas selection in favor of the mutant population would produce an upward bending curve (Kubitschek, 1970).

The mutation rate in glucose-limited chemostats is independent of growth rate for mutagens that act on the parental DNA (Kubitschek, 1970). Examples of this kind of mutagen are mutagenesis to T5 resistance (T5^R) in *E. coli* B/r by 254-nm radiation (Kubitschek and Bendigkeit, 1964*b*), and mutagenesis by acridine orange and visible light (photodynamic mutagenesis) (Webb and Kubitschek, 1963, 1965). On the other hand, with a mutagen that is incorporated in the DNA during replication, or acts at the DNA replication complex, causing replication errors to be made, the mutation rate is proportional to growth rate in glucose-limited chemostats (Kubitschek, 1970). Examples of mutagens acting at the replication fork include 2-aminopurine (Kubitschek and Bendigkeit, 1964*b*), 5-bromouracil (Witkin and Parisi, 1974), and caffeine (Kubitschek and Bendigkeit, 1964*a*; R. B. Webb and H. E. Kubitschek, unpublished observations).

Broad-spectrum near-UV mutagenesis to T5 resistance in glucose-limited continuous cultures was independent of growth rate (H. E. Kubitschek, personal communication), a result that is consistent with mutagenesis of duplex DNA rather than at the replication fork. Mutagenesis by broad-spectrum visible light (wavelengths below 400 nm were filtered out) under similar conditions also showed independence of growth rate (Webb and Malina, 1970).

Both near-UV and visible-light induction of resistance to phage T5 in continuous cultures were proportional to fluence rate over the tenfold range tested (Webb and Malina, 1970). This result is consistent with near-UV- and visible-light-induced mutagenesis occurring through a direct photochemical process. Oxygen dependence has been observed for both near-UV and visible-light induction of phage T5 resistance (Kubitschek, 1967; Webb and Malina, 1967, 1970).

Mutation to streptomycin resistance in continuous cultures of *E. coli* B/r also occurred at wavelengths longer than 320 nm (R. B. Webb, 1972, and unpublished data). The relative rate of induction of phage T5 resistance and streptomycin resistance (St^R) was significantly different in the far-UV, near-UV, and visible ranges. The ratio of the two mutation rates (T5^R/St^R) was 3 times greater for visible than for 254-nm radiation. The difference in the ratio of the two mutation rates in the three wavelength ranges suggests that different chromophores are involved in each of the three ranges.

4.2.2. Action Spectrum for Mutagenesis in Continuous Cultures

An action spectrum for the induction of T5 resistance in glucose-limited continuous cultures of *E. coli* B/r/1 *trp* shows a broad peak between

350 and 480 nm (Fig. 24). Above 320 nm, mutation induction is strongly oxygen dependent: under anoxic conditions, using nitrogen containing 5% CO_2, the induced mutation rate was not greater than the spontaneous mutation rate. Although the detail of this action spectrum is not great enough to make a selection among a number of possible chromophores, it shows a striking similarity to the absorption spectrum of riboflavin. However, the addition of riboflavin to the chemostat did not increase the mutation rate under continuous irradiation. This result is not conclusive as there is no evidence that exogenous riboflavin is taken up by the cell under the experimental conditions.

Additional evidence that there is a minimum in the action spectrum in the 340-nm range has been obtained using a strain of *E. coli* B/r that does not require tryptophan (R. B. Webb and M. S. Brown, unpublished data). Mutation rates with this strain were four times greater than mutation rates obtained with the tryptophan-requiring strain. The high mutation rates obtained eliminates exogenous tryptophan in the near-UV and visible-light mutational responses observed. An action spectrum for mutation to phage

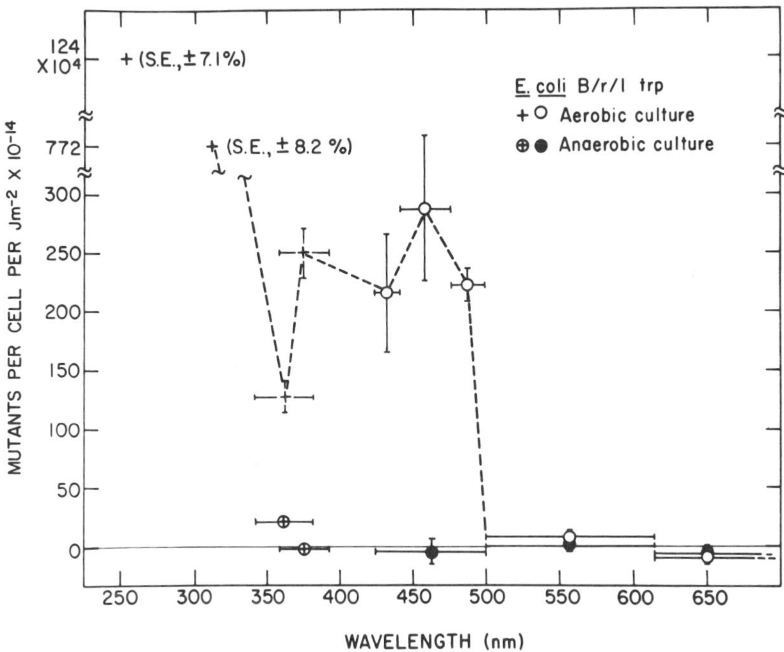

Fig. 24. Action spectrum for the induction of mutants resistant to phage T5 in aerobic and anaerobic continuous cultures (glucose limited) of *E. coli* B/r/1 *trp*. Standard error estimates are indicated by vertical bars. Horizontal bars indicate effective band width at half-maximum emission. From Webb and Malina (1970).

T5 resistance in *E. coli* B/r at monochromatic wavelengths over the range of 330–500 nm was similar to that shown in Fig. 24 except for a sharper minimum at 400 nm.

4.3. Mutagenesis at High Fluence Rates

4.3.1. Effects with Broad-Spectrum Radiation

Broad-spectrum near-UV radiation has been shown to produce genetic effects in several different systems. Kaplan (1956) and Kaplan and Kaplan (1956) observed the induction of color-sector and sulfonamide-resistant mutants in *S. marcescens* with near-UV radiation at wavelengths longer than 320 nm. The near-UV source was a high-pressure Hg lamp filtered through window glass, which removed wavelengths shorter than 320 nm. The mutant frequency response curve was approximately linear rather than the upward bending fluence-square-type curve characteristic of far-UV-induced mutagenesis. Fluences up to 15,000 J m^{-2} were delivered with no detectable cell killing, although mutants were readily produced. Kelner and Halle (1960; A Kelner, personal communication) found that radiation from a fluorescent blacklight lamp increased the mutation rate to streptomycin resistance in *E. coli* when the cells were irradiated continuously during growth. Other reports of genetic effects of broad-spectrum near-UV radiation from fluorescent BLB lamps or short-wavelength-filtered high-pressure Hg lamps include the inactivation of transforming DNA (Cabrera-Juarez, 1964), and the induction of *r* plaques in bacteriophage T4 (Ritchie, 1964). However, as the bacteriophage were suspended in a complete broth medium containing the cell lysate, the chromophore for the effect might have come from the suspending medium.

Cabrera-Juarez and Espinosa-Lara (1974) reported mutation induction by near-UV radiation (fluorescent blacklight, BLB) in *H. influenzae* at different stages of cell growth. Wavelengths shorter than 325 nm were removed by filtration. Mutation to both streptomycin resistance (12 µg/ml) and the ability to use protoporphyrin IX instead of hemin were reported after exposure to this near-UV source; however, the absence of dosimetry makes comparison to other work impossible. Generally, the mutant frequency curves showed a strong upward bend following the downward bend of the survival curves, except for the induction of protoporphyrin mutants in stationary-phase and early exponential-phase cells, in which there was little, if any, mutagenesis.

Mutagenesis by visible light (wavelengths longer than 400 nm) also has been observed in cultures of *S. typhimurium* by Speck and Rosenkranz

(1975). The effective wavelength range was identified as 400–500 nm through the use of filters and a continuous-spectrum blue-fluorescent source. Mutations from histidine dependence to histidine independence were identified as the base substitution type; frame-shift mutations were not induced. Since the cells were irradiated after plating on nutrient agar, the role of medium components and the possible requirement of growth for visible-light-induced mutagenesis (Webb and Malina, 1970) cannot be evaluated from these experiments.

An oxygen-dependent induction of inheritable variants by visible light mediated by endogenous pigments in *Euglena gracilis* was reported by Leff and Krinsky (1967). Although the white colony variants were very likely cytoplasmic rather than nuclear in nature, they were nevertheless comparable to those induced by far-UV and X radiation in this cell (Leff and Krinsky, 1967). Visible light from an incandescent source with wavelengths below 610 nm removed by filtration was effective for the production of white colony variants.

4.3.2. Effects with Monochromatic Radiation

Mutational responses to acute exposures at high fluence rates differ markedly from those after low-irradiance chronic exposures. In Fig. 25, the lethal and mutational response of *E. coli* B/r Hcr are shown at 254 and 365 nm at high fluence rates. Mutation induction roughly followed lethality and mutant frequency increase followed approximately the fluence-square relationship in both instances. This strain is approximately 7×10^5 times more sensitive at 254 nm than at 365 nm. The 365-nm mutational response of strains B/r and B/r Hcr is shown in Fig. 26. At fluences of 365-nm radiation up to 4×10^6 J m^{-2}, there were very few mutants induced in the wild-type strain B/r even though there was 70% inactivation. At greater fluences, there was a sharp upward bend in the mutation response curve. This result is consistent with differential near-UV damage to repair enzymes (see Section 2.4.3). If excision repair is more sensitive than *rec* repair to 365-nm inhibition, more of the premutational lesions would remain to be dealt with by the error-prone postreplication (*rec* and *lex*) repair system (Witkin, 1975), resulting in a large increase in mutant yield.

Oxygen dependence and photorepairability of lethal and mutational responses to 365-nm radiation are shown in Fig. 27. The large oxygen dependence of the lethal response was not seen in the mutational responses at acute fluence rates. The aerobic mutation curve showed the characteristic upward bend, but under anoxic conditions, the mutation curve was almost linear with fluence. At low 365-nm fluences, the anoxic mutation yield was actually greater than the yield under aerobic conditions.

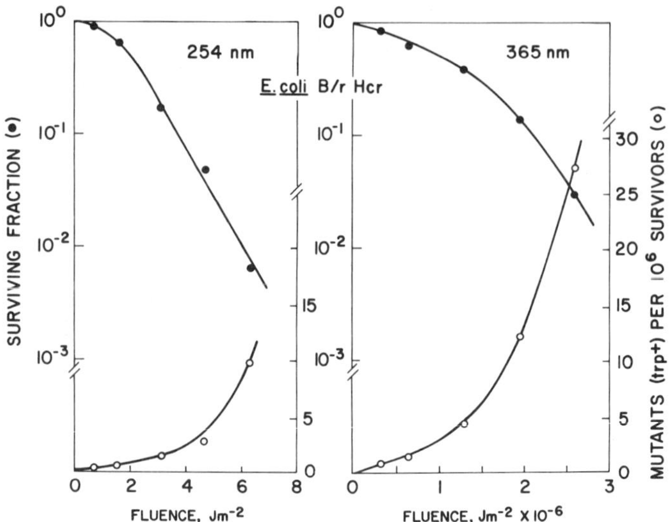

Fig. 25. Inactivation and mutagensis in stationary-phase cells of *E. coli* B/r Hcr (in buffer suspension) at 254 and 365 nm. Reversion to tryptophan independence ($trp^- \rightarrow trp^+$) was assayed on minimal plates supplemented with 3% nutrient broth. Fluence rates were 0.65 W m^{-2} at 354 nm and 1100 W m^{-2} at 365 nm. From Webb (1977*a*).

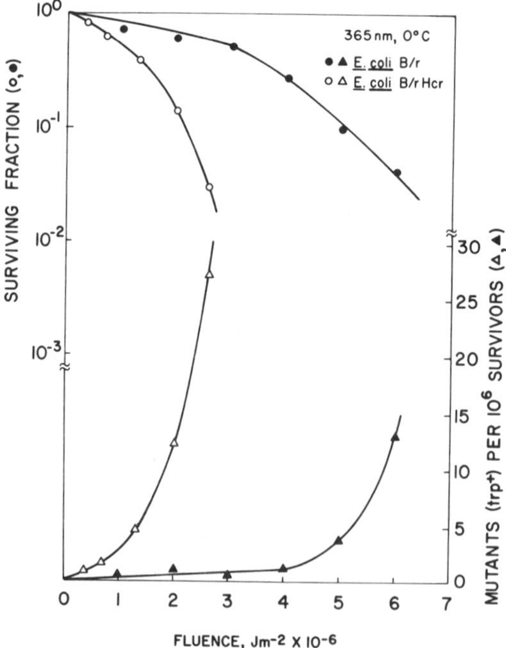

Fig. 26. Inactivation and mutagenesis ($trp^- \rightarrow trp^+$) at 365 nm in stationary-phase cells of *E. coli* B/r and *E. coli* B/r Hcr (in buffer suspension). The fluence rate was 1100 W m^{-2}. From Webb (1977*b*).

No photoreactivation of lethality has been seen for *E. coli* B/r Hcr (or any *rec*⁺ strain) after aerobic 365-nm irradiation (Brown, 1977; Webb *et al.*, 1976; R. B. Webb and M. S. Brown, unpublished data). However, after anoxic inactivation, a small but significant amount of photoreactivation could be demonstrated. In contrast to the absence of photorepair for aerobic lethality, mutational lesions induced under both aerobic and anerobic conditions were almost completely photoreactivable in *E. coli* B/r Hcr (Fig. 27). These data suggest that pyrimidine dimers are involved in the mutational response of this strain when irradiated at 365 nm at very high fluence rates (Webb, 1977*a*).

A comparison of lethality and mutagenesis at 254 and 365 nm is shown in Fig. 28. Mutation induction relative to killing was 6 times more efficient at 365 nm than at 254 nm. Under anoxic conditions, mutation induction at 365 nm was 18 times more efficient than at 254 nm. It is evident that the

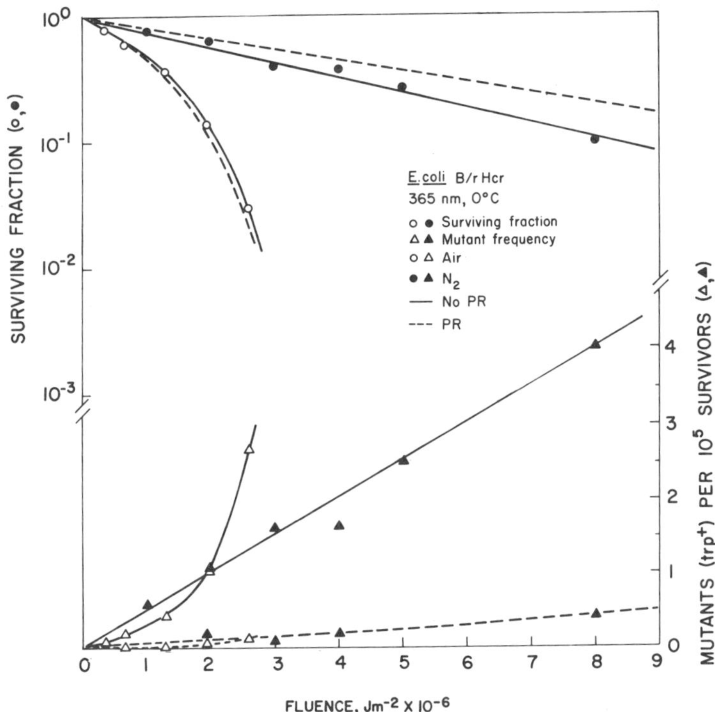

Fig. 27. Lethal and mutagenic (*trp*⁻ → *trp*⁺) effects at 365 nm (0°C) under aerobic and stringent anaerobic conditions in stationary-phase cells (in buffer suspension) of *E. coli* B/r Hcr. The effect of photoreactivating (PR) illumination (25°C) is indicated for both lethal and mutagenic effects. The fluence rate at 365 nm was 1100 W m⁻². The fluence rate of PR illumination (380–440 nm) was 75 W m⁻². From Webb (1977*a*).

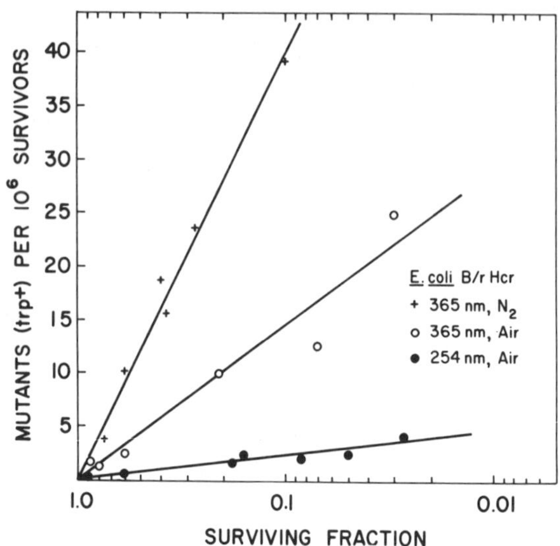

Fig. 28. Relationship between mutagenesis ($trp^-\rightarrow trp^+$) and inactivation in stationary-phase cells of *E. coli* B/r Hcr (in buffer suspension) at 365 nm (0°C) under aerobic and stringent anaerobic conditions, and at 254 nm (25°C) under aerobic conditions. The fluence rate was 0.65 W m^{-2} at 254 nm and 1100 W m^{-2} at 365 nm. From Webb (1977*a*).

mutagenic effect of dimers relative to their lethal action differs greatly under these three sets of conditions.

4.3.3. Mutagenesis in Partially Dehydrated Systems

S. J. Webb and Tai (1968, 1969, 1970) have reported highly provocative results indicating that high levels of forward mutations are produced by broad-spectrum near-UV radiation (fluorescent BLB) in semi-dry cells of *E. coli*. Wavelengths shorter than 320 nm were removed by a plastic filter. Cells were irradiated as aerosols in equilibrium with different relative humidities. In these experiments, forward mutations were assayed by the penicillin technique. In the strain used, *E. coli* K12S, almost all the mutants induced by near-UV radiation were histidine requiring, whereas a wide range of mutants were obtained with 254-nm radiation. Other strains were reported to yield different but highly specific mutations when exposed in stationary phase to near-UV radiation. In contrast, exponential-phase cultures showed a wide range of forward mutations.

This near-UV mutagenesis was found to be oxygen dependent, suggesting a photodynamic process. S. J. Webb and Tai (1970) and S. J. Webb

(1972) proposed that the chromophore, perhaps a part of the respiratory cytochrome chain, is associated with the DNA replication complex.

4.4. Comparison of Mutagenesis at High and Low Fluence Rates

There are striking differences between near-UV-induced mutagenesis in continuous cultures at low chronic exposures, and in stationary-phase cell suspensions subjected to large acute exposures. Mutation induction at low fluence rates above 330 nm was completely oxygen dependent (Fig. 24), in contrast to the complex oxygen independence at high fluence rates (inhibition at low fluences, enhancement at high fluences) (Fig. 27). Mutant induction at low levels of near-UV radiation involved procedures that should allow only very small numbers of dimers to be produced (Tyrrell, 1973), and these dimers should be largely monomerized by concomitant enzymatic photoreactivation at wavelengths in the range of the photoreactivation action spectrum (330–460 nm) (Jagger *et al.*, 1969). However, at high fluence rates, pyrimidine dimers play a major role in mutation induction (Fig. 27). Thus not only are the mutational lesions under the two conditions clearly different, but the chromophores and the nature of the sensitized processes concerned in their production also are of dissimilar types.

A preliminary action spectrum for the induction of tryptophan independence in *E. coli* B/r Hcr at 254, 313, 334, 365, and 405 nm for high fluence rates at O°C under aerobic conditions showed a generally declining mutation rate from 313 nm to the longer wavelengths, roughly parallel to aerobic lethality (see Fig. 2) in this wavelength range (R. B. Webb, unpublished data). These data are in striking contrast to the action spectrum for the induction of phage T5 resistance at low fluence rates in continuous cultures of *E. coli* B/r (Fig. 24), in which there was a minimum value at 340 nm, with mutation rates as great at 440 nm as at 370 nm (Webb and Malina, 1970; R. B. Webb and M. S. Brown, unpublished data).

5. SYNERGISTIC EFFECTS

5.1. Effects on Transforming DNA

Major advances in photobiology have been made using monochromatic or narrow-band radiation. Although preliminary action spectra can be obtained using broad-spectrum sources with sharp cutoff filters (e.g., Elkind and Sutton, 1957*b*; Webb and Malina, 1970), precise action spectra require monochromatic or narrow-band radiation if identification of the chromophore is to be achieved (Jagger, 1967; Giese, 1968; Smith and Hanawalt, 1969). While monochromatic radiation is required for the identification of chromophores, recent findings demonstrate that experiments using single wavelengths may fail to yield the full basis of the biological effects observed.

Transforming DNA of *B. subtilis* was observed to be severalfold more sensitive to inactivation by the radiant emission (mostly 365 nm) of a Philips HP 125-W lamp (Peak and Peak, 1973) than by 365-nm radiation from a monochromator using a stringent stray light filter (Peak *et al.,* 1973*a*). Through the use of a calibrated spectroradiometer, the Philips HP 125-W lamp was found to emit radiation at 313 nm (0.9%) and 334 nm (1.4%) in addition to the predominant 365-nm Hg line. Figure 29 depicts the inactivation of transforming DNA by 365-nm radiation at 750 W m^{-2} with and without the addition of 334-nm radiation in trace (0.6 W m^{-2}) amounts (Peak *et al.,* 1975). Inactivation at 334 nm at this low fluence rate was below detection for both the leucine and tryptophan markers up to the maximum fluence used (8.1 $\times$ 10^3 J m^{-2}). However, when this low fluence rate of 334-nm radiation was delivered simultaneously with 365-nm radiation at 750 W m^{-2}, the rate of inactivation of both markers was approximately 3 times the rate with 365-nm radiation alone (Peak *et al.,* 1975). A similar synergism was obtained when an even smaller fluence rate of 313-nm radiation (0.1 W m^{-2}) was added during 365-nm irradiation of transforming DNA. No inactivation by the 313-nm irradiation at the fluences involved could be detected. Prior irradiation of transforming DNA at 334 nm at the same fluence rate and same maximum fluence resulted in a similar threefold increase in the 365-nm sensitivity. This result indicates that a long-lived alteration in the DNA or bound chromophore is produced by the 334-nm irradiation. At present, no information has been obtained at other fluence rates or other wavelengths.

Mechanisms involved in this striking synergism cannot be identified from the present data. However, the following processes can be suggested. (1) An alteration by irradiation at 313 or 334 nm of a sensitizer bound to the DNA may occur, thereby increasing its absorbance at 365 nm. (2) Sublethal lesions may be induced in DNA by 313- or 334-nm irradiation that may interact with lesions induced by 365-nm radiation, thereby increasing the lethal potential of the 365-nm radiation.

5.2. Effects on Cell Lethality

Ramabhadran and Jagger (1975) reported that a strain of *E. coli* K12 K8 (wild type) was sensitized to moderate exposures at 405 nm by prior exposures to 334-nm radiation. This response prevented the accurate

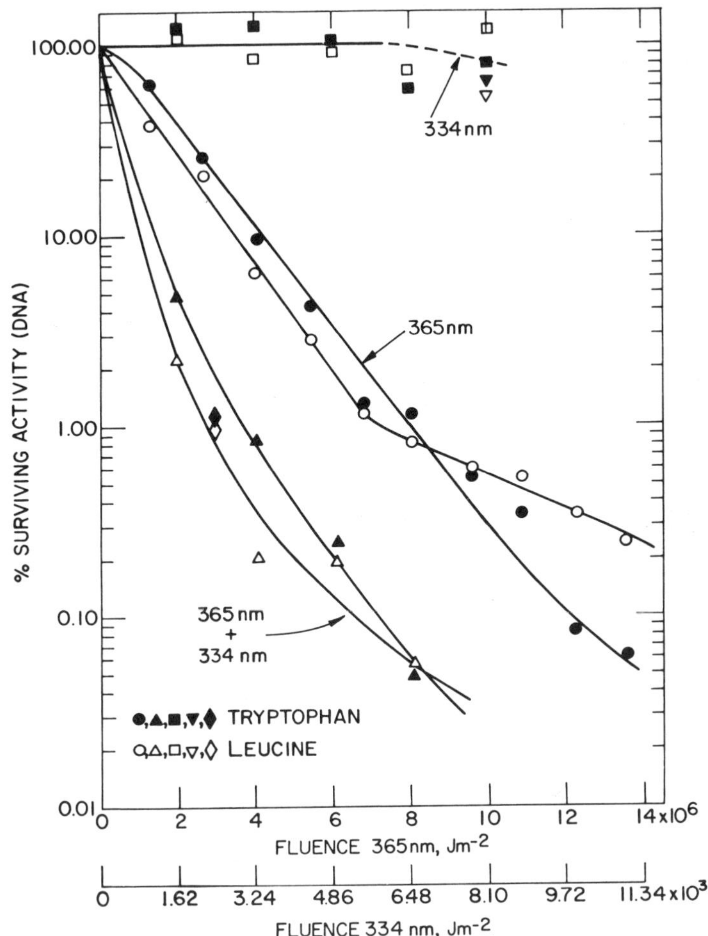

Fig. 29. Inactivation of *B. subtilis* transforming DNA by irradiation with 334-nm radiation alone at 0.6 W m⁻² (□, ■), with 365-nm radiation alone at 750 W m⁻² (○, ●), and with both wavelengths simultaneously (△, ▲) at the same fluence rates as before. In another experiment, the DNA was preirradiated with 334-nm radiation at 0.6 W m⁻² for a total of 8.1 × 10³ J m⁻² (▽, ▼) and then exposed to 3.0 × 10⁷ J m⁻² of 365-nm radiation (◇, ◆). Open symbols refer to the leucine marker and closed ones to the tryptophan marker. Modified from Peak *et al.* (1975).

determination of possible photoreactivation in this repair-proficient strain. A parallel effect was observed for another wavelength in *E. coli* K12 AB1157 (wild type) (Webb and Brown, 1977*a*). Irradiation of cells in stationary phase at 365 nm at a fluence of 2 × 10⁶ J m⁻² resulted in a twentyfold increase in the aerobic sensitivity of these cells to either monochromatic 405-nm or broad-band (380–440 nm) radiation. The pre-irradiated cells did not show the high sensitivity when they were irradiated at 405 nm under stringent anoxia.

Repair-deficient mutants of *E. coli* K12, AB1886 (*uvrA*), and AB2463 (*recA*) also were strongly sensitized to 405-nm radiation by a prior exposure to 365-nm radiation at low fluences. The highly sensitive double-mutant strain, *E. coli* K12 AB2480, was not sensitized to 405-nm radiation by prior irradiation at 334 nm (Ramabhadran and Jagger, 1975) or at 365 nm (Webb and Brown, 1977*a*). While the mechanism for this strong synergism at 334, 365, and 405 nm is unknown, the parallel with the synergism observed in transforming DNA (Peak *et al.*, 1975) is provocative. However, it should be noted that the synergism between 365-nm and 254-nm radiations clearly involves near-UV-induced inhibition to repair systems (Tyrrell and Webb, 1973).

Strains capable of repair are sensitized to far-UV (Fig. 11) (Tyrrell and Webb, 1973; Webb *et al.*, 1977*b*), X radiation (Tyrrell, 1974, 1976*a*), and mild heat (52°C) (Tyrrell, 1976*b*) by exposures to 365-nm radiation within the low fluence range (see Section 2.4.3). Synergism involving both far-UV and X radiation was clearly shown to be related to the disruption of DNA repair systems (Martignoni and Smith, 1973). Irradiation at 365 nm strongly inhibited the excision of pyrimidine dimers induced at 254 nm (Fig. 10). Furthermore, 365-nm irradiation strongly inhibited the repair of X-ray-induced DNA single-strand breaks by both the type II (buffer) and type III (full-medium) repair systems. The 365-nm-irradiated cells recovered their ability to perform the buffer type II repair upon holding for 30 min at 37°C before X irradiation; the cells did not recover their ability to perform the full-medium type III repair (Tyrrell, 1974).

Tyrrell (1976*a*) observed that near-UV radiation also strongly sensitized *E. coli* to inactivation by mild heat (52°C). A strong synergism was observed between radiation of 334, 365, or 405 nm and exposure to 52°C. Prior irradiation at 254 nm did not appreciably sensitize the cells to 52°C. Tyrrell also found that a 15-min heat treatment at 52°C caused little lethality, but strongly sensitized the cells to subsequent exposure to 365-nm but not 254-nm radiation. These data are consistent with synergistic damage to repair systems by both heat and near-UV radiation, a type of synergism that does not extend to far-UV radiation.

Synergism between far-UV radiation and X radiation occurred in buffer suspensions of cells of bacteria capable of dark repair whichever radiation was given first (Haynes, 1964, 1966; Davies *et al.*, 1967; Martignoni and Smith, 1972, 1973). In contrast, the near-UV and far-UV synergism of inactivation was not reciprocal (Webb *et al.*, 1977*b*); prior exposure to 254-nm radiation *decreased* the near-UV sensitivity of wild-type AB1157 and the excision-repair-deficient mutant AB1886 of *E. coli* K12 when the cells were held for 10 min or longer at 25°C between the 254-nm and 365-nm irradiations. This result is consistent with the induction of certain repair enzymes by 254-nm radiation that are not induced by exposure to 365-nm radiation alone.

Indirect evidence also exists for synergistic effects associated with inactivation by broad-spectrum radiation sources. Protection by carotenoid pigments against inactivation in *S. lutea* was much greater with broad-spectrum radiation such as natural sunlight (Mathews and Sistrom, 1959*b*), quartz-halogen incandescent lamps (Mathews, 1964*a*; Mathews-Roth and Krinsky, 1970), and fluorescent blacklight lamps (BLB) (Webb *et al.*, 1977*b*) than with monochromatic radiation in the near-UV and visible radiation ranges (see Section 2.6). This apparent synergism suggests that interaction of more than one chromophore is responsible for the near-UV inactivation of *S. lutea*. The location of carotenoid pigments on the cell membrane suggests that the chromophores and the sensitive targets may also occur on the membrane (or may be the membrane itself) in *S. lutea* (Mathews and Sistrom, 1959*a*). The chromophores responsible for the sensitivity of *S. lutea* to near-UV radiation have not been identified.

Eisenstark and co-workers observed that exponential-phase cells of *recA* strains of bacteria are much more sensitive to broad-spectrum near-UV sources (Eisenstark, 1970) than would be predicted from their sensitivity to monochromatic wavelengths in the same wavelength range (Fig. 3) (Mackay *et al.*, 1976; Yoakum *et al.*, 1977). It was suggested that synergism in *recA* strains may be caused by the interaction of near-UV-induced lesions in the DNA and by near-UV-induced photoproducts of endogenous tryptophan. Photoproducts of tryptophan are produced fiftyfold more efficiently by fluorescent blacklight than by any monochromatic wavelength within that wavelength range (Yoakum *et al.*, 1977). One of these tryptophan photoproducts has been shown to be hydrogen peroxide (McCormick *et al.*, 1975); however, hydrogen peroxide may be a breakdown product formed during purification (see Section 5.1).

Mutations in *E. coli* B/r (*trp*) (wild type for repair) to tryptophan independence (*trp*⁺) were readily induced at monochromatic wavelengths at 254, 313, 334, and 365 nm. However, no mutations to tryptophan independence were induced in this wild-type strain with broad-spectrum near-UV radiation (fluorescent BLB) (313–405 nm), although the cells were easily inactivated (R. B. Webb, unpublished data). In contrast, the excision-deficient strain *E. coli* B/r Hcr (*trp*) was readily mutated to tryptophan independence by both the monochromatic and broad-spectrum sources, but the broad-spectrum source was fourfold more efficient for both mutagenesis and lethality. The reason for the failure of broad-spectrum near-UV radiation to produce reversion to tryptophan independence in wild-type *E. coli* B/r has not been determined. It may be significant that a similar result was obtained in two strains of *N. crassa*: fluences of near-UV radiation (fluorescent BLB) that inactivated more than 99% of the conidia did not produce detectable nutritionally independent mutants at the *ad-3A* or *inl* loci. Mutants were easily detectable at similar survival levels after far-UV irradiation (Tuveson and Satterthwaite, 1976). In contrast to these results, mutation induction in wild-type strains by broad-spectrum near-UV radiation has been reported for color sector and sulfonamide resistance in *S. marcescens* (Kaplan, 1956), phage T5 resistance in continuous cultures of *E. coli* B/r/1 *trp* (Webb and Malina, 1967, 1970), and streptomycin resistance and protoporphyrin utilization in exponential-phase cells of *H. haemophilus* suspended in buffer (there were few or no mutants after irradiation of stationary-phase or early exponential-phase cells) (Cabrera-Juarez and Espinosa-Lara, 1974).

An explanation of the failure of broad-spectrum near-UV radiation to produce mutations in certain systems that are mutated strongly by far-UV radiation is the possibly greater sensitivity of the *rec*-repair system than the excision-repair system to broad-spectrum near-UV radiation. If there is such a differential sensitivity of the repair systems, a greater fraction of lesions must be repaired, if they are repaired at all, by the excision-repair system. As excision repair is largely error free and repair associated with the *rec* and *lex* gene products is error prone (Witkin, 1975), differential inhibition of *rec* or *lex* repair would be expected to reduce the induction of mutants.

A provocative observation in near-UV induction of human skin erythema may be related to some of the near-UV synergistic effects demonstrated in a variety of bacteria. Preirradiation of normal human skin with moderate fluences of broad-spectrum near-UV radiation (320–400 nm) caused that skin to be more sensitive to erythema induced by 280–320 nm radiation (Forbes, 1973; Pathak, 1972; Willis *et al.*, 1973). This result appears to be analogous to the findings of Peak *et al.* (1975), using transforming DNA, that 313- and 334-nm radiation act synergistically with 365-nm radiation. The biological significance of this photoaugmentation was further investigated by Parrish *et al.* (1974), who failed to find significant synergism, but observed a simple additive effect between the erythema-producing properties of near-UV and mid-UV radiations. The basis of the differences in the findings of the different groups of investigators is not known. However, the strong synergism for skin erythema between the mid-UV and near-UV ranges reported by Willis *et al.* (1973) is a result that should be investigated further in consideration of parallel findings in various bacteria.

6. MEDICAL ASPECTS OF NEAR-UV RADIATION

6.1. Induction of Skin Cancer

In accord with earlier assumptions (Blum, 1959; Cutchis, 1974), the action spectrum for skin cancer induction in albino mice was found to closely follow erythemal effectiveness for three monochromatic UV wavelengths, 300, 310, and 320 nm (Freeman, 1975). Mice exposed to the radiation in amounts proportional to its erythemal effectiveness developed skin tumors at approximately the same time and the same rate in each of the three groups. However, cancer induction departed sharply from erythemal effectiveness at 290 nm, where no tumors were observed. Although previous findings indicated that wavelengths longer than 320 nm were not effective in producing either erythema or cancer of the skin (Blum, 1943, 1959; Epstein, 1970), recent observations clearly demonstrate that high fluences of near-UV radiation (320–400 nm) can induce erythema (Parrish *et al.*, 1974). Similarly, carcinoma of the skin of hairless albino mice has been produced by continuous irradiation with a broad-spectrum near-UV source (330–400 nm) at a fluence rate of 10 W m^{-2}, which is somewhat lower than the near-UV component of direct sunlight (Forbes, 1973; Urbach *et al.*, 1974).

Synergism between mid-UV and near-UV wavelengths in the induction of skin cancer is suggested by provocative results obtained by Urbach and co-workers (Forbes, 1973; Urbach *et al.*, 1974). The two radiation sources emitted equal amounts of mid-UV radiation, but differed greatly in the amount of near-UV radiation emitted. Although both radiation sources were adjusted to produce the same amount of acute skin erythema in the mice, the prevalence of skin cancer was seven times greater in the animals that received both mid-UV and near-UV radiation. This result is consistent with the proposal of Setlow (1974), and clearly shows that the action spectrum for erythema, as traditionally defined, is significantly different from the action spectrum for carcinogenesis (Urbach *et al.*, 1974). Although the mechanism of interaction is not known, these results on skin cancer indicate that both mid-UV and near-UV radiation play significant roles in carcinogenesis. See Section 5 for additional examples of synergism.

6.2. Clinical Applications

With the advent of antibiotics and vitamin D supplementation more than 30 years ago, and the general recognition that phototherapy was not based on definitive clinical tests, the widespread use of phototherapy for a variety of health problems virtually ceased [see reviews by Luckiesh (1946) and Daniels (1974) for a discussion of early medical practices]. Recently, however, clinical applications of mid-UV, near-UV, and blue-visible radiations with and without added sensitizers have greatly expanded, concomitant with an increased concern in protection against harmful aspects of these radiations.

Some applications do not involve an added sensitizer. Treatment of hyperbilirubinemia (jaundice) in infants with a series of exposures to broad-spectrum near-UV–blue-visible radiation for a period of 1–4 days (Lucey *et al.*, 1968; Sisson, 1976) has become a standard practice in many medical centers. Although the exact *in vivo* mechanism of the reduction in bilirubin level is not known, the photoproducts of the *in vivo* reaction apparently have a low toxicity and are readily eliminated. Sisson (1970) observed a deleterious effect of blue light on the eyes of newborn piglets in the range of fluence rates used in the treatment of hyperbilirubinemia. Extensive damage to the retina of the irradiated piglets was evident several weeks after exposure to the light. Because of these findings, the eyes of newborn infants treated for hyperbilirubinemia are protected during exposure to the intense light.

In dentistry, high-intensity near-UV sources are in widespread use to polymerize UV-sensitive compounds for pit sealants, orthodontics, and tooth restoration (Council on Dental Materials and Devices, 1976; Roberts and Moffa, 1973). Possible damage to human oral mucosa produced by these high-irradiance near-UV sources has not been determined. The Nuva-Lite (L. D. Caulk Co., Milford, Delaware), for example, emits approximately 10^3 W m^{-2} of 365-nm radiation at the 4-mm working distance of the device (Birdsell *et al.*, 1977). Precautions were suggested because damage to the cornea and lens of the mouse eye and the possibility of damage to the human eye from exposure to large fluences of near-UV radiation have been well documented (Zigman, 1971; Zigman *et al.*, 1973, 1974; Grover and Zigman, 1972).

On the basis of observations that near-UV radiation can induce opacity in eye lenses, Zigman (1971) and Zigman *et al.* (1974) proposed that cataract of the eye lens can be a consequence of long term exposure to the near-UV component of natural sunlight. Zigman *et al.* (1973) also presented evidence that photoproducts of tryptophan may be involved in this process (see Section 2.5).

Some medical procedures utilize a sensitizer in combination with near-UV radiation or other wavelengths. Vitiligo has been treated for some time with a psoralen, applied locally, followed by exposure to natural sunlight (Kelly and Pinkus, 1955), or more recently, with a broad-spectrum near-UV source such as a bank of fluorescent blacklight (BLB) lamps with the psoralen given orally (Lerner *et al.*, 1969). At the doses employed (20–50 mg/kg), the oral application of psoralens is reported not to be carcinogenic (Pathak *et al.*, 1975).

Psoriasis, a common skin disease, has been refractory to the traditional therapeutic procedures. However, very recently, considerable success has been achieved in the treatment of this skin disorder by orally administered 8-methoxypsoralen (methoxalen) followed by exposure to near-UV radiation (320–390 nm) at a high fluence rate (56 W m^{-2}). In comparison, irradiation with a mid-UV–near-UV source (275–390 nm with a peak emission at 313 nm), produced much less improvement, even though the absorption peak for 8-methoxyposoralen is 304 nm. The initial treatments of psoriasis with 8-methoxypsoralen and near-UV radiation made use of topical application of the psoralen (Walter and Voorhees, 1973; Weber, 1974). However, the oral administering of psoralens may be of great importance (Parrish *et al.*, 1974), as skin cancer in mice can be produced with high efficiency with topical application of 8-methoxypsoralen followed by near-UV radiation (Hakim *et al.*, 1960; Ley *et al.*, 1977; Grube *et al.*, 1977). The differences in the number and kind of lesions formed and their repair between the oral and intraperitoneal application of psoralens are unknown.

Recent reports suggest that recurrent lesions of herpes simplex virus, both type 1 that occurs in the area of the mouth and type 2 that affects the genital region, can be successfully treated with photoactive dyes, such as neutral red or proflavine, applied locally followed by irradiation with visible light (Moore *et al.*, 1972; Felber *et al.*, 1973). In some cases, neutral red was combined with near-UV radiation, wavelengths not strongly absorbed by the dye (Rapp *et al.*, 1973). In this latter case, the dye may inhibit the repair of lesions produced in the viral nucleic acid by the near-UV radiation (see Section 2.3).

6.3. Phototoxicity and Protection Against Photosensitivity

In addition to the psoralens or dyes that are used because of their photoactive characteristics, the photosensitizing properties of a variety of drugs widely used in medicine are now recognized (for a discussion of phototoxic therapeutic agents, see review by Epstein, 1974). These photosensitizing drugs include sulfanilamide, tetracyclines, antihistamines, quinacrine, thiazide diuretics, phenathiazines, and halogenated salicylonides. The action spectrum of most of these agents fall into the near-UV range (Harber and Baer, 1972). Adequate studies of the consequences of extensive exposure to direct sunlight of patients under treatment with phototoxic agents have not been made.

Erythropoietic protoporphyria, a disease characterized by high levels of protoporphyrin in the blood and skin, results in a high sensitivity to near-UV and blue wavelengths (380–500 nm) (Magnus, 1968). Earlier observations that carotenoid pigments, when present, protect bacteria against the lethal effects of near-UV and violet radiation (Section 2.6) (Mathews and Sistrom, 1960), and that β-carotene could protect mice against lethal photosensitization by hematoporphyrin (Mathews, 1964*b*), were the basis of a clinical test on the effect of β-carotene on photosensitivity associated with high levels of protoporphyrin in the skin. These tests resulted in the successful amelioration of photosensitivity that accompanies erythropoietic protoporphyria by the oral intake of 180 mg per day of β-carotene (Mathews-Roth *et al.*, 1970, 1974*b*). The strong protective effect in many cases resulted in resistance to natural sunlight that approached the characteristics of normal subjects.

The colorless, UV-absorbing carotenoid phytoene was found to strongly protect guinea pigs against erythema from mid-UV irradiation (290–320 nm) (Mathews-Roth and Pathak, 1975). Since this carotenoid is widely distributed in edible plants, the possibility exists for the use of orally administered phytoene in the protection of human skin against sunburn.

7. SUMMARY

Previously, the injurious part of sunlight was assumed to be limited to wavelengths between 290 and 320 nm. Wavelengths shorter than 290 nm are effectively removed by the ozone layer in the upper atmosphere. Recently, it has been demonstrated that cellular DNA and repair systems in bacteria can be damaged by near-UV radiation (320–400 nm). Figure 30 presents a diagram of the lesions and interactions that appear to lead to lethality of *E. coli* by 365-nm radiation. Pyrimidine dimers and single-strand breaks (or alkali-labile bonds) are produced in bacterial DNA at approximately equal rates by 365-nm radiation. Dimers induced by 365-nm radiation at high fluence rates clearly account for most of the lethal effects in radiation-sensitive mutants (*E. coli* strains K12 AB2480 *uvrA recA* and B_{s-1} *uvrB lexA*) as shown by the large enzymatic photoreactivation (PR) sector. However, no PR of lethality has been demonstrated in wild-type and *uvrA* strains after aerobic near-UV inactivation. It is concluded that pyrimidine dimers, although present, are not significant lethal lesions for aerobic 365-nm inactivation of *rec*$^+$ strains after moderate fluences, because of their efficient repair. At very high fluences of 365-nm radiation capable of inhibiting dark repair and the PR enzyme, dimers may contribute to the lethality of *rec*$^+$ strains.

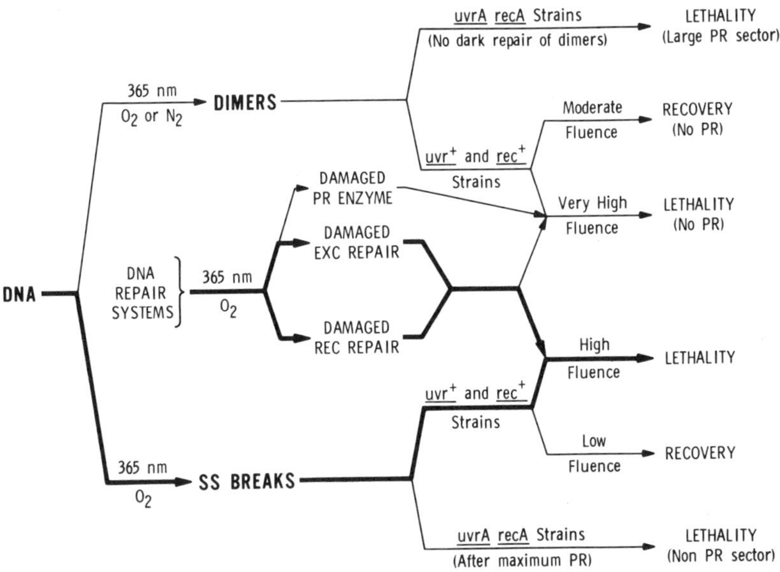

Fig. 30. Proposed pathways for the lethal effects of 365-nm radiation in stationary- or exponential-phase cells of *E. coli*. Single-strand (SS) breaks may be alkali-labile bonds or some other *oxygen-dependent* DNA photoproduct. For *E. coli*, low fluences are less than 1×10^6 J m^{-2}, moderate fluences are $1–3 \times 10^6$ J m^{-2}, high fluences are greater than 3×10^6 J m^{-2}, and very high fluences are greater than 5×10^6 J m^{-2}. PR, photoreactivation; EXC, excision repair; REC, postreplication repair; DIMERS, cyclobutane-type pyrimidine dimers; *uvrA*, excision-repair-deficient strain; *recA*, postreplication repair-deficient strain. The processes depicted here are discussed in Sections 2.4 and 2.9.

Single-strand breaks, which, like near-UV-induced lethality, are produced through strongly oxygen-dependent processes, are the strongest candidates for lethal DNA lesions. However, the possibility that some other oxygen-dependent photoproduct in DNA is the major lethal lesion has not been ruled out.

Near-UV radiation also damages both excision and postreplication repair systems, which enhances the lethal effects of DNA photoproducts produced by far-UV and X radiation in strains capable of dark repair. It is proposed that damage to the excision and postreplication repair systems increases the lethal effects of the DNA lesions produced by near-UV radiation.

Near-UV radiation also causes metabolic damage, including inhibition of respiration, membrane transport, and a variety of enzymes. However, these kinds of damage, except for repair enzymes, contribute little to lethality under the usual growth conditions. Furthermore, membrane damage does not appear to be lethal in most bacteria, although the site of lethal action in carotenoid-containing nonphotosynthetic cells clearly involves the cell membrane in some way. Carotenoids protect against near-UV radiation, showing protection in *S. lutea* at wavelengths between 300 nm and 415 nm only. Carotenoid protection is much greater with broad-spectrum near-UV sources than with monochromatic radiation.

The sensitivity of transforming DNA to near-UV radiation is 10–100 times too great to be accounted for by the formation of pyrimidine dimers. Furthermore, the sensitivity (which is oxygen dependent in *B. subtilis* DNA) is much too great to be accounted for by the direct absorption by DNA, suggesting the presence of chromophores bound tightly to the DNA.

Mutagenesis to phage T5 resistance at low fluence rates in continuous cultures shows a broad peak between 350 and 480 nm, with significant minima and 340 and 400 nm. This mutagenesis is oxygen dependent, indicating a photodynamic process mediated by an endogenous sensitizer. Although the absorption spectrum of riboflavin is consistent with the action spectrum for mutagenesis (Fig. 24), other evidence for this role is lacking.

Near-UV-induced mutagenesis at high, acute fluence rates shows a small complex oxygen dependence. The mutational lesions at 365 nm are subject to almost complete photoreactivation, clearly implicating pyrimidine dimers. However, interaction with other lesions likely occurs.

In partially dehydrated bacteria, both lethality and mutagenesis are induced much more efficiently than in fully hydrated system. In these dehydrated bacteria, unique patterns of mutation induction occur.

Striking synergism occurs between near-UV and mid-UV wavelengths in *B. subtilis* DNA. Synergism of various types occurs between mid-UV and near-UV wavelengths in exponential-phase *recA* strains of bacteria, in repair-proficient stationary-phase strains, and in carotenoid-containing *S. lutea*.

The high efficiency for the induction of genetic damage in bacteria relative to the expected absorbance of DNA at wavelengths between 320 and 400 nm clearly implicates indirect photosensitization. Furthermore, oxygen dependence for near-UV-induced effects in bacteria, bacterial DNA, and mammalian cells suggests that photodynamic reactions mediated by as yet unidentified endogenous chromophores are widespread natural processes.

ACKNOWLEDGMENTS

I wish to thank Drs. B. Hass, R. E. Krisch, H. E. Kubitschek, R. D. Ley, M. J. Peak, R. M. Tyrrell, G. H. Yoakum, and M. S. Brown for reading the manuscript and making helpful suggestions. I also wish to thank Drs. P. A. Cerutti, R. D. Ley, M. J. Peak, J. G. Peak, R. M. Tyrrell, G. H. Yoakum, and M. S. Brown for permission to use their unpublished work.

8. REFERENCES

Alper, T., 1963, Effects on irradiated micro-organisms of growth in the presence of acriflavine, *Nature (London)* **200**:534–536.

Alper, T., and Hodgkins, B., 1969, "Excision repair" and dose-modification: Questions raised by radiobiological experiments with acriflavine, *Mutat. Res.* **8**:15–23.

Alper, T., Forage, A. J., and Hodgkins, B., 1972, Protection of normal, lysogenic and pyocinogenic strains against ultraviolet radiation by bound acriflavine, *J. Bacteriol.* **110**:823–830.

Ananthaswamy, H. N., and Eisenstark, A., 1976, Near-UV-induced breaks in phage DNA sensitization by hydrogen peroxide (a tryptophan photoproduct), *Photochem. Photobiol.* **24**:439–442.

Ananthaswamy, H. N., and Eisenstark, A., 1977, Repair of hydrogen peroxide-induced single-strand breaks in *Escherichia coli* DNA, *J. Bacteriol.* **130**:187–191.

Anderson, T. F., 1948, The growth of T2 virus on ultra-violet killed host cells, *J. Bacteriol.* **56**:403–410.

Ashwood-Smith, M. J., Copeland, J., and Wilcockson, J., 1967, Sunlight and frozen bacteria, *Nature (London)* **214**:33–35.

Baptist, J. E., and Haynes, R. H., 1972, The UV-X-ray synergism in *E. coli*. I. Inhibition by the incorporation of 5-bromouracil and by purine starvation, *Photochem. Photobiol.* **16**:459–464.

Barran, L. R., Dooust, J. Y., Labelle, J. L., Martin, W. C., and Schneider, H., 1974, Differential effects of visible light on active transport in *E. coli, Biochem. Biophys. Res. Commun.* **56**:522–528.

Beukers, R., and Berends, W., 1960, Isolation and identification of the irradiation product of thymine, *Biochim. Biophys. Acta* **41**:550–551.

Birdsell, D. C., Bannon, P. J., and Webb, R. B., 1977, Harmful effects of near-ultraviolet radiation (365 nm) used for polymerizing a sealant and a composite resin, *J. Am. Dent. Assoc.* **94**:311–314.

Blanc, P. L., Tuveson, R. W., and Sargent, M. L., 1976, Inactivation of carotenoid-producing and albino strains of *Neurospora crassa* by visible light, blacklight, and ultraviolet radiation, *J. Bacteriol.* **125**:616–625.

Blum, H. F., 1941, *Photodynamic Action and Diseases Caused by Light,* Reinhold Publishing Co., New York.

Blum, H. F., 1943, Wavelength dependence of tumor induction by ultraviolet radiation, *J. Natl. Cancer Inst.* **3**:533–537.

Blum, H. F., 1959, *Carcinogenesis by Ultraviolet Light,* Princeton University Press, Princeton, New Jersey.

Bodmer, W. F., and Ganeson, A. T., 1964, Biochemical and genetic studies of integration and recombination in *Bacillus subtilis* transformation, *Genetics* **50**:717–738.

Boyce, R., and Setlow, R., 1963, The action spectra for ultraviolet-light inactivation of systems containing 5-bromouracil-substituted deoxyribonucleic acid, *Biochim. Biophys. Acta* **68**:446–454.

Bragg, P. D., 1971, Effect of near-ultraviolet light on the respiratory chain of *Escherichia coli, Can. J. Biochem.* **49**:492–495.

Brent, T. P., 1972, Repair enzymes suggested by mammalian endonuclease activity specific for ultraviolet-irradiated DNA, *Nature (London) New Biol.* **239:**172-173.

Bresler, S. E., Kalinin, V. L., and Perumov, D. A., 1964, Theory of inactivation and reactivation of transforming DNA, *Biopolymers* **2:**135-146.

Bresler, S. E., Kalinin, V. L., and Perumov, D. A., 1967, Inactivation and mutagenesis of DNA. I. Theory of inactivation of transforming DNA, *Mutat. Res.* **4:**389-398.

Brown, M. S., 1977, Biological evidence for the destruction of the photoreactivation enzyme by 365 nm radiation *Mutant. Res.* (submitted).

Brown, M. S., and Webb, R. B., 1972, Photoreactivation of 365 nm inactivation of *Escherichia coli, Mutat. Res.* **15:**348-352.

Brown, M. S., and Webb, R. B., 1977, Comparison of photoreactivation after inactivation of *E. coli* K12 AB2480 by 254 and 365 nm radiations (in preparation).

Bruce, A. K., 1958, Response of potassium retentivity and survival of yeast to far ultraviolet, near ultraviolet, visible, and X-radiation, *J. Gen. Physiol.* **41:**693-702.

Buchbinder, L., Solomey, M., and Phelps, E. B., 1941, Studies on microorganisms in simulated room environments. III. The survival rates of streptococci in the presence of natural daylight and sunlight, and artificial illumination, *J. Bacteriol.* **42:**353-366.

Burchard, R. P., and Dworkin, M., 1966, Light-induced lysis and carotenogenesis in *Myxococcus xanthus, J. Bacteriol.* **91:**535-545.

Burchard, R. P., Gordon, S. A., and Dworkin, M., 1966, Action spectrum for the photolysis of *Myxococcus xanthus, J. Bacteriol.* **91:**896-897.

Cabrera-Juarez, E., 1964, "Black light" inactivation of transforming deoxyribonucleic acid from *Haemophilus influenzae, J. Bacteriol.* **87:**771-778.

Cabrera-Juarez, E., and Espinosa-Lara, M., 1974, Lethal and mutagenic action of black light (325-400 nm) on *Haemophilus influenzae* in the presence of air, *J. Bacteriol.* **117:**960-964.

Cabrera-Juarez, E., and Swenson, P. A., 1975, Action spectrum for the oxygen independent inactivation of *Haemophilus influenzae* transforming DNA with near-UV light, *Photochem. Photobiol.* **21:**193-196.

Cabrera-Juarez, E., Setlow, J. K., Swenson, P. A., and Peak, M. J., 1976, Oxygen-independent inactivation of *Haemophilus influenzae* transforming DNA by monochromatic radiation: Action spectrum, effect of histidine and repair, *Photochem. Photobiol.* **23:**309-314.

Caldwell, M. M., 1971, Solar UV irradiation and growth and development of higher plants, in: *Photophysiology,* Vol. 6 (A. C. Giese, ed.), pp. 131-177, Academic Press, New York.

Claes, H., 1960, Interaction between chlorophyll and carotenes with different chromophoric groups, *Biochem. Biophys. Res. Commun.* **3:**585-590.

Claes, H., 1961, Energieiibertrangung non angeretem Chlorophyll auf C_{40}-Polyene mit verschiedenen Chromophoren Gruppen, *Naturforsch. Teil* **16B:**445-454.

Codd, G. A., 1972, The photoinhibition of malate dehydrogenase, *FEBS Lett.* **20:**211-214.

Coetzee, W. F., and Pollard, E. C., 1974, Near-UV effects on the induction of prophage, *Radiat. Res.* **57:**319-331.

Coetzee, W. F., and Pollard, E. C., 1975, Near ultraviolet inactivation studies on *Escherichia coli* tryptophanase and tryptophan synthetase, *Photochem. Photobiol.* **22:**29-32.

Cole, R. S., and Sinden, R. R., 1975, Psoralen cross-links in DNA: Biological consequences and cellular repair, in: *Radiation Research: Biochemical, Chemical, and Physical Perspectives* (O. F. Nygaard, H. I. Adler, and W. K. Sinclair, eds.), pp. 582-589, Academic Press, New York.

Council on Dental Materials and Devices, 1976, Guidelines on the use of ultraviolet radiation in dentistry, Reports of Councils and Bureaus, *J. Am. Dent. Assoc.* **92:**775-776.

Crounse, J. B., Feldman, R. P., and Clayton, R. K., 1963, Accumulation of polyene precursors of neurosporene in mutant strains of *Rhodopseudomonas spheroides, Nature (London)* **198:**1227-1228.

Cutchis, P., 1974, Stratospheric ozone depletion and solar ultraviolet radiation on earth, *Science* **184**:13–19.

Daniels, F., 1964, Sun exposure and skin aging, *N.Y. State J. Med.* **64**:2066–2069.

Daniels, F., 1974, Physiological and pathological extracutaneous effects of light on man and mammals, not mediated by pineal or other neuroendocrine mechanisms, in: *Sunlight and Man* (T. B. Fitzpatrick, M. A. Pathak, L. C. Harber, M. Seiji, and A. Kukita, eds.), pp. 247–258, University of Tokyo Press, Tokyo.

Danpure, H. J., and Tyrrell, R. M., 1975, The action of near UV (365 nm) radiation on two mammalian cell lines irradiated in inorganic buffer, *Abst. Am. Soc. Photobiol.*, pp. 93–94.

Danpure, H. J., and Tyrrell, R. M., 1976, Oxygen-dependence of near UV (365 nm) lethality and the interaction of near UV and X-rays in two mammalian cell lines, *Photochem. Photobiol.* **23**:171–177.

Davidoff-Abelson and Dubnau, D., 1973, Conditions affecting the isolation from transformed cells of *Bacillus subtilis* of high-molecular-weight single-stranded deoxyribonucleic acid of donor origin, *J. Bacteriol.* **116**:146–153.

Davies, D. R., Arlett, C. F., Munson, R. J., and Bridges, B. A., 1967, Interaction between ultraviolet light and γ-radiation damage in the induction of mutants of *Escherichia coli*: The response in strains with normal and reduced ability to repair ultraviolet damage, *J. Gen. Microbiol.* **46**:329–338.

Day, R. S., and Deering, R. A., 1968, Recovery of colony-forming ability and genetic marker activity by UV-damaged *Hemophilus influenzae, Biophys. J.* **8**:1119–1130.

Day, R. S., and Muel, B., 1974, Ultraviolet inactivation of the ability of *E. coli* to support the growth of phage T7: An action spectrum, *Photochem. Photobiol.* **20**:95–102.

Deering, R. A., and Setlow, R. B., 1963, Effects of ultraviolet light on thymidine dinucleotide and polynucleotide, *Biochim. Biophys. Acta* **63**:526–534.

Denniston, K. J., Webb, R. B., and Brown, M. S., 1972, Action spectrum for carotenoid protection against lethal photo-oxidation of *Sarcina lutea, Abst. Am. Soc. Microbiol.*, p. 184.

Doyle, R. J., and Kubitschek, H. E., 1976, Near ultraviolet light inactivation of an energy-independent membrane transport system in *Saccharomyces cerevisiae, Photochem. Photobiol.*, **24**:291–293.

Dubnau, D., and Cirigliano, C., 1972, Fate of transforming DNA following uptake by competent *Bacillus subtilis*. III. Formation and properties of products isolated from transformed cells which are derived entirely from donor DNA, *J. Mol. Biol.* **64**:9–29.

Duggar, B. M., 1936, Effects of radiation on bacteria, in: *Biological Effects of Radiation* (B. M. Duggar, ed.), pp. 1119–1149, McGraw-Hill Book Co., New York.

Dulbecco, R., and Weigle, J. J., 1952, Inhibition of bacteriophage development in bacteria illuminated with visible light, *Experientia* **8**:386–389.

Durham, N. M., and Wyss, O., 1956, An example of non-inherited radiation resistance, *J. Bacteriol.* **72**:95–100.

Dworkin, M., 1958, Endogenous photosensitization in a carotenoidless mutant of *Rhodopseudomonas speroides, J. Gen. Physiol.* **41**:1099–1112.

Eisenstark, A., 1970, Sensitivity of *Salmonella typhimurium* recombinationless (*rec*) mutants to visible and near-visible light, *Mutat. Res.* **10**:1–6.

Eisenstark, A., 1971, Mutagenic and lethal effects of visible and near-ultraviolet light on bacterial cells, in: *Advances in Genetics*, Vol. 12 (E. W. Caspari, ed.), pp. 167–198, Academic Press, New York.

Eisenstark, A., 1973, Tryptophan photoproduct as a genetic probe: Effects on bacteria, in: *Fifth Stadler Genetics Symposium* (G. Kimber and G. P. R. Redei, eds.), pp. 49–60, University of Missouri, Columbia.

Eisenstark, A., and Ruff, D., 1970, Repair in phage and bacteria inactivated by light from fluorescent and photo lamps, *Biochem. Biophys. Res. Commun.* **38**:244–248.

Elkind, M. M., and Sutton, H., 1957a, Lethal effect of visible light on a mutant strain of haploid yeast. I. General dependencies, *Arch. Biochem. Biophys.* **72**:84–95.

Elkind, M. M., and Sutton, H., 1957b, Lethal effect of visible light on a mutant strain of haploid yeast. II. Absorption and action spectrum, *Arch. Biochem. Biophys.* **72**:96–111.

Epstein, J. H., 1970, Ultraviolet carcinogenesis, in: *Photophysiology,* Vol. 5 (A. C. Giese, ed.), pp. 235–273, Academic Press, New York.

Epstein, J. H., 1974, Phototoxicity and photoallergy: Clinical syndromes, in: *Sunlight and Man* (T. B. Fitzpatrick, M. A. Pathak, L. C. Harber, M. Seiji, and A. Kukita, eds.), pp. 459–477, University of Tokyo Press, Tokyo.

Felber, T. D., Smith, E. B., Knox, J. M., Wallis, C., and Melnick, J. L., 1973, Photodynamic inactivation of herpes simplex, *J. Am. Med. Assoc.* **223**:289–292.

Ferron, W. L., Eisenstark, A., and Mackay, D., 1972, Distinction between far- and near-ultraviolet light killing of recombinationless (*recA*) *Salmonella typhimurium, Biochim. Biophys. Acta* **277**:651–658.

Fong, F., Peters, J., Pauling, C., and Heath, R. L., 1975, Two mechanisms of near-ultraviolet lethality in *Saccharomyces cerevisiae.* Respiratory capacity-dependent and an irreversible inactivation, *Biochim. Biophys. Acta* **387**:451–460.

Foote, C. S., 1976, Photosensitized oxidation and singlet oxygen: Consequences in biological systems, in: *Free Radicals in Biology* Vol. 2 (W. A. Pryor, ed.), pp. 85–124, Academic Press, New York.

Foote, C. S., and Denny, R. W., 1968, Chemistry of singlet oxygen. VII. Quenching by β-carotene, *J. Am. Chem. Soc.* **90**:6233–6235.

Foote, C. S., Chang, Y. C., and Denny, R. W., 1970, Chemistry of singlet oxygen. X. Carotenoid quenching parallels biological protection, *J. Am. Chem. Soc.* **92**:5216–5218.

Forbes, P. D., 1973, Influence of long wave UV on photocarcinogenesis, *Abst. Am. Soc. Photobiol.,* p. 136.

Forbes. P. D., Davies, R. E., D'Aloisio, L. C., and Cole, C., 1976, Emission spectrum differences in fluorescent blacklight lamps. *Photochem. Photobiol.* **23**:613–615.

Freeman, R. G., 1975, Data on the action spectrum for ultraviolet carcinogenesis, *J. Natl. Cancer Inst.* **55**:1119–1121.

Freeman, R. G., Owens, D. W., and Knox, J. M., 1966, Relative energy requirements for an erythemal response of skin to monochromatic wavelengths of ultraviolet present in the solar spectrum, *J. Invest. Dermatol.* **47**:586–592.

Freese, E. B., Gerson, J., Taber, H., Rhaese, H.-J., and Freese, E., 1967, Inactivating DNA alterations induced by peroxides and peroxide-producing agents, *Mutat. Res.* **4**:517–531.

Fuks, Z.. and Smith, K. C., 1971, Effect of quinacrine on X-ray sensitivity and repair of damaged DNA in *Escherichia coli* K-12, *Radiat. Res.* **48**:63–73.

Ganesan, A. K., and Smith, K. C., 1972, Requirement for protein synthesis in *rec*-dependent repair of deoxyribonucleic acid in *Escherichia coli* after ultraviolet or x irradiation, *J. Bacteriol.* **111**:575–585.

Gates, F. L., 1930, A study of the bactericidal action of ultraviolet light. III. The absorption of ultraviolet light by bacteria, *J. Gen. Physiol.* **14**:31–42.

Giese, A. C., 1946, An intracellular photodynamic sensitizer in *Blepharisma, J. Cell. Comp. Physiol.* **28**:119–127.

Giese, A. C., 1953, Some properties of a phytodynamic pigment from *Blepharisma, J. Gen. Physiol.* **37**:259–269.

Giese, A. C., 1968, Ultraviolet action spectra in perspective: With special reference to mutation, *Photochem. Photobiol.* **8**:527–546.

Giese, A. C., 1971, Photosensitization by natural pigments, in: *Photophysiology*, Vol. 6 (A. C. Giese, ed.), pp. 77–129, Academic Press, New York.

Giese, A. C., 1976, *Living with Our Sun's Ultraviolet Rays*, Plenum Press, New York.

Ginsberg, D. M., and Jagger, J., 1965, Evidence that initial ultraviolet lethal damage in *Escherichia coli* is independent of growth phase, *J. Gen. Microbiol.* **40**:171–184.

Glatzer, L., 1977, Radioactive near-ultraviolet photoproducts of L-tryptophan solutions (in preparation).

Grover, D., and Zigman, S., 1972, Coloration of human lenses by near-UV photooxidized tryptophan, *Exp. Eye Res.* **13**:70–76.

Grube, D. D., Ley, R. D., and Fry, R. J. M., 1977, Photosensitizing effects of 8-methoxypsoralen on the skin of hairless mice. II. Strain and spectral difference for tumorigenesis, *Photochem. Photobiol.* **25**:271–278.

Hakim, R. E., Griffin, A. C., and Knox, J. M., 1960, Erythema and tumor formation in methoxsalen treated mice exposed to fluorescent light, *Arch. Dermatol.* **82**:572–577.

Hanawalt, P. C., 1966, The UV sensitivity of bacteria: Its relationship to the replication cycle, *Photochem. Photobiol.* **5**:1–12.

Harber, L. C., and Baer, R. L., 1972, Pathogenic mechanisms of drug-induced photosensitivity, *J. Invest. Dermatol.* **58**:327–342.

Hariharan, P. V. and Cerutti, P. A., 1974, The incision and strand rejoining step in the excision repair of 5,6-dihydroxy-dihydrothymine by crude *E. coli* extracts, *Biochem. Biophys. Res. Commun.* **61**:375–379.

Hariharan, P. V., and Cerutti, P. A., 1976, Excision of ultraviolet and gamma ray products of the 5,6-dihydroxydihydrothymine-type by nuclear preparations of xeroderma pigmentosum cells, *Biochem. Biophys. Acta* **477**:375–378.

Hariharan, P. V., and Cerutti, P A., 1977, Formation of products of the 5,6-dihydroxy-dihydrothymine type by ultraviolet light in HeLa cells, *Biochemistry* **16**:2791–2795.

Hariharan, P. V., Achey, P. M., and Cerutti, P. A., 1977, Biological effect of thymine ring-saturation in coliphage ϕX174-DNA, *Radiat. Res.* **69**:375–378.

Harm, W., 1966, Repair effects in phage and bacteria exposed to sunlight, *Radiat. Res. Suppl.* **6**:215–216.

Harm, W., 1967, Differential effects of acriflavine and caffeine on various ultraviolet-irradiated *Escherichia coli* strains and T1 phage, *Mutat. Res.* **4**:93–110.

Harm, W., 1969, Biological determination of the germicidal activity of sunlight, *Radiat. Res.* **40**:63–69.

Harrison, A. P., 1967, Survival of bacteria; harmful effects of light, with some comparisons with other adverse physical agents, *Annu. Rev. Microbiol.* **21**:143–156.

Hausser, K. W., and Vahle, W., 1927, Sunburn and suntanning, *Wiss. Veroeff. Siemens Konzern* **6**:101–120.

Haynes, R. H., 1964, Molecular localization of radiation damage relevant to bacterial inactivation, in: *Physical Processes in Radiation Biology* (L. Augenstein, R. Mason, and B. Rosenberg, eds.), pp. 51–72, Academic Press, New York.

Haynes, R. H., 1966, The interpretation of microbial inactivation and recovery phenomena, *Radiat. Res. Suppl* **6**:1–29.

Haynes, R. H., 1975, The influence of repair processes on radiobiological survival curves, in: *Cell Survival after Low Doses of Radiation: Theoretical and Clinical Implications* (T. Alper, ed.), pp. 197–208, Wiley, New York.

Hill, R. F., 1956, Effects of illumination on plaque formation by *Escherichia coli* infected with T1 bacteriophage, *J. Bacteriol.* **71**:231–235.

Hollaender, A., 1943, Effect of long ultraviolet and short visible radiation (3500–4900 Å) on *Escherichia coli*, *J. Bacteriol.* **46**:531–541.

Hollaender, A., and Emmons, C. W., 1946, Induced mutations and speciation in fungi, *Cold Spring Harbor Symp. Quant. Biol.* **11**:78–84.

Howard-Flanders, P., and Boyce, R. P., 1966, DNA repair and genetic recombination: Studies on mutants of *Escherichia coli* defective in these processes, *Radiat. Res. Suppl.* **6**:156–184.

Hutchinson, F., and Hales, H. B., 1970, Mechanism of the sensitization of bacterial transforming DNA to ultraviolet light to the incorporation of 5-bromouracil, *J. Mol. Biol.* **50**:59–69.

Igali, S., Bridges, B. A., Ashwood-Smith, M. J., and Scott, B. R., 1970, Mutagenesis in *Escherichia coli.* IV. Photosensitization to near ultraviolet light by 8-methoxypsoralen, *Mutat. Res.* **9**:21–30.

Jagger, J., 1964, Photoprotection from far ultraviolet effects in cells, in: *Advances in Chemical Physics,* Vol. 7 (J. Duchesne, ed.), pp. 584–601, Interscience, New York.

Jagger, J., 1967, *Introduction to Research in Ultraviolet Photobiology,* Prentice-Hall, Englewood Cliffs, N.J.

Jagger, J., 1972, Growth delay and photoprotection induced by near-ultraviolet light, in: *Research Progress in Organic, Biological and Medicinal Chemistry,* Vol. 3 (U. Gallo and L. Santamaria, eds.), Part 1, pp. 383–401, American Elsevier, New York.

Jagger, J., 1973, Realm of the ultraviolet, *Photochem. Photobiol.* **18**:353–354.

Jagger, J., 1975, Inhibition by sunlight of the growth of *Escherichia coli* B/r, *Photochem. Photobiol.* **22**:67–70.

Jagger, J., and Stafford, R. S., 1962, Biological and physical ranges of photoprotection from ultraviolet damage in microorganisms. *Photochem. Photobiol.* **1**:245–257.

Jagger, J., Wise, W. C., and Stafford, R. S., 1964, Delay in growth and division induced by near ultraviolet radiation in *Escherichia coli* B and its role in photoprotection and liquid holding recovery, *Photochem. Photobiol.* **3**:11–24.

Jagger, J., Stafford, R. S., and Snow, J. M., 1969, Thymine-dimer and action-spectrum evidence for indirect photoreactivation in *Escherichia coli, Photochem. Photobiol.* **10**:383–396.

Kaplan, R. W., 1956, Dose-effect curves of S-mutation and killing in *Serratia marcescens, Arch. Microbiol.* **24**:60–79.

Kaplan, R. W., and Kaplan, C., 1956, Influence of water content on UV-induced S-mutation and killing in *Serratia, Exp. Cell Res.* **11**:378–392.

Kashket, E. R., and Brodie, A. F., 1962, Effects of near-ultraviolet irradiation on growth and oxidative metabolism of bacteria. *J. Bacteriol.* **83**:1094–1100.

Kelly, E. W., Jr., and Pinkus, H., 1955, Local application of 8-methoxypsoralen in vitiligo, *J. Invest. Dermatol.* **25**:453–456.

Kelner, A., and Halle, S., 1960, Mutagenesis by visible light in a mutable strain of *Escherichia coli, Bacteriol. Proc.,* p. 67.

Knowles, A., 1975, The effects of photodynamic action involving oxygen upon biological systems, in: *Radiation Research: Biochemical, Chemical, and Physical Perspectives* (O. F. Nygaard, H. I. Adler, and W. K. Sinclair, eds.), pp. 612–622, Academic Press, New York.

Koch, A. L., Doyle, R. J., and Kubitschek, H. E., 1976, Inactivation of membrane transport in *Escherichia coli* by black light, *J. Bacteriol.* **126**:140–146.

Koller, L. R., 1939, Bactericidal effects of ultraviolet radiation produced by low pressure mercury vapour lamps, *J. Appl. Phys.* **10**:621–630.

Krinsky, N. I., 1976, Cellular damage initiated by visible light, in: *The Survival of Vegetative Microbes* (T. R. G. Gray and J. R. Postgate, eds.), pp. 209–230, Cambridge University Press, Cambridge.

Kubitschek, H. E., 1967, Mutagenesis by near-visible light, *Science* 155:1545–1546.

Kubitschek, H. E., 1970, *Introduction to Research with Continuous Cultures* (A. Hollaender, ed.), Prentice-Hall, Englewood Cliffs, N.J.

Kubitschek, H. E., and Bendigkeit, H. E., 1964a, Mutation in continuous cultures. I. Dependence of mutational response upon growth limiting factors, *Mutat. Res.* 1:113–120.

Kubitschek, H. E., and Bendigkeit, H. E., 1964b, Mutation in continuous cultures. II. Mutations induced with ultraviolet and 2-aminopurine, *Mutat. Res.* 1:209–218.

Kubitschek, H. E., and Doyle, R. J., 1977, Near-UV induced inhibition of succinate transport in *E. coli B/r* (in preparation).

Kubitschek, H. E., Nance, S. L., and Doyle, R. J., 1975, Induction of growth delay by inactivation of membrane transport after exposure to near-UV, *Abst. Am. Soc. Photobiol.*, p. 118.

Leff, J., and Krinsky, N. I., 1967, A mutagenic effect of visible light mediated by endogenous pigments in *Euglena gracilis, Science* 158:1332–1334.

Lerner, A. B., Denton, C. A., and Fitzpatrick, T. B., 1969, Clinical and experimental studies with 8-methoxypsoralen in vitiligo, *Arch. Dermatol.* 100:224–229.

Ley, R. D., Grube, D. D., and Fry, R. J. M., 1977, Photosensitizing effects of 8-methoxypsoralen on the skin of hairless mice. I. Formation of interstrand cross-links in epidermal DNA, *Photochem. Photobiol.* 25:265–270.

Lucey, J. F., Ferreiro, M., and Hewitt, J., 1968, Prevention of hyperbilirubinemia of prematurity by phototherapy, *Pediatrics* 41:1047–1056.

Luckiesh, M., 1946, *Applications of Germicidal, Erythemal, and Infrared Energy,* Van Nostrand, New York.

Luria, S. E., 1955, Radiation and viruses, in: *Radiation Biology,* Vol. 2 (A. Hollaender, ed.), pp. 333–364, McGraw-Hill, New York.

Mackay, D., Eisenstark, A., Webb, R. B., and Brown, M. S., 1976, Action spectra for lethality in recombinationless strains of *Salmonella typhimurium* and *Escherichia coli, Photochem. Photobiol.* 24:337–344.

Magnus, I. A., 1968, Photobiological aspects of porphyria, *Proc. Roy. Soc. Med.* 61:196–198.

Maguire, B. H., 1960, Lethal effect of visible light on cavernicolous astracods, *Science* 132:226–227.

Marmur, J., Anderson, W. F., Mathews, L., Berns, K., Gajewski, E., and Doty, P., 1961, The effects of ultraviolet light on the biological and physical chemical properties of deoxyribonucleic acids, *J. Cell. Comp. Physiol.* 58:33–55 (Suppl. 2).

Martignoni, K. D., and Smith, K. C., 1972, The synergistic interaction of UV and X-radiation in mutants of *E. coli* K12, *Radiat. Res.* 51:487–488.

Martignoni, K. D., and Smith, K. C., 1973, The synergistic action of ultraviolet and X-radiation on mutants of *Escherichia coli* K-12, *Photochem. Photobiol.* 18:1–8.

Mathews, M. M., 1964a, The effect of low temperature on the protection by carotenoids against photosensitization in *Sarcina lutea, Photochem. Photobiol.* 3:75–77.

Mathews, M. M., 1964b, Protective effect of β-carotene against lethal photosensitization by haematoporphyrin, *Nature (London)* 203:1092.

Mathews, M. M., and Krinsky, N. I., 1965, The relationship between carotenoid pigments and resistance to radiation in non-photosynthetic bacteria, *Photochem. Photobiol.* 4:813–817.

Mathews, M. M., and Sistrom, W. R., 1959a, Intracellular location of carotenoid pigments and some respiratory enzymes in *Sarcina lutea, J. Bacteriol.* 78:778–789.

Mathews, M. M., and Sistrom, W. R., 1959b, Function of carotenoid pigments in non-photosynthetic bacteria, *Nature (London)* 184:1892–1893.

Mathews, M. M., and Sistrom, W. R., 1960, The function of the carotenoid pigments of *Sarcina lutea, Arch. Mikrobiol.* 35:139–146.

Mathews-Roth, M., 1967, The photosensitizing ability of prodigiosin, *Photochem. Photobiol.* **6:**923–926.

Mathews-Roth, M. M., and Krinsky, N. I., 1970, Studies on the protective function of the carotenoid pigments of *Sarcina lutea, Photochem. Photobiol.* **11:**419–428.

Mathews-Roth, M. M., and Pathak, M. A., 1975, Phytoene as a protective agent against sunburn (> 280 nm) radiation in guinea pigs, *Photochem. Photobiol.* **21:**261–263.

Mathews-Roth, M. M., Pathak, M. A., Fitzpatrick, T. B., Harber, L. C., and Kass, E. H., 1970, Beta-carotene as a photoprotective agent in erythropoietic protoporphyria, *N. Engl. J. Med.* **282:**1231–1234.

Mathews-Roth, M. M., Wilson, T., Fujimori, E., and Krinsky, N. I., 1974*a*, Carotenoid chromophore length and protection against photosensitization, *Photochem. Photobiol.* **19:**217–222.

Mathews-Roth, M. M., Pathak, M. A., Fitzpatrick, T. B., Harbor, L. C., and Kass, E. H., 1974*b*, Beta-carotene as an oral photoprotective agent in erythropoietic protoporphyria, in: *Sunlight and Man* (T. B. Fitzpatrick, M. A. Pathak, L. C. Harber, M. Seiji, and A. Kukita, eds.), pp. 659–668, University of Tokyo Press, Tokyo.

Mathews-Roth, M. M., Pathak, M. A., Fitzpatrick, T. B., Harber, L. C., and Kass, E. H., 1974*c*, β-Carotene as an oral photoprotective agent in erythropoietic protoporphyria. *J. Am. Med. Assoc.* **228:**1004–1008.

McCormick, J. P., Fischer, J. R., Pochlatko, J. P., and Eisenstark, A., 1975, Characterization of a cell-lethal product from the photooxidation of tryptophan: Hydrogen peroxide, *Science* **191:**468–469.

McGrath, R. A., and Williams, R. W., 1966, Reconstruction *in vivo* of irradiated *Escherichia coli* deoxyribonucleic acid; the rejoining of broken pieces, *Nature (London)* **212:**534–535.

Moore, C., Wallis, C., Melnick, J. L., and Kuns, M. D., 1972, Photodynamic treatment of herpes keratitis, *Infect. Immunol.* **5:**169–171.

Moroson, H., and Alexander, P., 1961, Changes produced by ultraviolet light in the presence and in the absence of oxygen on the physical chemical properties of deoxyribonucleic acid, *Radiat. Res.* **14:**29–49.

Morton, R. A., and Haynes, R. B., 1969, Changes in the ultraviolet sensitivity of *Escherichia coli* during growth in batch cultures, *J. Bacteriol.* **97:**1379–1385.

Musajo, L., and Rodighiero, G., 1972, Mode of photosensitizing action of furocoumarins, in: *Photophysiology,* Vol. 7 (A. C. Giese, ed), pp. 115–147, Academic Press, New York.

Notani, N., and Goodgal, S. H., 1966, On the nature of recombinants formed during transformation in *Haemophilus influenzae, J. Gen. Physiol.* **49:**197–209.

Parrish, J. A., Ying, C. Y., Pathak, M. A., and Fitzpatrick, T. B., 1974, Erythemogenic properties of long-wave ultraviolet light, in: *Sunlight and Man* (T. B. Fitzpatrick, M. A. Pathak, L. G. Harber, M. Seiji, and A. Kukita, eds.), pp. 131–141, University of Tokyo Press, Tokyo.

Pathak, M. A., 1972, Biochemical changes in epidermal nucleic acids following UV irradiation, *Abst. Int. Congr. Photobiol.,* p. 44.

Pathak, M. A., and Stratton, K., 1969, Effects of ultraviolet and visible radiation and production of free radicals in skin, in: *The Biologic Effects of Ultraviolet Radiation* (F. Urbach, ed.), pp. 207–222, Pergamon Press, New York.

Pathak, M. A., Riley, F. C., and Fitzpatrick, T. B., 1962, Melanogensis in human skin following exposure to long-wave ultraviolet light and visible light, *J. Invest. Dermatol.* **39:**435–443.

Pathak, M. A., Kramer, D. M., and Fitzpatrick, T. B., 1975, Photobiology and photochemistry of furocoumarins (psoralens), in: *Sunlight and Man* (T. B. Fitzpatrick, M. A.

Pathak, L. C. Harber, M. Seiji, and A. Kukita, eds.), pp. 335–368, University of Tokyo Press, Tokyo.

Patrick, M., and Rupert, C. S., 1967, The effect of host-cell reactivation on assay of UV-irradiated *Haemophilus influenza* transforming DNA, *Photochem. Photobiol.* **6**:1–20.

Peak, M. J., 1970, Some observations on the lethal effects of near-ultraviolet light on *Escherichia coli,* compared with the lethal effects of far-ultraviolet light, *Photochem. Photobiol.* **12**:1–8.

Peak, M. J., and Peak, J. G., 1973, Protection by histidine against the inactivation of DNA transforming activity by near-ultraviolet light, *Photochem. Photobiol.* **18**:525–527.

Peak, M. J., and Peak, J. G., 1974, Protection of transforming DNA against X-rays by histidine, glycerol, AET and the *uvr⁻* genotype, *Radiat. Res.* **59**:288.

Peak, M. J., and Peak, J. G., 1975, Protection by AET against inactivation of transforming DNA by near-ultraviolet light, action spectrum. *Photochem. Photobiol.* **22**:147–148.

Peak, M. J., Peak, J. G., and Webb, R. B., 1973*a*, Inactivation of transforming DNA by ultraviolet light. I. Near-UV action spectrum for marker inactivation, *Mutat. Res.* **20**:129–135.

Peak, M. J., Peak, J. G., and Webb, R. B., 1973*b*, Inactivation of transforming DNA by ultraviolet light. II. Protection by histidine against near-UV irradiation: Action spectrum. *Mutat. Res.* **20**:137–141.

Peak, M. J., Peak, J. G., and Webb, R. B., 1973*c*, Inactivation of transforming DNA by ultraviolet light. III. Further observations on the effects of 365 nm radiation, *Mutat. Res.* **20**:143–148

Peak, M. J., Peak, J. G., and Webb, R. B., 1975, Synergism between different near-ultraviolet wavelengths in the inactivation of transforming DNA, *Photochem. Photobiol.* **21**:129–131.

Piechowska, M., and Fox, M. S., 1971, Fate of transforming deoxyribonucleate in *Bacillus subtilis, J. Bacteriol.* **108**:680–689.

Pollard, E. C., 1974, Cellular and molecular effects of solar ultraviolet radiation, *Photochem. Photobiol.* **20**:301–308.

Rahn, R. O., 1973, Denaturation in ultraviolet-irradiated DNA, in: *Photophysiology,* Vol. 8 (A. C. Giese, ed.), pp. 231–255, Academic Press, New York.

Rahn, R. O., Landry, L. C., and Carrier, W. L., 1974, Formation of chain breaks and thymine dimers in DNA upon photosensitization at 313 nm with acetophenone, acetone, or benzophenone, *Photochem. Photobiol.* **19**:75–78,

Ramabhadran, T. V., 1975, Effects of near-ultraviolet and violet radiations (313–405 nm) on DNA, RNA, and protein synthesis in *E. coli* B/r: Implications for growth delay, *Photochem. Photobiol.* **22**:117–123.

Ramabhadran, T. V., and Jagger, J., 1975, Evidence against DNA as the target for 334 nm-induced growth delay in *Escherichia coli, Photochem. Photobiol.* **21**:227–233.

Ramabhadran, T. V., and Jagger, J., 1976, Mechanism of growth delay induced in *Escherichia coli* by near ultraviolet radiation, *Proc. Natl. Acad. Sci. USA* **73**:59–63.

Ramabhadran, T. V., Fossum, T., and Jagger, J., 1976, *In vivo* induction of 4-thiouridine-cytidine adducts in tRNA of *E. coli* B/r by near-ultraviolet radiation, *Photochem. Photobiol.* **23**:315–321.

Randolph, M. L., and Setlow, J. K., 1971, Mechanism of inactivation of transforming deoxyribonucleic acid by X rays, *J. Bacteriol.* **106**:221–226.

Rapp, F., Lui-Lien, H. L., and Jerkofsky, M., 1973, Transformation of mammalian cells by DNA containing viruses following photodynamic inactivation, *Virology* **55**:339–346.

Resnick, M. A., 1970, Sunlight-induced killing in *Saccharomyces cerevisiae, Nature (London)* **226**:377–378.

Ritchie, D. A., 1964, Mutagenesis with light and proflavine in phage T4, *Genet. Res. Camb.* **5**:168–169.

Robb, F. T., Hauman, J. H., and Peak, M. J., 1977, Similar spectra for the inactivation by monochromatic light of two distinct leucine transport systems in *Escherichia coli, Photochem. Photobiol.* (in press).

Roberts, M. W., and Moffa, J. P., 1973, Repair of fractured incisal angles with an ultraviolet light-activated fissure sealant and a composite resin: Two-year report of 60 cases, *J. Am. Dent. Assoc.* **75**:121–128.

Rupert, C. S., 1968, Shapes of the UV inactivation curves for single and double linked markers of *Haemophilus influenzae* transforming DNA, *Photochem. Photobiol.* **7**:437–449.

Rupert, C. S., and Goodgal, S. H., 1960, Shape of ultraviolet inactivation curves of transforming deoxyribonucleic acid, *Nature (London)* **185**:556–557.

Rupert, C. S., and Harm, W., 1966, Reactivation after photobiological damage, in: *Advances in Radiation Biology,* Vol. 2 (L. G. Augenstein, R. Mason, and M. Zelle, eds.), pp. 1–81, Academic Press, New York.

Rupp, W. D., and Howard-Flanders, P., 1968, Discontinuities in the DNA synthesized in an excision-defective strain of *Escherichia coli* following ultraviolet irradiation, *J. Mol. Biol.* **31**:291–304.

Scotto, J., Kopf, A. W., and Urbach, F., 1974, Non-melanoma skin cancer among Caucasians in four areas of the United States, *Cancer* **34**:1333–1338.

Setlow, J. K., 1967, The effects of ultraviolet radiation and photoreactivation, in: *Comprehensive Biochemistry, Photobiology, Ionizing Radiations* (M. Florkin and E. H. Stotz, eds.), pp. 157–203, Elsevier, New York.

Setlow, J. K., and Boling, M. E., 1965, The resistance of *Micrococcus radiodurans* to ultraviolet radiation. II. Action spectra for killing, delay in DNA synthesis, and thymine dimerization, *Biochim. Biophys. Acta* **108**:259–265.

Setlow, R. B., 1964, Physical changes and mutagenesis, *J. Cell Comp. Physiol.* **64**:51–68 (Suppl. 1).

Setlow, R. B., 1966, Cyclubutane-type pyrimidine dimers in polynucleotides, *Science* **153**:379–386.

Setlow, R. B., 1968, Photoproducts in DNA irradiated *in vivo, Photochem. Photobiol.* **7**:643–649.

Setlow, R. B., 1974, Wavelengths in sunlight effective in producing skin cancer: A theoretical analysis, *Proc. Natl. Acad. Sci. USA* **71**:3363–3366.

Setlow, R. B., and Carrier, W. L., 1966, Pyrimidine dimers in ultraviolet-irradiated DNA's, *J. Mol. Biol.* **17**:237–254.

Setlow, R. B., and Hart, R. W., 1975, Direct evidence that damaged DNA results in neoplastic transformation—a fish story, in: *Radiation Research, Biochemical, Chemical, and Physical Perspectives* (O. F. Nygaard, H. I. Adler, and W. K. Sinclair, eds.), pp. 879–884, Academic Press, New York.

Setlow, R. B., and Setlow, J. K., 1972, Effects of radiation on polynucleotides, *Ann. Rev. Biophys. Bioeng.* **1**:293–346.

Sisson, T. R. C., 1976, Visible light therapy of neonatal hyperbilirubinemia, in: *Photochemical and Photobiological Reviews,* Vol. 1 (K. C. Smith, ed.), pp. 241–268, Plenum Publishing Co., New York.

Sisson, T. R. C., Glauser, S. C., Glauser, E. M., and Kumabara, T., 1970, Retinal changes produced by phototherapy, *J. Pediatr.* **77**:221–227.

Sistrom, W. R., Griffiths, M., and Stanier, R. Y., 1956, The biology of photosynthetic bacterium which lacks colored carotenoids, *J. Cell. Comp. Physiol.* **48**:473–515.

Smith, K. C., 1971, Roles of genetic recombination and DNA polymerase in repair of damaged DNA, in: *Photophysiology,* Vol. 6 (A. C. Giese, ed.), pp. 209–278, Academic Press, New York.

Smith, K. C., 1974*a*, Molecular changes in nucleic acids produced by ultraviolet and visible radiation in: *Sunlight and Man* (T. B. Fitzpatrick, M. A. Pathak, L. C. Harber, M. Seiji, and A. Kukita, eds.), pp. 57–66, University of Tokyo Press, Tokyo.

Smith, K. C., 1974*b*, Cellular repair and radiation damage, in: *Sunlight and Man* (T. B. Fitzpatrick, M. A. Pathak, L. C. Harber, M. Seiji, and A. Kukita, eds.), pp. 67–77, University of Tokyo Press, Tokyo.

Smith, K. C., and Hanawalt, P. C., 1969, *Molecular Photobiology: Inactivation and Recovery,* Academic Press, New York.

Speck, W. T., and Rosenkranz, H. S., 1975, Base substitution mutations induced in *Salmonella* strains by visible light, *Photochem. Photobiol.* **21:**369–371.

Spikes, J. D., 1968, Photodymatic action, in: *Photophysiology,* Vol. 3 (A. C. Giese, ed.), pp. 33–64, Academic Press, New York.

Spikes, J. D., and Livingston, R., 1969, The molecular biology of photodynamic action, in: *Advances in Radiation Biology,* Vol. 3 (L. G. Augenstein, R. Mason, and M. Zelle, eds.), pp. 29–121, Academic Press, New York.

Sprott, G. D., Dimock, K., Martin, W. G., and Schneider, H., 1975, Coupling of glycine and alanine transport to respiration in cells of *Escherichia coli, Can. J. Biochem.* **53:**262–268.

Stanier, R. Y., 1959, Formation and function of photosynthetic pigment system in purple bacteria, *Brookhaven Symp. Biol.* **11:**43–53.

Stanier, R. Y., and Cohen-Bazire, G., 1957, The role of light in the microbial world: Some facts and speculation, *Symp. Soc. Gen. Microbiol.* **7:**56–89.

Stanier, R. Y., and Straight, R., 1967, Sensitized photochemical processes in biological systems, *Annu. Rev. Phys. Chem.* **18:**405–436.

Stoien, J. D., and Wang, R. J., 1974, Effect of near-ultraviolet and visible light on mammalian cells in culture. II. Formation of toxic photoproducts in tissue culture medium by blacklight, *Proc. Natl. Acad. Sci. USA* **71:**3961–3965.

Stuy, J. H., 1965, Fate of transforming DNA in *Haemophilus influenzae* transforming system, *J. Mol. Biol.* **13:**554–570.

Stuy, J. H., 1974, Acid-soluble breakdown of homologous DNA absorbed by *Haemophilus influenzae*: Its biological significance, *J. Bacteriol.* **120:**917–922.

Swenson, P. A., 1976, Physiological responses of *Escherichia coli* to far-ultraviolet radiation, in: *Photochemical and Photobiological Reviews,* Vol. 1 (K. C. Smith, ed.), pp 269–387, Plenum Publishing Co., New York.

Swenson, P. A., and Schenley, R. L., 1970, Role of pyridine nucleotides in the control of respiration in ultraviolet-irradiated *Escherichia coli* B/r cells, *J. Bacteriol.* **104:**1230–1235.

Swenson, P. A., and Schenley, R. L., 1974, Evidence relating cessation of respiration, cell envelope changes, and death in ultraviolet-irradiated *Escherichia coli* B/r cells, *J. Bacteriol.* **117:**551–559.

Swenson, P. A., and Setlow, R. B., 1970, Inhibition of the induced formation of tryptophanase in *Escherichia coli* by near-ultraviolet radiation, *J. Bacteriol.* **102:**815–819.

Szybalski, W., and Opara-Kubinska, L., 1965, Radiobiological and physiochemical properties of 5-bromodeoxyuridine-labelled transforming DNA as related to the nature of the critical radiosensitive structures, in: *Cellular Radiation Biology,* pp. 223–240, Williams and Wilkins, Baltimore.

Town, C. D., Smith, K. C., and Kaplan, H. S., 1973, Repair of X-ray damage to bacterial DNA, *Curr. Top. Radiat. Res. Q.* **8:**351–399.

Tuveson, R. W., and Satterthwaite, M. A., 1976, Comparison of ultraviolet and blacklight for the induction of nutritional independence at two loci in *Neurospora crassa, Mutat. Res.* **36:**165–170.

Tyrrell, R. M., 1973, Induction of pyrimidine dimers in bacterial DNA by 365 nm radiation *Photochem. Photobiol.* **17:**69–73.

Tyrrell, R. M., 1974, The interaction of near-UV (365 nm) and X-radiations on wild-type and repair-deficient strains of *Escherichia coli* K12: Physical and biological measurements, *Int. J. Radiat. Biol.* **25:**373–390.

Tyrrell, R. M., 1976a, *RecA*$^+$-dependent synergism between 365 nm and ionizing radiation in log-phage *Escherichia coli*: A model for oxygen-dependent near-UV inactivation by disruption of DNA repair, *Photochem. Photobiol.* **23:**13–20.

Tyrrell, R. M., 1976b, Synergistic lethal action of ultraviolet–violet radiations and mild heat in *Escherichia coli, Photochem. Photobiol.* **24:**345–352.

Tyrrell, R. M., and Davies, D. J. G., 1974, The kinetics of photoreactivation in the ultraviolet sensitive mutant *Escherichia coli* K12 AB2480, *Mutat. Res.* **23:**151–161.

Tyrrell, R. M., and Webb, R. B., 1973, Reduced dimer excision following near ultraviolet (365 nm) radiation, *Mutat. Res.* **19:**361–364.

Tyrrell, R. M., Moss, S. H., and Davies, D. J. G., 1972a, The variation in UV sensitivity of four K12 strains of *Escherichia coli* as a function of their stage of growth, *Mutat. Res.* **16:**1–12.

Tyrrell, R. M., Moss, S. H., and Davies, D. J. G., 1972b, The variation in photoreactivating enzyme activity as a function of stage of growth of three K12 strains of *Escherichia coli, Mutat. Res.* **16:**345–352.

Tyrrell, R. M., Webb, R. B., and Brown, M. S., 1973, Destruction of photoreactivating enzyme by 365 nm radiation, *Photochem. Photobiol.* **18:**249–254.

Tyrrell, R. M., Ley, R. D., and Webb, R. B., 1974, Induction of single-strand breaks (alkali-labile bonds) in bacterial and phage DNA by near-UV (365 nm) radiation, *Photochem. Photobiol.* **20:**395–398.

Urbach, F., Epstein, J. H., and Forbes, P. D., 1974, Ultraviolet carcinogenesis: Experimental, global, and genetic aspects, in: *Sunlight and Man* (T. B. Fitzpatrick, M. A. Pathak, L. C. Harber, M. Seiji, and A. Kukita, eds.), pp. 260–283, University of Tokyo Press, Tokyo.

Van der Schueren, E., and Smith, K. C., 1974, Inhibition of the *exrA* gene-dependent branch of the DNA excision repair system in *Escherichia coli* K-12 by 2,4-dinitrophenol, *Photochem. Photobiol.* **19:**95–102.

Van der Schueren, E., Smith, K. C., and Kaplan, H. S., 1973, Modification of DNA repair and survival of X-irradiated *pol, rec,* and *exr* mutants of *Escherichia coli* K-12 by 2,4-dinitrophenol, *Radiat. Res.* **55:**346–355.

Venema, G., Pritchard, R. H., and Venema-Schröder, T., 1965, Fate of transforming deoxyribonucleic acid in *Bacillus subtilis, J. Bacteriol.* **89:**1250–1255.

Wahl, R., 1946, Quelques précisions au sujet de l'action de la lumière sur les bactériophages, *Ann. Inst. Pasteur* **72:**284–286.

Wahl, R., and Latarjet, R., 1947, Near-UV inactivation of several bacteriophages maximum in the near UV and violet regions, *Ann. Inst. Pasteur* **73:**957–971.

Walter, J. F., and Voorhees, J. J., 1973, Psoriasis improved by psoralen plus black light, *Acta Derm. Venereol. (Stockholm)* **53:**469–472.

Wang, R. J., Stoien, J. D., and Landa, F., 1974, Lethal effect of near-ultraviolet irradiation on mammalian cells in culture, *Nature (London)* **247:**43–45.

Webb, R. B., 1972, Photodynamic lethality and mutagenesis in the absence of added sensitizers, in: *Organic, Biological and Medicinal Chemistry*, Vol. 3 (U. Galo and L. Santamaria, eds.), Part 2, pp. 511–530, American Elsevier, New York.

Webb, R. B., 1977*a*, Near-ultraviolet mutagenesis. I. Photoreactivation of mutational lesions induced by 365 nm radiation in *Escherichia coli* WP2$_8$, *J. Bacteriol.* (submitted).

Webb, R. B., 1977*b*, Near-ultraviolet mutagenesis, II. Mutation induction by 365 nm radiation in strains of *Escherichia coli* that differ in repair capability, *Mutat. Res.* (submitted).

Webb, R. B., and Brown, M. S., 1976, Sensitivity of strains of *Escherichia coli* differing in repair capability to far UV, near UV and visible radiations, *Photochem. Photobiol.* **24**:425–432.

Webb, R. B., and Brown, M. S., 1977*a*, Strong synergism between 365 nm and longer wavelength radiation in repair-proficient strains of *E. coli* (in preparation).

Webb, R. B., and Brown, M. S., 1977*b*, Oxygen dependence of sensitization to 254 nm radiation by prior exposures to 365 nm radiation (in preparation).

Webb, R. B., and Kubitschek, H. E., 1963, Mutagenic and antimutagenic effects of acridine orange in *Escherichia coli, Biochem. Biophys. Res. Commun.* **13**:90–94.

Webb, R. B., and Kubitschek, H. E., 1965, Photodynamic mutation in continuous cultures, in: *Argonne National Laboratory Biology and Medicine AEC Research and Development Report,* ANL-7136, pp. 145–148.

Webb, R. B., and Lorenz, J. R., 1970, Oxygen dependence and repair of lethal effects of near ultraviolet and visible light, *Photochem. Photobiol.* **12**:283–289.

Webb, R. B., and Malina, M. M., 1967, Mutagenesis in *Escherichia coli* by visible light, *Science* **156**:1104–1105.

Webb, R. B., and Malina, M. M., 1970, Mutagenic effects of near ultraviolet and visible radiant energy on continuous cultures of *Escherichia coli, Photochem. Photobiol.* **12**:457–468.

Webb, R. B., Brown, M. S., and Tyrrell, R. M., 1976, Lethal effects of pyrimidine dimers induced at 365 nm in strains of *E. coli* differing in repair capability, *Mutat. Res.* **37**:163–172.

Webb, R. B., Brown, M. S., and Hass, B. S., 1977*a*, Action spectrum for carotenoid protection in *Sarcina lutea,Radiat. Res.* (submitted).

Webb, R. B., Tyrrell, R. M., and Brown, M. S., 1977*b*, Synergism between 365 and 254 nm radiation (in preparation).

Webb, R. B., Ley, R. D., and Hass, B. S., 1977*c*, The role of oxygen-dependent lesions and damage to repair processes in the oxygen-dependent inactivation of *E. coli* by 365 nm radiation, in preparation.

Webb, S. J., 1963, The effect of relative humidity and light on air-dried organisms, *J. Appl. Bacteriol.* **26**:307–313.

Webb, S. J., 1972, Semi-dehydration and the action of ultraviolet light, in: *Research Progress in Organic, Biological and Medicinal Chemistry*, Vol. 3 (U. Gallo and L. Santamaria, eds.), Part 2, pp. 737–753, American Elsevier, New York.

Webb, S. J., and Bhorjee, J. S., 1967, The effect of 3000–4000 Å light on the synthesis of B. galactosidase and bacteriophages by *Escherichia coli B, Can. J. Microbiol.* **13**:69–79.

Webb, S. J., and Tai, C. C., 1968, Lethal and mutagenic action of 3200–4000 Å light, *Can. J. Microbiol.* **14**:727–735.

Webb, S. J., and Tai, C. C., 1969, Physiological and genetic implications of selective mutation by light at 320–400 nm, *Nature (London)* **224**:1123–1125.

Webb, S. J., and Tai, C. C., 1970, Differential, lethal and mutagenic action of 254 nm and 320–400 nm radiation on semi-dried bacteria, *Photochem. Photobiol.* **12**:119–143.

Weber, G., 1974, Combined 8-methoxypsoralen and black light therapy of psoriasis: Technique and results, *Br. J. Dermatol.* **90:**317–323.

Wells, W. F., 1955, *Air Hygiene and Air Contagion,* Harvard University Press, Boston.

Wells, W. F., and Wells, N. W., 1936, Air-borne infection, *J. Am. Med. Assoc.* **107:**1698–1703.

Willis, I., Kligman, A., and Epstein, J., 1973, Effects of long ultraviolet rays on human skin: Photoprotective or photoaugmentative, *J. Invest. Dermatol.* **59:**416–420.

Witkin, E. M., 1963, The effect of acriflavine on photoreversal of lethal and mutagenic damage produced in bacteria by ultraviolet light, *Proc. Natl. Acad. Sci. USA* **50:**425–430.

Witkin, E. M., 1975, Elevated mutability of *polA* and *uvrA polA* derivatives of *Escherichia coli* B/r at sublethal doses of ultraviolet light: Evidence for an inducible error-prone repair system ("SOS repair") and its anomalous expression in these strains, *Genetics* **79:**199–213.

Witkin, E. M., and Parisi, E. C., 1974, Bromouracil mutagenesis: Mispairing or misrepair? *Mutat. Res.* **25:**407–409.

Wright, L. J., and Rilling, H. C., 1963, The function of carotenoids in a photochromogenic bacterium, *Photochem. Photobiol.* **2:**339–342.

Wulff, D. L., and Fraenkel, G., 1961, On the nature of thymine photoproduct, *Biochim. Biophys. Acta* **51:**332–339.

Wurtman, R. J., 1975, The effects of light on man and other animals, *Annu. Rev. Physiol.* **37:**467–483.

Yoakum, G. H., 1975, Tryptophan photoproduct(s): Sensitized induction of strand breaks (or alkali-labile bonds) in bacterial deoxyribonucleic acid during near-ultraviolet irradiation, *J. Bacteriol.* **122:**199–205.

Yoakum, G., and Eisenstark, A., 1972, Toxicity of L-tryptophan photoproduct on recombinationless (*rec*) mutants of *Salmonella typhimurium, J. Bacteriol.* **112:**653–655.

Yoakum, G., Ferron, W., Eisenstark, A., and Webb, R. B., 1974, Inhibition of replication gap closure in *Escherichia coli* by near-ultraviolet light photoproducts of L-tryptophan, *J. Bacteriol.* **119:**62–69.

Yoakum, G., Eisenstark, A., and Webb, R. B., 1975, Near-UV photoproduct(s) of L-tryptophan: An inhibitor of medium-dependent repair of X-ray-induced single-strand breaks in DNA which also inhibits replication-gap closure in *Escherichia coli* DNA; molecular mechanisms for repair of DNA, in: *Molecular Mechanism for Repair of DNA* (P. C. Hanawalt and R. B. Setlow, eds.), Part B, pp. 453–458, Plenum Press, New York.

Yoakum, G., Webb, R. B., and Eisenstark, A., 1977, Toxic photoproducts of L-tryptophan: A role in near-UV sensitivity of *Salmonella typhimurium recA* strains, *Photochem. Photobiol.* (submitted).

Youngs, D. A., Van der Schueren, E., and Smith, K. C., 1974, Separate branches of the *uvr* gene-dependent excision repair process in ultraviolet-irradiated *Escherichia coli* K-12 cells; their dependence upon growth medium and the *polA, recA, recB* and *exrA* genes, *J. Bacteriol.* **117:**717–725.

Zelle, M. R., and Hollaender, A., 1955, Effects of radiation on bacteria, in: *Radiation Biology.* Vol. 2 (A. Hollaender, ed.), pp. 365–430, McGraw-Hill, New York.

Zetterberg, G., 1964, Mutagenic effects of ultraviolet and visible light, in: *Photophysiology,* Vol. 3 (A. C. Giese, ed.), pp. 247–281, Academic Press, New York.

Zigman, S., 1971, Eye lens color: Formation and function, *Science* **171:**807–809.

Zigman, S., and Hare, J. D., 1976, Inhibition of cell growth by near ultraviolet light photoproducts of tryptophan, *Mol. Cell. Biochem.* **10:**131–135.

Zigman, S., Schultz, J., Yulo, T., and Griess, G., 1973, The binding of photo-oxidized tryptophan to a lens gamma-crystallin, *Exp. Eye Res.* **17:**209–217.

Zigman, S., Yulo, T., and Schultz, J., 1974, Cataract induction in mice exposed to near UV light, *Ophthalmic Res.* **6**:259–270.

Zirenberg, B. E., Krämer, D. M., Geisert, M. G., and Kirste, R. G., 1971, Effects of sensitized and unsensitized longwave UV-irradiation on the solution properties of DNA, *Photochem. Photobiol.* **14**:515–520.

5

DNA Repair Enzymes in Mammalian Cells

Errol C. Friedberg, Kem H. Cook, James Duncan, and
Kristien Mortelmans

*Laboratory of Experimental Oncology, Department of Pathology, Stanford University School
of Medicine, Stanford, California 94305*

1. INTRODUCTION

In the past two decades, considerable strides have been made in understanding the phenomenon of DNA damage and its repair at all levels of biological organization. Studies on living cells using prokaryote models have been extensively reviewed by a number of authors (Town *et al.,* 1973; Setlow and Setlow, 1972; Smith, 1971; Witkin, 1969; Hanawalt, 1968; Howard-Flanders, 1968; Strauss, 1968). In recent years, considerable emphasis has been placed on DNA repair in mammalian systems, and this area of investigation too has been the subject of a number of review articles (Cleaver and Bootsma, 1975; Cleaver, 1974*a*; Strauss, 1974; Painter, 1970). Closely paralleling these biological studies, there have been significant efforts made to identify, purify, and characterize the numerous enzymatic and nonenzymatic components involved in the molecular mechanisms of DNA repair. These efforts have enjoyed particular success in prokaryote models such as *Micrococcus luteus, Escherichia coli,* and phage T4-infected *E. coli* (for recent reviews, see Grossman *et al.,* 1975; Grossman, 1974). The present chapter addresses itself to a review of mammalian cell enzymes that may be significant in DNA repair. Where appropriate, we have discussed aspects of prokaryote enzymology that have not received much attention in the literature, and that are, or may be, directly relevant to mammalian repair systems.

2. PHOTOREACTIVATING ENZYME (PRE) AND ENZYMATIC PHOTOREACTIVATION (EPR) IN PLACENTAL MAMMALS

Enzymatic photoreactivation (EPR) is one of a number of distinct light-dependent reactivation phenomena that have been described in prokaryote and eukaryote forms, but not until recently in cells or tissues derived from placental mammals (Rupert, 1975; Paterson *et al.,* 1974; Cook, 1970; Cook and McGrath, 1967). Aside from its intrinsic interest as a ubiquitous DNA repair mechanism, the phenomenon has particular usefulness as an experimental probe because of its extraordinary specificity for one particular form of DNA damage, *viz.,* cyclobutyl pyrimidine dimers in DNA. Thus any effect of ultraviolet (UV) irradiation on an organism or on DNA that is reversed or prevented by EPR can be ascribed specifically to pyrimidine dimers. Elegant use has been made of this property of photoreactivating enzyme (PRE) in a number of studies (Kondo and Jagger, 1966; Witkin, 1964). One of the more recent studies is that by Hart and Setlow (1975) strongly implicating pyrimidine dimers in the pathogenesis of neoplastic transformation by UV radiation in fish cells that are deficient in excision repair but are capable of EPR. These workers have irradiated tissue homogenates (comprising a mixture of liver, heart, and thyroid cells) from the gynogenetic fish *Poecilia formosa* and injected them into isogenic recipients. Hart and Setlow (1975) have reported that thyroid carcinomas were present in 100% of the fish injected with 5×10^5 cells exposed to an average incident fluence of 10–20 J m^{-2}. Exposure of the tissue homogenates to photoreactivating light conditions prior to UV irradiation had no effect on tumor incidence. However, such treatment applied after UV irradiation effected a highly significant reduction in the number of fish with thyroid carcinoma. Qualitatively similar results were reported using liver cells from these fish (Hart and Setlow, 1973).

The studies reported during the past 7 years on the human disease xeroderma pigmentosum suggest that UV radiation is also carcinogenic in humans (Cleaver, 1974*a*; Cleaver *et al.*, 1975; Robbins *et al.*, 1974; Robbins and Burk, 1973; Epstein *et al.*, 1971). In this disease, a potentially attractive tissue culture model exists with which to explore the relationship between DNA damage and carcinogenesis. In this model system, the capability for specifically implicating cyclobutyl pyrimidine dimers in neoplastic trans-formation in humans would be enhanced by the type of photoreactivation phenomenon described above. Thus the recent assertion by Sutherland and her colleagues of the existence of PRE activity in cell-free extracts of human and mouse tissues (Sutherland and Sutherland, 1975; Sutherland, 1974; Sutherland *et al.*, 1974), as well as the reported light-dependent disap-pearance of thymine dimers from the acid-insoluble fraction of living cells (Sutherland *et al.*, 1975, 1976), is of great potential significance.

As suggested by Cook (1970), in order to establish that EPR occurs in a given biological system, the following should be demonstrable.

1. PRE activity in a cell-free system.
2. A light-dependent loss of thymine dimers from the DNA of UV-irradiated living cells.
3. A light-dependent recovery (or reactivation) from the injurious effects of UV radiation in living cells.

Additionally, if possible, mutants should be isolated that are defective in PRE activity. Such mutants should not be capable of (2) and should show a significant reduction of (3).

In 1974, Sutherland reported the identification and purification of a PRE from extracts of peripheral blood leukocytes. The enzyme (purified to apparent homogeneity) has a molecular weight of 40,000. It has the properties expected of a PRE in that it requires UV-irradiated DNA as substrate, and photoreactivating light for activity. In addition, Sutherland *et al.* (1974) have shown that the kinetics of disappearance of cytosine (measured as uracil) in cytosine–thymine and cytosine–cytosine dimers from the acid-insoluble fraction of DNA correlate with the appearance of cytosine monomer (also measured as uracil) in the acid-soluble fraction following incubation of irradiated DNA with leukocyte PRE.

The human leukocyte enzyme has been shown to have two distinctive properties that differentiate it from previously studied PREs: (1) It is maximally active at an ionic strength of 0.05 μ, and is inhibited at higher salt concentrations. (2) It has an action spectrum for photoreactivation that extends to a wavelength of about 600 nm. Ignorance of these properties may well have contributed to the failure of earlier attempts to demonstrate PRE activity in tissues from placental mammals. The critical ionic strength dependence is self-explanatory. With respect to the action spectrum, if "dark" controls were incubated under yellow light (which eliminates photo-reactivating wavelengths required by the yeast and *E. coli* enzymes), the human enzyme would still be active, and no differences between the results of control and test reactions would be observed (Sutherland and Sutherland, 1975).

Studies on the distribution of PRE in various human tissues reveal curious results (Sutherland *et al.*, 1974). The activity has not been detected in erythrocyte extracts, extracts of spleen, or serum. When leukocytes were fractionated on Ludox gradients, activity was confined to extracts of monocytes and polymorphonuclear forms. Lymphocytes are apparently lacking in activity. The enzyme has also been detected in extracts of bovine bone marrow, mouse 3T3 cells in culture, and human fibroblasts in culture (Sutherland *et al.*, 1974). In addition, Harm (1974) has reported preliminary results indicating the presence of PRE activity in crude extracts of rabbit, bovine, and human leukocytes, and in extracts of rabbit liver, lymph nodes, and bone marrow. This rather wide distribution of activity in placental mammalian tissues makes it unlikely that the presence of PRE in human phagocytic white cells is an artifact associated with the engulfment of bacteria. Nonetheless, since tissue culture studies can be complicated by cryptic mycoplasma contamination, it would be reassuring to have a demonstration of enzyme activity in fresh human tissues other than leukocytes.

A highly interesting observation by Sutherland *et al.* (1975) is that levels of PRE are lower in skin fibroblasts from patients with the disease xeroderma pigmentosum (XP) than in control subjects (Table 1). These differences are apparently not related to the age of the skin biopsy donor or to the cell passage number. In addition, there is no evidence that the enzyme from XP fibroblasts has different pH, metal, or ionic strength requirements. Also, mixing experiments provide no evidence for an inhibitor in extracts of

TABLE 1. Levels of Photoreactivating Enzyme in Normal and Xeroderma Cells

Line	Description	Age at biopsy (yr)	Xeroderma complementation group	Photoreactivating enzyme activity	
				Units	Percent
NHF	Human foreskin fibroblasts	Neonatal	—	627[a]	100
HESM	Human skin and muscle fibroblasts	Embryonic	—	625	99
CRL 1199 (Po Co)	Xeroderma pigmentosum	28	B	0	0
CRL 1200 (Te Ko)	Xeroderma pigmentosum	11	D	50	8
CRL 1161 (Ge Ar)	Xeroderma pigmentosum	12	C	99.6	15.8
CRL 1223 (Jay Tim)	Xeroderma pigmentosum	7	A	227	36.3
CRL 1259 (XP 2)	Xeroderma pigmentosum	34	E	311.1	49.5

[a] Typical ranges of enzyme activity were $\pm 5\%$ for samples with high activity and $\pm 10\%$ for those of lower activity. From Sutherland *et al.* (1975).

XP cells. There is no apparent relationship between the residual levels of PRE activity and of excision repair ability determined by unscheduled DNA synthesis in cells from the various complementation groups of XP. Indeed, three of four patients with the so-called XP variant group (in which levels of excision repair are normal) had between 4% and 11% residual PRE activity (Sutherland and Oliver, 1975). However, it should be noted that with the exception of the XP variant cases, Sutherland's comparisons are based on single cases from each complementation group. Furthermore, there is some controversy as to the residual levels of excision repair in the XP complementation groups (Huang and Vincent, 1975; Day, 1975).

In light of these observations, it is clearly of interest to examine PRE levels in other disease states characterized by abnormal photosensitivity. Certainly, the observation that the one human disease clearly associated with defective excision repair (XP) is also apparently defective in a second DNA repair mode is intriguing. As pointed out by Sutherland *et al.* (1975), the reduction of PRE might result from mutation in a single gene controlling both PRE and the excision enzymes. Alternatively, "the low levels of both PRE and the excision enzyme(s) might also result from independent mutations leading to inadequate repair of UV-induced lesions."

The presence of active PRE *in vitro* would imply that cells containing the enzyme can carry out the phenomenon of EPR *in vivo*. Support for the existence of this phenomenon would be provided by fulfilling the second and third of the three criteria listed above for the establishment of PRE in a living system. Sutherland *et al.* (1975) have reported the light-dependent loss of thymine dimers from the acid-insoluble fraction of cells labeled with [³H]thymidine. For this experiment, XP cells severely defective in excision repair were chosen with the rationale that dark repair (dimer excision) might mask the detection of a light-dependent loss of dimers. As shown in Fig. 1, the response reported was dramatic. In view of the extraordinarily rapid kinetics of this phenomenon compared to the kinetics of dimer excision in cultured human fibroblasts, the light-dependent loss of dimers should be readily observable in normal cells. This is indeed the case as evidenced by recent studies with a strain of normal cells called HESM (Sutherland *et al.*, 1976). Additional support for the phenomenon of EPR in living cells is provided by the following results reported by Sutherland *et al.* (1976).

1. The radioactive fraction identified by chromatography as the substrate for photoreactivation was reported to be thymine dimers as evidenced by its apparent conversion to thymine by irradiation at 248 nm.

2. The action spectrum for EPR *in vivo* was reported to be essentially

similar to that of PRE *in vitro* in XP12BE (Fig. 2). Similar results have been reported for normal human fibroblasts (Sutherland *et al.*, 1976).

The experiments reported by Sutherland and her colleagues represent a potentially important breakthrough in photobiology, and have raised a number of interesting questions. Perhaps the most significant of these is whether the presence of PRE in extracts of human and other higher mammalian tissues has any real biological significance. Cleaver *et al.* (1976) have tried unsuccessfully to reproduce the light-dependent loss of dimers from living cells. In these experiments, cells were grown in a variety of tissue culture media, including Eagle's modified essential medium (MEM), Ham's medium, and commercially obtained Dulbecco's medium (Grand Island Biological Company). More recently, Sutherland and Oliver (1976) have reported that EPR *in vivo* is not clearly demonstrable in cells grown in MEM, but is when cells were maintained in Dulbecco's medium prepared in their laboratory (Fig. 3). Correlating with these observations, these authors

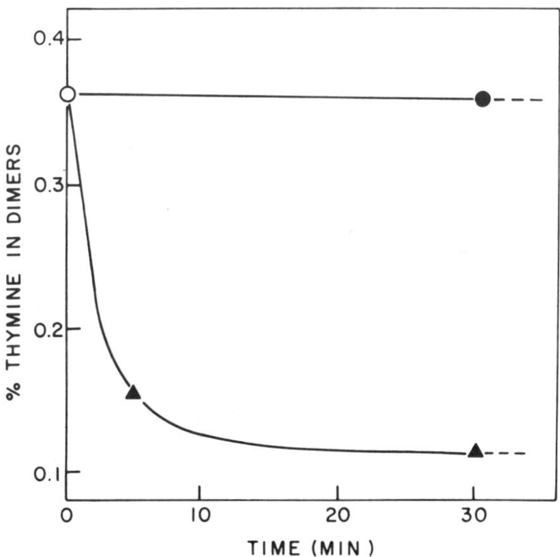

Fig. 1. Photoreactivation of pyrimidine dimers in xeroderma pigmentosum cells. Strain CRL 1223 cells (XP group A) were grown in monolayers on coverslips, labeled with [³H]thymidine, rinsed, and exposed to 100 J m⁻² of 254-nm radiation. One sample was harvested immediately after UV irradiation (O), two were exposed to photoreactivating light (▲), and one was kept in the dark (●). The samples were harvested, and the dimer content in the DNA was determined. In a parallel experiment (not shown), samples were exposed to photoreactivating light for 1 or 24 h; no further reduction in dimer content could be detected. In the dark, the dimer content of the DNA was only slightly decreased after 24 h. From Sutherland *et al.* (1975).

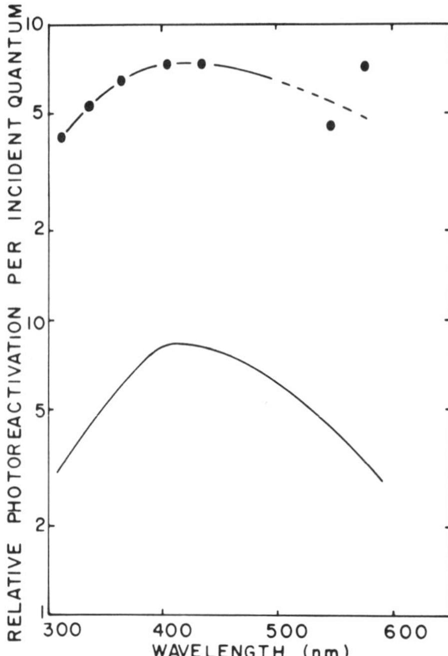

Fig. 2. Action spectrum for dimer monomerization in XP12BE cells (upper curve) and the action spectrum of the photoreactivating enzyme from human leukocytes (lower curve). In the cellular photoreactivation experiments, the fluence entering the irradiation chamber was 10^{16} photons/s and the irradiation time was 2 min. During this time, no significant excision of dimers occurred. From Sutherland *et al.* (1976).

have indicated that when cells grown in their medium were transferred to MEM there was a significant reduction in the specific activity of PRE in crude extracts of the cells (Fig. 4). Using medium and cells obtained from Sutherland, as well as medium made up in our laboratory according to the formula used by Sutherland and Oliver (1976), K. Mortelmans and E. C. Friedberg (unpublished observations) have been able to qualitatively reproduce the result shown in Fig. 1. However, the extent of the light-dependent loss of thymine dimers from the DNA of living cells was significantly less in our hands; only about 20% of the thymine dimers disappear from DNA in 20 min of illumination.

At the time of writing, it would appear that the phenomenon of EPR in human cells has an exquisite dependence on the medium in which the cells are grown. Table 2 lists the identifiable differences between the MEM and Dulbecco's media used by Sutherland and Oliver (1976). Aside from the fact that the Dulbecco's medium is somewhat richer than the MEM in all components they share in common, the latter medium is deficient in glycine and serine, while the former contains no added leucine. Parenthetically, it should be noted that the formula of the Dulbecco's medium used by Sutherland and Oliver (1976) is different from that adopted by the Tissue Culture Standards Committee (Morton, 1970). The latter includes leucine and $Fe(NO_3)_3 \cdot 9H_2O$, neither of which is present in the formula used by Sutherland and Oliver (1976). Sutherland and Oliver (1976) have pointed out that previous studies in prokaryotes have indicated that single compounds can exert an extraordinary effect on the specific activity of PRE. Thus Nishioka and Harm (1972) observed that in *E. coli* strain B_{8-1}-160 (an adenine-requiring mutant) photoreactivation increased by about 500-fold in the absence of added adenine. It would be highly enlightening to determine whether the current dependency of EPR on the growth of cells in Dulbecco's medium is, in fact, a function of the presence or absence of a single component in the medium.

Wagner *et al.* (1975) have reported an apparent biological role of EPR in living human cells by showing a light-dependent reduction in UV-induced killing of herpes simplex virus (Fig. 5). However, if, as indicated by Sutherland *et al.* (1975), there are significant variations in the level of PRE in different XP cells, this is not reflected in the survival of irradiated virus. For

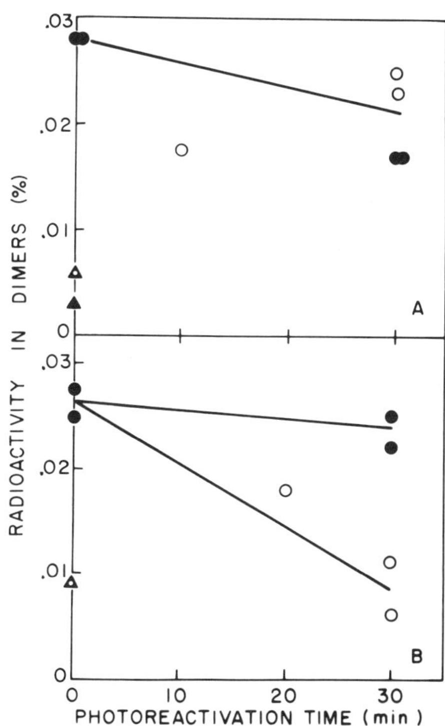

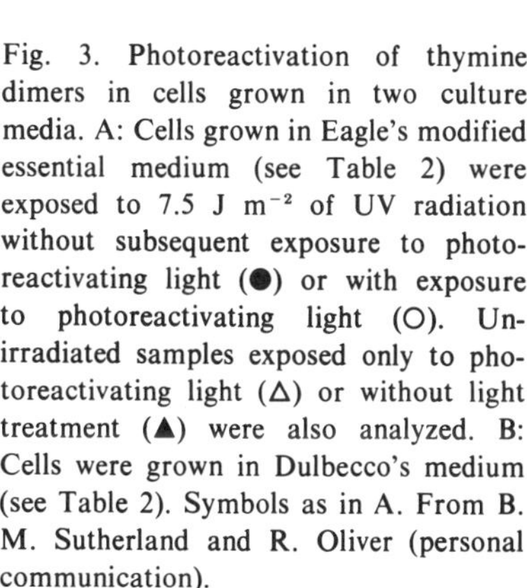
Fig. 3. Photoreactivation of thymine dimers in cells grown in two culture media. A: Cells grown in Eagle's modified essential medium (see Table 2) were exposed to 7.5 J m⁻² of UV radiation without subsequent exposure to photoreactivating light (●) or with exposure to photoreactivating light (O). Unirradiated samples exposed only to photoreactivating light (△) or without light treatment (▲) were also analyzed. B: Cells were grown in Dulbecco's medium (see Table 2). Symbols as in A. From B. M. Sutherland and R. Oliver (personal communication).

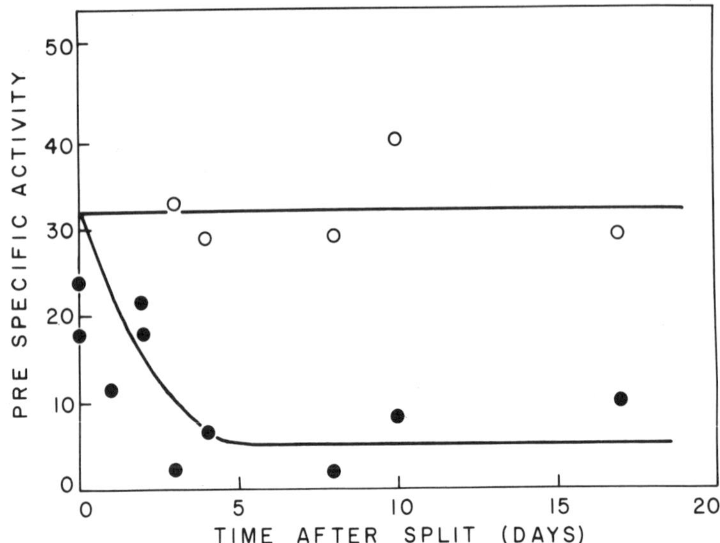

Fig. 4. Photoreactivating enzyme levels in XP12BE cells. Cells were either maintained in Dulbecco's medium (O) (see Table 2) or transferred to Eagle's modified essential medium (●) (see Table 2). The ordinate indicates the measured specific activity of photoreactivating enzyme in cell-free extracts expressed in terms of percent of that in normal cells. The abscissa indicates the times after cells were subcultured into fresh medium. From B. M. Sutherland and R. Oliver (personal communication).

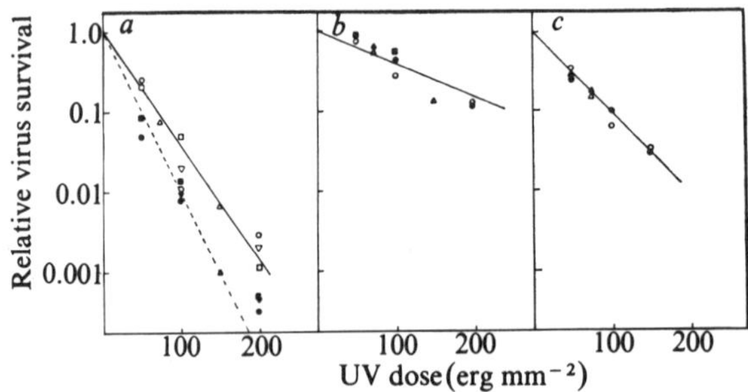

Fig. 5. Photoreactivation of UV-irradiated herpes simplex virus by normal and XP human cells. (a) XP12BE cells, (b) HESM (normal) cells, (c) XP-1 cells. Virus survival was calculated from the relative plaque titers obtained by diluting unirradiated virus and carrying out a parallel infection in illuminated or dark cells. The open symbols represent titers on photoreactivated cells, and the closed symbols represent titers on cells kept in the dark. The different symbols represent different experiments. From Wagner *et al.* (1975).

TABLE 2.   Comparison of Constituents Present in Eagle's Modified Essential Medium (EMEM) and Dulbecco's Medium[a]

Component	EMEM (mg/liter)	Dulbecco's Modified EMEM (mg/liter)
Arginine	126.4	73.7
Cystine	24.0	42.0
Glutamine	300.0	525.0
Glycine	—	30.0
Histidine	41.9	36.75
Isoleucine	52.5	91.8
Leucine	52.5	—
Lysine	73.0	127.85
Methionine	14.9	26.0
Phenylalanine	33.0	57.8
Serine	—	42.0
Threonine	47.6	83.37
Tryptophan	10.2	14.28
Tyrosine	36.2	63.4
Valine	46.9	82.0
Ca-pantothenate	1.0	3.5
Choline	1.0	3.5
Folic acid	1.0	3.5
Inositol	2.0	7.0
Nicotinamide	1.0	3.5
Pyridoxal	1.0	3.5
Riboflavin	1.0	3.5
Thiamin	1.0	3.5

[a] Data calculated from Flow Laboratories' formulation for their amino acid and vitamin solution. From B. M. Sutherland and R. Oliver (personal communication).

instance, it is difficult to explain the observation that XP-1 cells with less than 1% residual PRE (Wagner *et al.*, 1975) are more efficient at host-cell reactivation of herpes simplex virus than are Jay Tim cells with 36.3% PRE activity.

In the final analysis, much more work needs to be done on the question of photoreactivation in the cells of placental mammals. Aside from the problems of the apparent specific medium dependency at this stage, it remains to be shown that the phenomenon *in vivo* is really an enzyme-catalyzed event rather than a nonenzymatic photochemical effect. It is also imperative to determine whether the observations on cells in culture have any relevance to living placental mammals.

3. THE ENZYMOLOGY OF EXCISION REPAIR IN MAMMALIAN CELLS

3.1. Classification of DNA Damage

As pointed out by Cerutti (1975), DNA repair phenomena are often classified according to the exogenous agents by which the damage was induced: ultraviolet repair, X-ray repair, repair of alkylation damage, etc. This classificatory viewpoint stemmed fairly naturally out of the fact that early studies in DNA repair were largely biological in character. More recent years have witnessed a blooming of the biochemistry of DNA repair, notably the discovery and characterization of a large number of deoxyribonucleases specific for damaged DNA. Cerutti (1975) has pointed out that many agents that interact with DNA produce a heterogeneous set of reaction products, and it is likely that some products produced by different agents are structurally related. It has thus become increasingly apparent that the enzymology of DNA repair can be more meaningfully considered in terms of the *products* of interaction of given chemical and physical agents with DNA than in terms of the nature of the damaging agents themselves. One classification recently proposed by Cerutti (1975) distinguishes three major classes of DNA base damage as follows:

Class I: Monofunctional base lesions causing negligible helix distortion.
Class II: Monofunctional base lesions causing minor helix distortion.
Class III: Monofunctional and difunctional base lesions causing major helix distortion.

This classification is significant in terms of the general principles stated above. However, in our view, there is at present insufficient evidence that endonucleases that attack damaged DNA specifically recognize varying degrees of helix distortion as a major determinant of substrate specificity. Therefore, for the purposes of the present discussion, we have adopted the simpler classification of Grossman *et al.* (1975), who have recognized two general categories of base damage:

1. *Diadduct damage,* defined as that type of damage "in which more than one nitrogenous base is involved in the final chemical product."

2. *Monoadduct damage,* defined as "single-base modifications in which an addition reaction, adduct formation or a chemical transformation leads to the alteration of a single nitrogenous base."

3.2. Enzymatic Repair of Diadduct Damage to DNA

The most common model of diadduct damage to DNA is that of cyclobutyl pyrimidine dimers resulting from UV irradiation. The enzymology of the repair of this form of DNA damage in prokaryotes has been intensively studied and thoroughly reviewed (for recent reviews, see Friedberg, 1975; Grossman *et al.*, 1975; Grossman, 1974). The following comments are confined to a consideration of enzymes from higher eukaryote forms.

At about the time that the pioneering studies of Boyce and Howard-Flanders (1964), Setlow and Carrier (1964), and Riklis (1965) established the phenomenon of thymine dimer excision in *E. coli,* Rasmussen and Painter (1964) laid claim to an analogous event in mammalian cells in culture. In their experiments, HeLa cells were exposed to UV radiation and labeled with [³H]thymidine for various periods. Autoradiographic analyses revealed that, whereas in cultures of unirradiated cells about 25% of the cells showed tritium uptake into nuclei (reflecting the fraction of cells in S phase), in irradiated cultures 100% of the cells contained visible grains. Additionally, a linear relationship was observed between the number of nuclear grains/cell not in S phase and the UV fluence. The uptake of [³H]thymidine into DNA by cells not in S phase is referred to as unscheduled DNA synthesis (UDS), and is generally considered to reflect repair synthesis in cells undergoing excision repair. Subsequent studies in a number of laboratories have shown that UV-irradiated mammalian cells that carry out UDS also carry out repair replication (as defined by the prototype experiment of Pettijohn and Hanawalt, 1964) (Cleaver, 1968; Rasmussen and Painter, 1966), as well as excision of thymine-containing pyrimidine dimers from DNA (Duncan *et al.*, 1975; Paterson *et al.*, 1973; Wilkins, 1973; Cleaver and Trosko, 1970; Setlow *et al.*, 1969).

Whether or not UDS is a quantitatively or even a qualitatively valid index of excision repair remains, in our minds, a vexing question. Nonetheless, the direct demonstration of a selective loss of dimers from the DNA of mammalian cells *in vivo* (Duncan *et al.*, 1975; Cleaver and Trosko, 1970; Setlow *et al.*, 1969) provides excellent evidence that such cells carry out excision repair. Therefore, in cell-free extracts one would expect to identify enzymes whose catalytic functions are appropriate to current models of excision repair.

3.2.1. Endonucleases in Normal and Xeroderma Pigmentosum Cells

All models of enzymatic excision of pyrimidine dimers from DNA require the endonucleolytic cleavage of at least one phosphodiester bond in close proximity to each pyrimidine dimer. A number of laboratories have identified nuclease activities "specific" for UV-irradiated DNA (Table 3). In 1971, Burt and Brent reported that crude extracts of HeLa cells effected the degradation of UV-irradiated DNA at a significantly greater rate than of unirradiated DNA. The activity was found to bear a linear relationship to the UV fluence, and was not competitively inhibited by unirradiated double-stranded or heat-denatured DNA. In this study, enzyme activity was determined by measuring the acid-soluble radioactive product, and it was not established whether the activity was endonucleolytic, exonucleolytic, or both. In a subsequent study, Brent (1972) identified an endonuclease activity specifically. He observed that incubation with HeLa cell extracts at pH 7.0 in the presence of 1.0 mM EDTA effected the conversion of about 40% of form I PM2 DNA molecules irradiated at 1700 J m^{-2} to the form II state. Unirradiated DNA was unaffected under these incubation conditions. Similar results were reported by Bacchetti *et al.* (1972). These workers incubated crude extracts of HeLa cells with unirradiated ^{14}C-labeled and UV-irradiated (500 J m^{-2}) ^{3}H-labeled adenovirus 5 DNA in the presence of 1.0 mM EDTA. Sedimentation of the DNA in alkaline sucrose density gradients revealed preferential degradation of the irradiated DNA. Qualitatively comparable results were obtained by monitoring the conversion of RF I phage ϕX174 DNA to the RF II form.

The primary characteristic one might expect of an endonuclease involved in the excision of pyrimidine dimers is that it would be specific for dimers, as opposed to other photoproducts in UV-irradiated DNA. When this property of the human cell UV endonuclease was investigated by Bacchetti *et al.* (1972), a negative correlation was obtained. They treated UV-irradiated adenovirus DNA with a purified preparation of PRE from *Streptomyces griseus*. Subsequent incubation of this DNA with HeLa cell extract revealed no detectable loss of endonuclease sensitive sites (Fig. 6). Bacchetti *et al.* (1972) concluded that the endonucleolytic activity was not acting at dimers, but suggested that it may nevertheless be involved in the repair of some other UV-induced damage.

Studies in our laboratory (K. Minton and E. C. Friedberg, unpublished observations) have also detected endonucleolytic activity from mammalian cells (HeLa and WIL II lymphocytes) that preferentially attacks UV-irradiated DNA. We were struck by the observations that (1) the number of endonucleolytic incisions determined from the observed reduction in sedimentation velocity of DNA in alkaline sucrose gradients was significantly less than the calculated number of pyrimidine dimers/DNA strand, and (2) endonucleolytic activity was not detected using DNA irradiated at fluences of less than about 100 J m^{-2}. These observations led us to conclude that the endonuclease activity in question was not specific for pyrimidine dimers in DNA.

TABLE 3. Endonuclease Activities That Attack UV-Irradiated DNA

Authors	Source	Substrate specificity	Metal requirements	pH optimum	Inhibitors	Molecular weight
Brent (1972)	HeLa cells	UV-irradiated DNA (1700 J m^{-2}) γ-irradiated DNA (3300 rads)	Nil Nil	— —	— —	— —
Bacchetti *et al.* (1972)	HeLa cells	UV-irradiated DNA (500 J m^{-2}) Substrate nonphotoreactivable	Nil	—	—	—
Bacchetti and Benne (1975)	Calf thymus	UV-irradiated DNA (300 J m^{-2}) Substrate nonphotoreactivable Does not degrade denatured DNA Does not degrade apurinic DNA Does not degrade 8-methoxy-psoralen cross-linked DNA	Acts with equal efficiency in EDTA, MG^{2+} or Ca^{2+}	7.0–7.5	Heat 50°C NaCl 50 mM tRNA	~30,000
Brent (1975)	Human lymphoblasts fraction VCL	Predominantly active with depurinated DNA but also attacks UV-irradiated DNA (500 J m^{-2}) and unirradiated DNA	Works in EDTA but stimulated by 3 mM Mg^{2+}	—	—	—
	Fraction VD2	UV-irradiated DNA predominantly (500 J m^{-2})	Nil	—	Total loss of activity in 2 days at 4°C	—
Van Lancker and Tomura (1974)	Rat liver	UV-irradiated DNA (2% thymine dimers) Acetylaminofluorene-DNA X-irradiated DNA (1500–3000 rads)	Acts without added metal Stimulated about fivefold by 5–10 mM Mg^{2+}	~7.5	Not inhibited by caffeine	15,000–20,000
Duker and Tabor (1975)	HeLa cells WI-38 cells	UV-irradiated DNA (40–60 dimers/molecule)	Nil	—	—	—

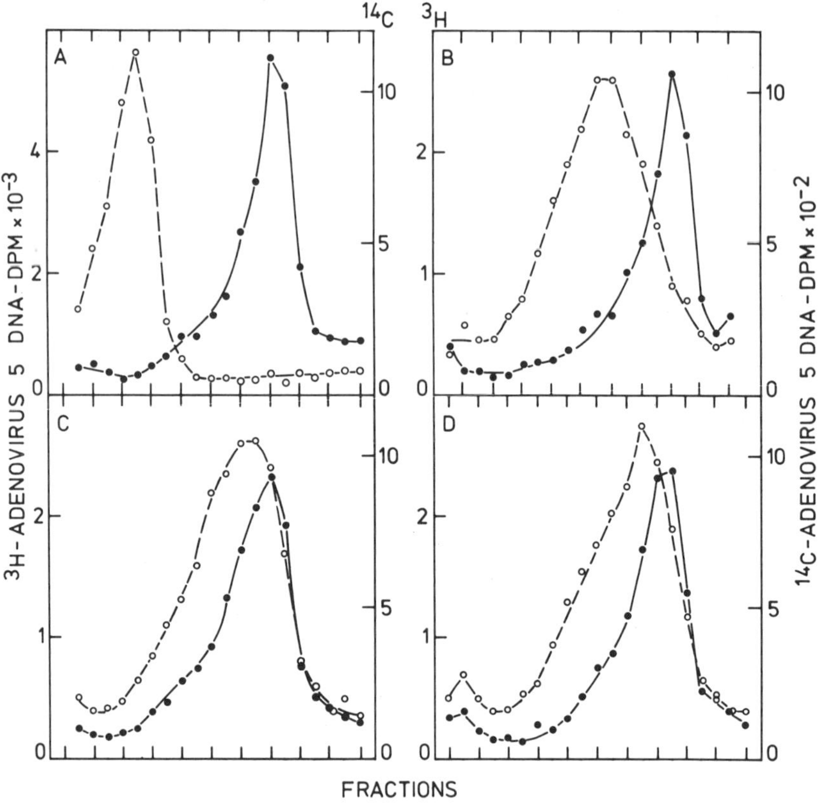

Fig. 6. Sedimentation profiles of adenovirus 5 [³H]DNA showing lack of photoreactivation of endonuclease-sensitive sites. The DNA was UV irradiated at 500 J m⁻² and treated as follows. A: UV endonuclease from *Micrococcus luteus*. B: PR enzyme and UV endonuclease from *M. luteus* (note the loss of endonuclease-sensitive sites in this control experiment). C: Extract from XP10 cells. D: PR enzyme and extract from XP10 cells. ●, Unirradiated [¹⁴C]DNA; O, UV-[³H]DNA. The direction of sedimentation in alkaline sucrose gradients is from left to right. From Bacchetti *et al.* (1972).

A possible explanation for the results obtained by Brent (1972), Bacchetti *et al.* (1972), and Minton and Friedberg (unpublished observations) is provided by studies of Ljungquist *et al.* (1974). These workers have recently described an endonuclease from mammalian cells that attacks apurinic and possibly apyrimidinic sites in DNA. This enzyme will be discussed in detail in Section 3.3.7a. Of relevance to the present discussion, however, is their observation that PM2 DNA irradiated with a fluence of 6–7 J m⁻² was insensitive to attack by their enzyme. However, on exposure of PM2 DNA to a UV fluence of 270 J m^{-2}, alkali-labile and endonuclease-sensitive sites were introduced. These authors point out that the glycosidic bond in pyrimidine nucleotides is very greatly labilized by saturation of the pyrimidine ring, and "a plausible explanation of our results is that some of the pyrimidine radiation products were hydrolyzed off during the incubation period after irradiation, leaving apyrimidinic sites in the DNA." Even if such lesions in heavily UV-irradiated DNA are not recognized by an endonuclease activity in cell-free preparations, depyrimidinated and depurinated sites are highly alkali labile. If the assay of endonuclease activity involves the sedimentation of DNA in alkaline solutions, a spurious result will be obtained unless the specific control of sedimenting UV-irradiated DNA without enzyme incubation is included in the experiment. Furthermore, as indicated in later discussion, N-glycosidase activities have been detected in prokaryote and mammalian cell sources that can remove certain bases from DNA. It is distinctly possible that these and/or other N-glycosidase activities can recognize and remove damaged bases in UV-irradiated DNA. Thus an apparently "UV-specific" endonuclease activity could possibly result from the action of an N-glycosidase followed by enzymatic and/or alkali-catalyzed phosphodiester bond cleavage.

Further pitfalls associated with the use of heavily UV-irradiated DNA as a substrate emerge from the studies of Slor and Lev (1973). They detected a nuclease activity in extracts of HeLa cells, XP cells, and normal peripheral blood lymphocytes. The activity in their hands degraded single-stranded DNA very effectively; indeed, the latter substrate was a potent competitive inhibitor of the activity on heavily irradiated duplex DNA. They showed that the UV irradiation of HeLa cell DNA resulted in the generation of single-stranded regions (defined by susceptibility to S_1 nuclease) in a UV-fluence-dependent fashion. They point out that the use of large fluences of UV radiation results in both single-strand breaks and areas of denaturation that render the substrate susceptible to attack by non-specific nucleases.

Further studies from the laboratories of Brent and Bacchetti have reinforced the notion that the endonucleases detected in crude extracts of HeLa cells are not pyrimidine dimer specific. Bacchetti and Benne (1975) have purified an endonuclease activity from calf thymus about 8000-fold with a final yield of 2% of that in the crude extract. Sephadex gel filtration studies provided a molecular weight estimate of about 30,000. Electrophoresis of the purified enzyme in SDS gels revealed two bands of protein with molecular weights of 28,000 and 32,000. The purified preparation of enzyme had no detectable endonuclease activity on unirradiated PM2 DNA. However, when PM2 DNA was irradiated with either UV or γ-rays, the enzyme catalyzed single-strand breaks in a dose- and enzyme-concentration-dependent manner. The enzyme apparently worked to a very limited extent on either unirradiated or UV-irradiated alkali-denatured phage T7 DNA. As was true of the activity present in crude extracts of HeLa cells, substrate sites for the purified calf thymus enzyme were not photoreactivable with PRE.

To determine whether the endonucleolytic activity on UV- and γ-irradiated DNA reflects activities of a single enzyme, several parameters of the enzyme assay were varied, and the activity on both substrates was compared. The pH optimum curves were found to vary similarly, and endonucleolytic activity on both substrates was observed in the presence of EDTA. When divalent cations such as Mg^{2+} and Ca^{2+} were substituted for EDTA, no significant variation of the enzymatic activities relative to that in EDTA was observed for concentrations of Mg^{2+} up to 3.5 mM, and of Ca^{2+} below 1 mM. Addition of NaCl (above 50 mM) or tRNA to reaction mixtures resulted in pronounced inhibition of both enzyme activities. These observations, together with the fact that the two activities cochromatographed on phosphocellulose, DNA-agarose, and Sephadex G200, led Bacchetti and Benne (1975) to conclude that both activities are associated with a single protein. Analysis of the chemistry of the termini produced by endonucleolytic incision of either UV-or γ-irradiated DNA revealed the presence of 5′ phosphoryl groups.

In considering the nature of the specific substrate created by either UV- or γ-irradiation of DNA, Bacchetti and Benne (1975) indicate that the enzyme had no activity on depurinated DNA or on DNA cross-linked by treatment with 8-methoxypsoralen and visible light. It is to be hoped that continued studies on this interesting endonuclease will result in the identification of the specific substrate sites recognized in irradiated DNA.

Brent (1973) also observed that the endonuclease activity he detected in crude extracts of HeLa cells attacks γ-irradiated DNA in addition to UV-irradiated DNA. He partially purified three endonuclease activities by fractionating crude extracts of a human lymphoblast line grown in suspension culture (Brent, 1975). Crude extracts from these cells, like those of HeLa, contained an activity that extensively nicked phage PM2 DNA irradiated at a fluence of 500 J m^{-2} but did not significantly affect untreated DNA in the presence of 1 mM EDTA. On fractionation, an endonuclease (referred to as fraction VC3) was isolated and found to attack both UV-irradiated and unirradiated DNA to about the same extent. A second endonuclease (fraction VC1) was found to be predominantly active on partially depurinated DNA but also slightly active on UV-irradiated and native DNA. The activity on all three substrates was stimulated ten- to twentyfold by Mg^{2+}. Finally, a third chromatographically distinct peak of endonuclease activity was isolated as fraction VD2. This fraction was several times more active with UV-irradiated DNA than with native DNA, and was not a significantly active on depurinated DNA.

With the limited information available, it seems reasonable to conclude, at least tentatively, that fraction VD2 of Brent (1975) resembles the enzyme purified by Bacchetti and Benne (1975), while fraction VC1 resembles the apurinic endonuclease purified from calf thymus by Ljungquist *et al.* (1974) (see later discussion). Since fraction VC3 attacks native DNA as effectively as UV-irradiated DNA, it has no obvious features that distinguish it as an activity relevant to damaged DNA.

Other laboratories have reported the results of a search for UV-specific endonuclease activities. Van Lancker and Tomura (1974) reported the purification to apparent homogeneity of an enzyme from rat liver. The purified enzyme has a specific activity 10,000 times greater than that in crude extracts, and a molecular weight of 15,000–20,000, estimated by gel filtration. Sedimentation velocity analyses in neutral and alkaline sucrose density gradients indicated that the purified enzyme catalyzed the formation of single-strand breaks in duplex UV-irradiated or acetylaminofluorene (AAF)-treated DNA (Fig. 7). (The UV-irradiated DNA contained 2% thymine dimers, which we estimate would require about 500 J m^{-2} of irradiation.) The enzyme was apparently active without added $MgCl_2$, but was optimally stimulated at concentrations between 5 and 10 mM $MgCl_2$. The pH optima and thermosensitivity of the activity on both substrates were identical.

The question of the substrate specificity of this endonuclease and its possible relationship to the enzymes studied by Bacchetti and Benne (1975) and Brent (1975) is complicated. Unfortunately, Van Lancker and Tomura (1974) have not reported the effect of the enzyme on UV-irradiated DNA in which pyrimidine dimers have been monomerized by PRE. In addition to UV-irradiated and AAF-treated DNA, it has been reported that 7-bromo-benzanthracene-treated (Maher *et al.,* 1974) and X-irradiated DNA (Tomura and Van Lancker, 1975) are attacked by the enzyme. Van Lancker and Tomura (1974) concluded that the specificity of the enzyme is not directed toward thymine dimers or the guanosine acetylaminofluorine complex, but rather toward the distortion in the double helix brought about by the base alterations.

In a recent study, Duker and Tabor (1975) also used closed circular DNA as a substrate for the detection of "UV endonuclease" activity. They irradiated form I phage PM2 DNA to produce about 50 pyrimidine dimers per molecule. Based on the determination that 0.1 J m^{-2} produces an average of five dimers/*E. coli* genome equivalent (Friedberg and Clayton, 1972), we calculated the UV fluence used by Duker and Tabor (1975) to be about 454 J m^{-2}. These workers circumvented the possible artifact of con-

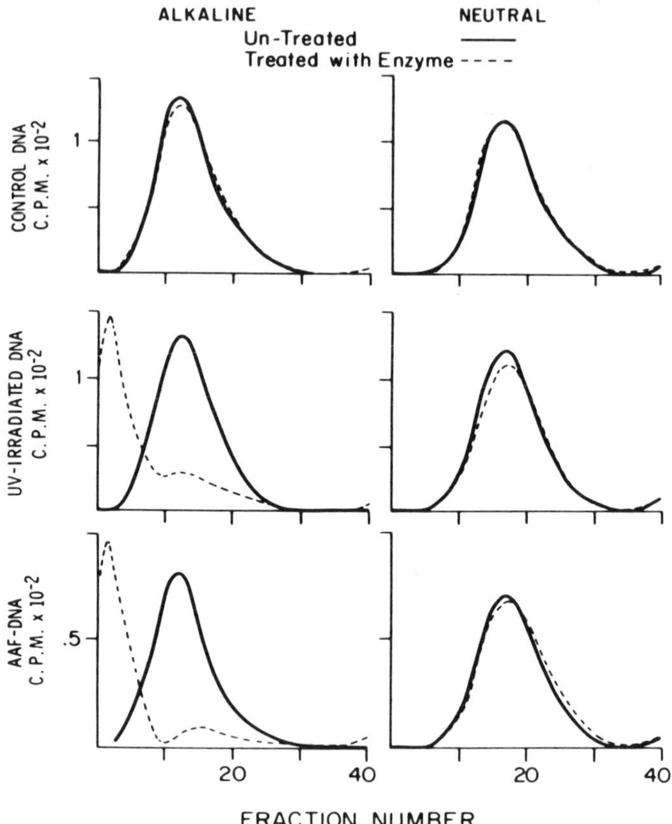

Fig. 7. Enzymatic degradation of UV-irradiated and AAF-treated DNA. The differential centrifugation in alkaline and neutral sucrose gradients indicates the introduction of single-strand breaks in both UV-irradiated and AAF-treated DNA incubated with enzyme. Sedimentation is from left to right. From Van Lancker and Tomura (1974).

fusing alkali-labile bonds with endonucleolytic incisions by comparing the results of sedimentation of the DNA in alkaline sucrose gradients to those obtained with neutral cesium chloride gradients. They detected an endonuclease activity in extracts of HeLa cells and of XP skin fibroblasts that attacked UV-irradiated and not unirradiated DNA when incubations were performed under the conditions described by Brent (1972). Comparison of the specific activity of the endonuclease in a variety of cell types showed the greatest activity in HeLa and WI-38 cell extracts. The activity in extracts of four different XP cell strains and in extracts of normal fetal and adult skin fibroblasts was about half that in HeLa cells. While there is no direct experimental evidence indicating the dimer specificity of this endonuclease activity, the use of relatively low fluences of UV radiation in this study is significant. The results obtained with extracts of XP cells led these authors to suggest that this activity is not the UV repair endonuclease. However, they do raise the very pertinent possibility that several so-called UV endonucleases may exist in extracts of human cells, and that in crude extracts the absence of one of these enzymes may not be detectable.

In a number of the studies discussed above, the demonstration of an endonuclease activity in extracts of XP cells has led to the conclusion that the activity is not involved in the excision repair of pyrimidine dimers in normal cells. This conclusion is based on the results of a number of studies on living cells which are consistent with a DNA incision defect in UV-irradiated XP cells. For instance, Cleaver (1969, 1971) carried out experiments in which the extent of repair replication in XP cells was measured after two kinds of radiation damage, *viz.*, base damage without chain breakage (induced by UV radiation) and chain breakage (induced by ionizing or by UV radiation after bromodeoxyuridine incorporation into DNA). The rationale for these experiments was that repair of damage to DNA bases presumably requires endonucleolytic incision of the polynucleotide chain, whereas repair of strand breaks does not. His results showed that fibroblasts from homozygous XP patients can perform repair replication after chain breakage, but not after damage that requires enzymatic polynucleotide chain incision. These results are consistent with a defective pyrimidine dimer-specific endonuclease (or endonuclease-related function), or a defect in a step in excision repair that precedes endonucleolytic incision in XP cells. Studies from a number of other laboratories have examined the capacity of XP cells to carry out repair synthesis of DNA *in vivo* after different types of DNA damage (Stich, 1975; Regan and Setlow, 1974; Cleaver, 1971, 1973; Slor, 1973; Jacobs *et al.*, 1972; Setlow and Regan, 1972; Kleijer *et al.*, 1970). They have confirmed the contention that the repair of damage requiring endonucleolytic incision is defective in XP cells, whereas damage that itself causes strand breakage is associated with normal levels of repair replication.

Essentially similar conclusions were arrived at by Setlow *et al.* (1969) on the basis of sedimentation velocity studies in alkaline sucrose gradients. They reported that during the 24 h following irradiation of normal cells, single-strand breaks were demonstrable, following which the DNA was restored to high molecular weight. Breaks in the DNA of XP cells were not observed after irradiation.

Cleaver *et al.* (1975) attempted to show a DNA incision defect in XP cells by demonstrating that phage T4 UV endonuclease action would restore normal levels of DNA repair synthesis in XP cells that were incubated with phage T4. The negative results obtained are open to a variety of interpretations, since actual uptake of phage, release of naked DNA, and synthesis of phage proteins were not monitored in these experiments. More recently, Tanaka *et al.* (1975) obtained positive results based on a different experimental approach to the same problem. By treating XP cells with inactivated Sendai virus, they were able to effect uptake by the cells of purified T4 UV endonuclease. Such cells carried out normal levels of UDS following UV irradiation, suggesting that the enzyme circumvented an incision defect in XP cells.

While these results can be interpreted as being consistent with a DNA incision defect in excision repair of UV radiation damage, they do not indicate that the defect specifically involves an endonuclease. In recent years, our laboratory has attempted to identify an excision-repair-specific UV endonuclease activity in extracts of normal cells, and to determine whether such an activity is defective in XP cells. In order to avoid selecting for a non-dimer-specific endonuclease, we have measured the excision of thymine dimers from UV-irradiated DNA by cell-free extracts as the assay for endonuclease activity. Initial experiments were performed with HeLa and WIL II cells grown in suspension culture (Duncan *et al.*, 1975). Recently, more definitive experiments have been carried out with WI-38 fibroblasts growing in monolayer and with human lymphocytes harvested from fresh peripheral blood (Mortelmans *et al.*, 1976). Extracts of these cells were prepared by sonication. The supernatants were incubated in the presence of 1.0 mM $MgCl_2$, and labeled DNA purified from either prokaryote or mammalian sources and irradiated at a fluence of 100 J m^{-2}. The excision of dimers was monitored by measuring the thymine dimer content of the acid-precipitable fraction of DNA (Cook and Friedberg, 1976), while general degradation of DNA was determined from the radioactivity released into the acid-soluble fraction. Using 1.5–5.0×10^6 cell equivalents of crude extract per incubation, we were able to observe the loss of between 50% and 90% of the thymine dimers in the DNA, with the degradation of only 0.1–2.0% of the substrate (Fig. 8A). This extent of DNA degradation is a highly reproducible result, suggesting that nonspecific deoxyribonuclease activity is limited under our incubation conditions.

The highly selective loss of thymine dimers from DNA implies the presence of a specific endonuclease activity in these cell-free preparations. At present, little information on this activity is available. If either the cells from which extracts were obtained or the extracts themselves were frozen at $-20°C$ without any special precautions, dimer-excising activity was lost. When purified T4 UV endonuclease was added to the frozen and thawed extracts, excising activity was restored, suggesting that it is the endonucleave (or some cofactor that it requires) that is sensitive to freezing (Table 4). Similar results were obtained when the UV-irradiated substrate was chromatin rather than purified DNA (Fig. 8A). In these experiments, cells were prelabeled with [³H]thymidine, sonicated, and exposed to UV radiation. The entire sonicate was incubated in the presence of $MgCl_2$, and dimer excision from the labeled chromatin was measured.

Direct comparisons between normal and XP cell extracts have revealed a highly interesting distinction. Whereas XP cell extracts from complementation group A readily effected the excision of thymine dimers from

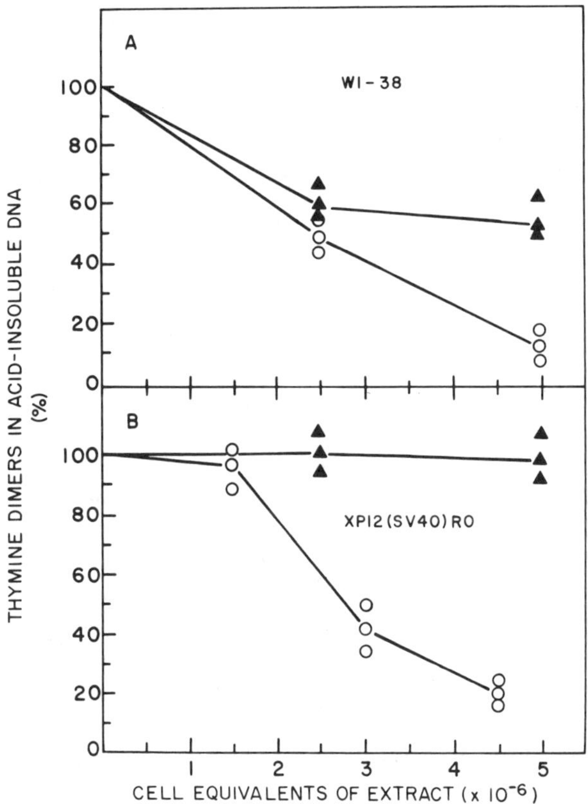

Fig. 8. Excision of thymine dimers from UV-irradiated DNA or chromatin by cell-free extracts. Crude extracts of unlabeled WI-38 cells (A) or XP12(SV40)RO (group A cells) (B) were incubated with [³H]thymine-labeled *Escherichia coli* DNA (O). The data plotted show the percent thymine dimers remaining in the acid-insoluble fraction of the DNA following incubation with increasing amounts of cell extract. In separate experiments, extracts of [³H]thymine-prelabeled cells were incubated in order to determine excision from the DNA (presumably present as chromatin) in the cell-free extract (▲). From Mortelmans et al. (1976).

TABLE 4.   Effect of Freezing HeLa Cells on Thymine Dimer
Excision *in Vitro*[a]

Cells frozen	Enzyme addition to cell-free extract	Thymine as thymine dimers in acid-precipitable T7 DNA (%)	
		0 min	60 min
−	Nil	100	57.3
+	Nil	100	111.9
+	T2 UV exonuclease	100	100.0
+	T4 UV endonuclease	100	54.4
+	T2 exonuclease T4 endonuclease	100	57.4

[a] From Duncan *et al.* (1975).

purified DNA, they were not able to effect excision from their own labeled chromatin (Fig. 8B). The excision defect does not appear to reside in the chromatin itself, since, when extracts of WI-38 cells were mixed with extracts of prelabeled XP cells, a normal loss of dimers from the XP chromatin was observed. These results indicate that, in cell-free preparations, the molecular mechanisms of thymine dimer excision from naked DNA and from chromatin differ, and that XP group A cells are defective in the latter process. This defect possibly resides at the level of endonucleolytic incision of chromatin since the addition of purified T4 UV endonuclease to extracts of prelabeled group A cells stimulates the excision of dimers. However, such an interpretation is based on the assumption that the human and T4 UV endonuclease attack pyrimidine dimers at the same site. If, for instance, the human endonuclease attacks dimers in chromatin on their 3′ side, the 5′ site could still be susceptible to endonucleolytic attack by the T4 enzyme. Another possibility to be considered is that the XP cells are defective in a step that precedes endonucleolytic incision of chromatin, and that the T4 UV endonuclease is not dependent on this step. Nonetheless, the establishment of cell-free assays that can measure both endonuclease activity (using purified DNA) and an additional necessary factor(s) (using chromatin) is encouraging, and should facilitate attempts to purify and characterize these components from normal and XP cells.

3.2.2.   Pyrimidine Dimer-Excising Nucleases

It is theoretically possible that pyrimidine dimers could be excised from DNA by the concerted action of two distinct endonucleases that incise the polynucleotide chain on both sides of the dimer. Studies with prokaryote systems have not borne out such a model, however. Rather, it appears that a single endonucleolytic incision, by creating free ends in the polynucleotide chain, provides the substrate for exonucleases that recognize free termini, and effect dimer excision by DNA degradation from these ends.

Experience with such enzymes in prokaryote systems suggests that the capacity of a given exonuclease to effect the excision of dimers is determined by the lack of substrate specificity, which would otherwise preclude such a role. Thus exonucleases that specifically recognize damaged nucleotides as being "different" are inhibited in their action when such sites are encountered in the course of processive DNA degradation (Grossman, 1974; Grossman *et al.*, 1968). A second factor that selects for dimer-excising activity is determined by the polarity of the endonucleolytic incision. Incisions 5′ to dimers require exonucleolytic degradation in the 5′ → 3′ direction, while incisions 3′ to dimers require degradation in the reverse direction. At the present time, no mammalian dimer-specific endonuclease has been characterized with respect to the polarity of DNA incision. Nonetheless, a number of studies have identified exonuclease activities in extracts of mammalian tissues that are able to degrade UV-irradiated DNA, and in some instances these activities have been shown to effect the excision of thymine dimers from DNA previously incised with prokaryote UV endonucleases.

3.2.2a.   DNases III and IV from Rabbit Tissues

Lindahl *et al.* (1969*a,b*) have reported the presence of two new deoxyribonuclease activities in extracts of rabbit myeloid and lymphoid tissues. In accord with the numbering system established by Cunningham and Laskowski (1953) for mammalian DNases, they referred to these two enzymes as DNases III and IV. A number of reports of activities resembling DNase III had been made prior to 1969 (Eron and McAuslan, 1966; Morrison and Keir, 1966; Georgatsos and Symeonidis, 1965).

DNase III has been purified 680-fold from normal rabbit bone marrow (Lindahl *et al.*, 1969*b*). The enzyme was shown to have a molecular weight of about 52,000, a pH optimum at 8.5, and a requirement for either Mg^{2+} or Mn^{2+}. Support for an exonucleolytic mode of action was obtained from studies using sedimentation in alkaline sucrose. After 2% of the DNA had been digested to acid-soluble products, the remaining material showed no detectable decrease in sedimentation coefficient. Thus no endonucleolytic activity was detected in the purified preparation. Studies on the kinetics of hydrolysis of double-labeled polymers indicated that DNase III degrades both single- and double-stranded polydeoxyribonucleotides preferentially from the 3′ end (Lindahl *et al.*, 1969*b*). When DNase III was incubated with UV-irradiated [³H]poly(dT) or [³²P]DNA, thymine dimers were liberated in the acid-soluble fraction; however, the presence of photoproducts in the DNA significantly inhibited the rate of degradation by DNase III. Conceivably, an enzyme like DNase III could subserve a role in pyrimidine dimer excision from DNA in which incisions were present on the 3′ side of the dimers (Lindahl, 1971).

DNase IV was partially purified from rabbit bone marrow (Lindahl *et al.*, 1969*a*), and subsequently from rabbit lung (Lindahl, 1970). The properties of the enzyme from both sources are identical. The enzyme from bone marrow showed optimal activity at pH 8.5 and required a divalent cation for activity. At an optimal concentration of 2 mM, Mn^{2+} was twice as effective as Mg^{2+}. The enzyme was completely inactivated by 0.3 mM *p*-hydroxymercuribenzoate, and 50% inhibition was observed in the presence of 80 mM NaCl or 33 mM potassium phosphate. After 5 min at 52°C, less than 1% of the activity was detectable. The molecular weight calculated from the sedimentation coefficient and the Stokes radius was about 42,000, while that measured by gel filtration was about 43,000.

Lindahl *et al.*, (1969*a*) have reported that after 35% hydrolysis of dA-[³H]:dT by DNase IV, the major radioactive acid-soluble product was 5′ dAMP. Acid-soluble radioactivity in dinucleotide was less than 10% that in mononucleotide, and in oligonucleotide it was less than 3% that in mononucleotide. The results indicate that the enzyme acts exonucleolytically, releasing mononucleotides as the product. Further support for an exonucleolytic action was provided by showing that when 5% of [³H]dAT was degraded to an acid-soluble form, only a very small reduction in the sedimentation coefficient was detected. The kinetics of degradation of terminally labeled polydeoxyribonucleotides support an exonucleolytic mechanism operating from the 5′ end exclusively. Although initial experiments apparently failed to detect the degradation of single-stranded synthetic polymers, subsequent reports indicate that denatured DNA was degraded at a slow rate (Lindahl, 1972). Moreover, an apparently very similar enzyme obtained from rat lung hydrolyzed denatured DNA 4 times faster than native DNA. Surprisingly, both the rabbit and rat enzymes degraded double-stranded DNA much more slowly than synthetic polymers.

In contrast to DNase III, DNase IV was not significantly inhibited by photoproducts in DNA or poly(dA):poly(dT). About 40% of the products of the degradation of the latter UV-irradiated substrate were oligonucleotides containing two to eight nucleotides. Thymine dimers were found in oligonucleotides five to eight residues in size (Lindahl, 1971). Lindahl (1972) also incubated DNase IV with a preparation of *Bacillus subtilis* DNA that had been UV-irradiated and preincubated with a UV endonuclease preparation from *M. luteus*. Analysis of the acid-soluble fraction indicated an enrichment of the ratio of thymine dimers to thymine monomers, suggesting that dimers were preferentially excised from the DNA. Since all known *M. luteus* UV endonucleases attack irradiated DNA on the 5′ side of pyrimidine dimers (Grossman, 1975), these results indicate that rabbit DNase IV can excise thymine dimers *in vitro* by exonucleolytic degradation in the 5′ → 3′ direction.

DNase IV bears a striking resemblance to the 5′ → 3′ exonuclease moiety of *E. coli* DNA polymerase I. As pointed out by Lindahl (1971), the similar properties between the two activities include:

1. The preference for double-stranded substrates.
2. The sites of attack (5′ ends with either phosphate or hydroxy end groups).
3. The ability to excise pyrimidine dimers from DNA effectively.
4. The more rapid degradation *in vitro* of double-stranded synthetic polydeoxyribonucleotides of simple repeating base composition compared to DNA from natural sources.

This comparison is of more than passing interest, since the exonuclease activity in *E. coli* is normally associated with a DNA-polymerizing moiety from which it can be readily separated by limited proteolysis. The question then arises as to whether a similar association exists in living mammalian cells that is particularly susceptible to proteolysis during cell extraction. Lindahl (1971) has pointed out indirect evidence against such a possibility (including the absence of detectable DNA polymerase activity in purified preparations of DNase IV), but the definitive proof against such an enzyme association is, of course, difficult to obtain.

3.2.3b. Human Thymine Dimer-Excising Activities

Activities resembling DNase IV of rabbit tissues have been identified from human sources. As indicated in the discussion on UV endonucleases, our laboratory has established a human cell-free system that preferentially excises thymine dimers from UV-irradiated DNA or chromatin (Mortelmans *et al.*, 1976). Although freezing and thawing either the cells or the extracts from which they are derived inactivates endonuclease activity, dimer-excising activity is retained under these conditions. This can readily be demonstrated by using UV-irradiated DNA preincubated with purified T4 UV endonuclease as the substrate (Duncan *et al.*, 1975). It should be noted that the use of such a substrate specifically selects for 5′ → 3′ exonuclease activities, since the T4 endonuclease creates endonucleolytic incisions 5′ with respect to dimers (Friedberg and Lehman, 1974).

The activity is present in crude extracts of human WI-38 fibroblasts, the human lymphocyte line WIL 2, HeLa cells, and KB cells (Duncan *et al.*, 1975; Friedberg *et al.*, 1974). The activity selectively excises thymine dimers from specifically incised DNA (Fig. 9). Incubation of unnicked UV-irradiated DNA with extracts of frozen cells does not result in the loss of thymine dimers from acid-insoluble DNA (Fig. 9). Like DNase IV from rabbit tissues, the human activity is sensitive to SH inhibitors; 1.6 mM *p*-chloromercuriphenylsulfonic acid completely inactivates the enzyme(s) (Duncan *et al.*, 1975; Friedberg *et al.*, 1974). In addition, the activity has an absolute requirement for either Mg^{2+} or $Mn,^{2+}$ the former being twice as effective a cofactor at a concentration of 3.0 mM. Mg^{2+} or Mn^{2+} cannot be replaced by Ca^{2+} as a cofactor for the activity (Duncan *et al.*, 1975; Friedberg *et al.*, 1974).

In considering the molecular mechanism of pyrimidine dimer excision by enzymes such as these exonucleases, the intriguing question arises as to how $5' \rightarrow 3'$ exonucleolytic degradation is limited and controlled. As indicated by Lindahl (1972), after removal of a damaged residue and a few adjacent mononucleotides in DNA by an exonuclease, further nuclease action in some way must be prevented in order to avoid extensive degradation of the DNA. A number of plausible theoretical schemes merit brief consideration:

1. The combined effects of a pyrimidine dimer and an endonucleolytic incision adjacent to it may destabilize the H-bonding in that seg-

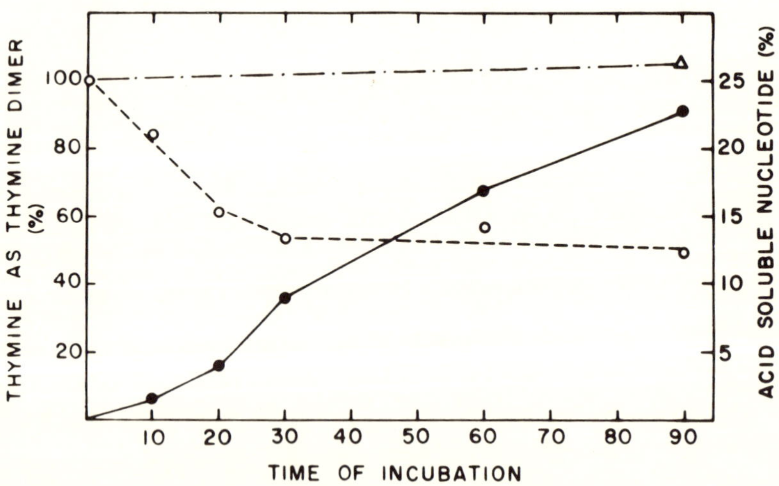

Fig. 9. Excision of thymine dimers by a crude extract of WIL II lymphocytes. O, Thymine dimer content of prenicked DNA; ●, acid-soluble nucleotide; Δ, thymine dimer content of unnicked DNA. From Duncan *et al.* (1975).

ment of the polynucleotide chain so that a significant region exists in a single-stranded type of conformation. Exonucleases that excise pyrimidine dimers may specifically recognize such areas, and terminate DNA degradation where normal H-bonding becomes reestablished in the DNA duplex.

2. It is possible that repair synthesis and exonucleolytic degradation take place concurrently, and that factors regulating the extent of strand displacement during DNA synthesis determine the extent of exonucleolytic degradation.

3. Lindahl (1972) has postulated that the kinetics of exonucleolytic degradation by these enzymes may provide for a "partly processive" or "multiple" mode of attack, in which, on the average, ten monomeric residues are released as the result of a single enzyme–substrate encounter.

4. Finally, it is possible that direct inhibitors of exonuclease activity exist in living cells, although no evidence for such molecules has been reported.

3.2.2c. *Exonuclease Defectiveness in Human DNA Repair*

The demonstration of multiple complementation groups in the disease xeroderma pigmentosum leads to the obvious possibility that the clinical syndrome may arise from defects in different steps of the excision repair pathway. With this hypothesis in mind, Cook *et al.* (1975) measured the specific activity of $5' \rightarrow 3'$ exonuclease present in crude extracts of human cells, using UV-irradiated DNA previously incised with purified phage T4 UV endonuclease as the substrate. Comparisons between a variety of control and XP cells, including cells from complementation groups A, C, D, and E, revealed no significant differences (Table 5). However, more recent experiments (K. Cook and E. C. Friedberg, unpublished observations) have shown that phosphocellulose chromatography of crude extracts of HeLa or KB cells resolves a number of distinct peaks of $5' \rightarrow 3'$ dimer-excising activity (Fig. 10). This result does not appear to be the result of proteolysis in the extracts, since the same results are obtained in the presence of inhibitors of proteolysis. Similar results have been observed by Grossman and his colleagues using extracts of human placenta (L. Grossman, personal communication). If these chromatographically separable peaks of exonuclease activity indeed represent distinct enzymes, it is possible that one or more are specifically defective in some of the XP complementation groups. Studies aimed at investigating this possibility are currently in progress in our laboratory.

One $5' \rightarrow 3'$ exonuclease identified in human placenta has been purified by Doniger and Grossman (1975). A preliminary report indicates that the enzyme is able to degrade duplex DNA that has extending single-stranded termini. It hydrolyzes at both $3'$ and $5'$ termini with equal facility, releasing oligonucleotides (predominantly four to seven residues in length) with 5′ phosphorylated ends. The exonuclease specifically excises pyrimidine dimers from UV-irradiated DNA incised with a UV-specific endonuclease from *M. luteus*.

Besides XP, recent observations on a number of other human diseases are suggestive of defects in DNA repair. One of these, Fanconi's anemia (FA), was first described as a clinical entity about 50 years ago (Fanconi, 1927). As indicated by Fanconi (1967), the disorder is rare and is characterized by hematological abnormalities consisting of pancytopenia and bone marrow hypoplasia, symptoms that usually appear between the fourth and twelfth years of life. The course of the disease is chronic and progressive; death usually occurs in childhood as a result of hemorrhage or some other manifestation of failure of the marrow. The syndrome often affects siblings, and this, along with an increased incidence of parental consanguinity, is quoted as evidence for its being genetic in origin and transmitted as an autosomal recessive. It is generally accepted that individuals with this syndrome have an increased incidence of leukemia and other malignancies. In addition, Todaro *et al.* (1966) have demonstrated an increased susceptibility to transformation by SV40 virus of fibroblasts in tissue culture from Fanconi's anemia patients.

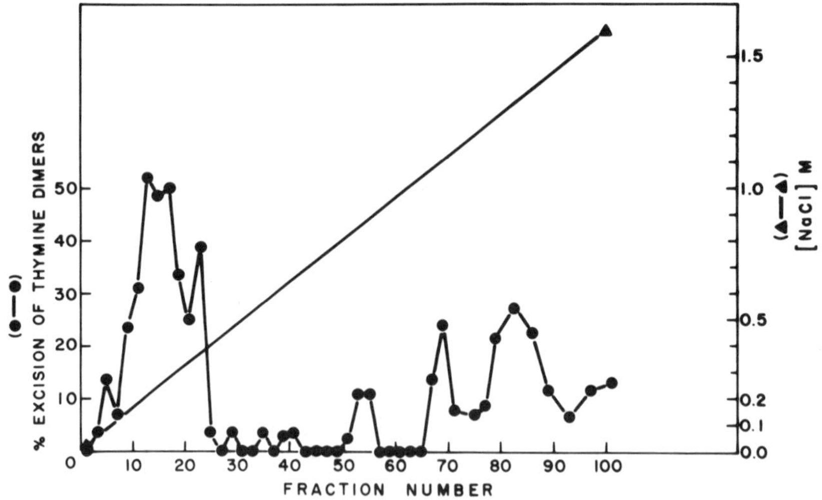

Fig. 10. Chromatographic separation of exonuclease activities on phosphocellulose. A cytoplasmic fraction of KB cells freed of nuclei was loaded onto a phosphocellulose column (Bio-Rad Cellex-P) and eluted with a linear gradient of NaCl. Fractions were assayed for the ability to excise thymine dimers from ³H-labeled *E. coli* DNA preincised with T4 UV endonuclease. From K. H. Cook and E. C. Friedberg (unpublished data).

TABLE 5. Thymine Dimer-Excision Activity in Normal and XP Cell Extracts[a]

Source of extract	Complementation group	Thymine dimers excised in 30 min (%)	Total acid-soluble nucleotides (%)
197 P (normal)	—	22	5.4
XP25RO	A	20	3.4
XP1PW	C	19	4.8
XP2NE	D	32	2.8
XP25SF	E	25	5.2

[a] From Cook *et al.* (1975).

Schroeder *et al.* (1964) are credited with the first reported observation of spontaneously occurring chromosomal breakage in the lymphocytes of patients with this disease. Since then, only a few patients have been reported not to show this phenomenon (Bloom *et al.,* 1966), which is also demonstrable in cultured skin fibroblasts (German, 1972). Schuler *et al.* (1969) showed that the cells of an FA patient were abnormally sensitive to chromosomal aberrations induced by the polyfunctional alkylating agent tetramethanesulfonil-*d*-mannit, and Sasaki and Tonomura (1973) reported similar findings with difunctional alkylating agents such as nitrogen mustard and mitomycin C, as well as with 8-methoxypsoralen treatment in the presence of long wavelength UV irradiation. In contrast to these agents, all of which can cross-link DNA, Sasaki and Tonomura (1973) observed no particular sensitivity to monofunctional alkylating agents, 4-nitroquinoline-1-oxide, γ-irradiation (^{60}Co), or short wavelength UV irradiation. These authors suggested the possibility that FA cells may be defective in the repair of interstrand cross-links; however, they found normal levels of UDS in FA fibroblasts and in phytohemagglutinin-stimulated FA peripheral blood lymphocytes treated with mitomycin C or nitrogen mustard.

Direct studies on the repair of damaged DNA in FA cells have been reported by Regan *et al.* (1970) and Poon *et al.* (1974). The former authors reported normal kinetics of thymine dimer excision in UV-irradiated skin fibroblasts growing in tissue culture. In contrast, the latter group reported a significant reduction in the ability of cells to excise thymine dimers compared to normal cells. Of particular relevance to the present discussion was the speculation that FA cells may be defective in the exonucleolytic step of dimer excision. Sedimentation velocity analysis of UV-irradiated cells showed a reduction in the molecular weight of the DNA of both normal and FA fibroblasts, suggesting normal endonucleolytic incision in the latter. Additionally, levels of UDS in both cell types bore the same relationship to UV fluence, consistent with normal DNA repair synthesis. On the other

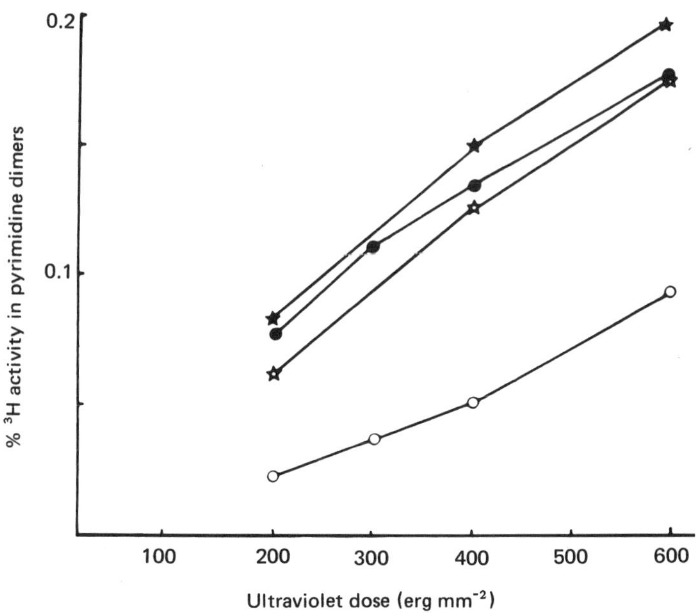

Fig. 11. Comparison of excision of thymine-containing pyrimidine dimers from UV-irradiated, [³H]thymidine-labeled normal human fibroblasts and fibroblasts from a patient with FA. ●, Normal human fibroblasts immediately after irradiation; O, 24 h after irradiation. ★, FA cells immediately after irradiation; ☆, 24 h after irradiation. From Poon *et al.* (1974).

hand, measurement of the thymine dimer content of acid-precipitable DNA as a function of UV fluences between 20 and 60 J m⁻² showed a significant failure of FA cells to remove dimers effectively during a 24-h period (Fig. 11). On the basis of these observations, Poon *et al.* (1974) suggested that FA cells may be defective in an exonuclease that excises thymine dimers.

3.2.3. The Role of DNA Polymerases in Excision Repair

All models of excision repair postulate a DNA synthesis step to replace the excised nucleotides. Models based on prokaryote studies have suggested that DNA synthesis may follow the completion of excision (cut and patch) or may be the driving force for excision by strand displacement during DNA synthesis (patch and cut) (Haynes, 1966; Howard-Flanders and Boyce, 1966).

In prokaryotes, a number of DNA polymerases have an associated $5' \rightarrow 3'$ exonuclease function that makes them highly attractive candidates for subserving the roles of both excision and repair synthesis during DNA repair (Grossman, 1975). In mammalian cells, the DNA polymerases of nuclear and cytoplasmic origin thus far identified have no demonstrable $5' \rightarrow 3'$ or $3' \rightarrow 5'$ exonuclease activities (Weissbach, 1975), and there is no obvious basis for identifying interesting enzymes that may function in repair synthesis. For the sake of completeness, we offer the following brief survey of current knowledge on mammalian cell DNA polymerases. The interested reader is advised to consult a number of recent reviews on this topic (Weissbach, 1975; Fansler, 1974; Kornberg, 1974; Loeb, 1974).

DNA Polymerase α. Even though DNA polymerase α is often the major activity in cytoplasmic extracts of growing cells, its actual location *in vivo* may be in the nucleus. The α-polymerase, according to Weissbach (1975), is a high-molecular-weight enzyme with an often observed sedimentation coefficient of approximately 6–8 S and an estimated molecular weight of 110,000–220,000. The enzyme has a pronounced tendency to exist as aggregate forms of higher molecular weight under conditions of low ionic strength (Byrnes *et al.,* 1973). As with all the known DNA polymerases, the α-polymerase requires $3'$ OH sites as primers for initiation of synthesis. It is especially active on primed duplex DNA containing gaps, and from this point of view would certainly fit the requirements of a cut-and-patch mechanism of repair synthesis. The enzyme has no detectable nuclease activity *in vitro,* nor can it catalyze pyrophosphate exchange with deoxynucleoside triphosphates (Chang and Bollum, 1973). The enzyme is strongly inhibited by SH-group inhibitors (Byrnes *et al.,* 1973).

DNA Polymerase β. DNA polymerase β is recovered almost entirely in nuclear extracts, but its presence has also been reported in the cytoplasm (Chang and Bollum, 1972). It comprises a single polypeptide chain, and has a molecular weight of about 45,000 (Chang and Bollum, 1972). However, under conditions of low ionic strength the enzyme forms large aggregates having apparent molecular weights of over 250,000 (Wang *et al.,* 1975). Like the α-enzyme, this polymerase has no associated nuclease activity, and utilizes "gapped" DNA very effectively as a substrate (Chang and Bollum, 1972). It is insensitive to SH-group inhibitors (Wang *et al.,* 1975).

DNA Polymerase γ. DNA polymerase γ has the unique ability to copy synthetic ribohomopolymers such as $(rA)_n \cdot dT_{15}$ at a significantly greater rate than deoxyribohomopolymer templates, or "activated" (gapped) DNA itself (Fridlender *et al.,* 1972). It represents only 1–2% of the total cellular DNA polymerase. In HeLa cells it has a molecular weight of 110,000 and may exist in multiple forms (Spadari and Weissbach, 1974). In a recent study, Bertazzoni *et al.* (1976) found that following stimulation of human lymphocytes with phytohemagglutinin, two waves of induction of DNA polymerase activity were observed. The first wave, between the third and fifth days, occurred in parallel with an increase in DNA synthesis rate, and mainly involved DNA polymerase α. Later, when DNA synthesis had declined to almost resting levels, a wave of mainly DNA polymerase β activity was observed that coincided with a peak in the capacity of cells to perform repair synthesis. These authors have suggested that DNA polymerase β may be specifically required for DNA repair, rather than for semiconservative DNA synthesis.

3.2.4. DNA Ligases in Mammalian Cells

Any model of excision repair in which new nucleotides are inserted into DNA necessarily requires a biochemical step involving the joining of the last newly inserted nucleotide to the adjacent nucleotide of the extant DNA. The joining of nucleotides by phosphodiester linkage in DNA is a reaction carried out by an enzyme called polynucleotide (DNA) ligase. This enzyme has been extensively purified from a number of prokaryote sources, and characterized in detail (Gellert, 1967; Olivera and Lehman, 1967; Weiss and Richardson, 1967). Its role in DNA repair in these organisms has been established mainly by the use of mutants defective in ligase function (Morse and Pauling, 1975; Sugimoto *et al.*, 1968). In eukaryotes, the luxury of available mutants that can complement biochemical studies is not yet at hand. Nonetheless, it is reasonable to assume that at this level of biological organization, DNA ligase is also an important component of the enzymology of DNA repair.

The first report of polynucleotide ligase activity in extracts of mammalian cells was by Lindahl and Edelman (1968). These workers identified and purified this activity about 200-fold from extracts of rabbit bone marrow. The enzyme, estimated to have a molecular weight of about 95,000, is dependent for its activity on a double-stranded DNA substrate with adjacent 3′OH and 5′P termini, ATP, and Mg^{2+}. DPN could not be substituted for ATP. The enzyme was shown to catalyze an ATP-dependent pyrophosphate exchange in the absence of DNA. ATP was also hydrolyzed by the enzyme in the absence of DNA. This reaction was stimulated by the presence of DNA nicked with pancreatic DNase I, suggesting that an enzyme–adenylate complex is a reaction intermediate. In subsequent studies with an enzyme purified from calf thymus (Söderhäll and Lindahl, 1973*a*), a covalently linked enzyme–adenylate complex was isolated. The isolated complex was shown to catalyze the joining of adjacent 3′OH and 5′P termini in DNA in the absence of added ATP, with release of the adenylate residue. Furthermore, on incubation of ligase–[14C]adenylate complex with pyrophosphate, the radioactivity was recovered as ATP. These data suggest that the mammalian polynucleotide ligase functions in the same manner as the enzymes obtained from prokaryote sources.

Studies on subcellular distribution indicated that 60% of the activity was in the nuclear fraction, 35% in the cytoplasm, and 5% in a crude mitochondrial fraction (Lindahl and Edelman, 1968). However, in view of the well-known propensity for nuclear enzymes to leach into the cytoplasm during aqueous extraction of nuclei (Wang, 1967), these distributions may have limited significance. DNA ligase activity requiring ATP as a cofactor was identified in extracts of spleen, thymus, liver, and small intestine, but not in rabbit kidney, brain, heart, or lung extracts (Lindahl and Edelman, 1968). No enzyme activity was detected that could utilize DPN instead of ATP. Struck by the observation that the rabbit enzyme resembles the phage T4 rather than the *E. coli* enzyme in terms of its cofactor requirement, Lindahl and Edelman (1968) appropriately raised the possibility of virus-induced ligase activity in mammalian cells.

Polynucleotide ligase activity has now been isolated from a number of other mammalian cell sources (Table 6). Sambrook and Shatkin (1969) demonstrated ATP-dependent ligase activity in extracts of chicken, hamster, mouse, monkey, and human cells grown in culture. After infection of mouse embryo, monkey kidney, and HeLa cells with polyoma virus, SV40, and vaccinia virus, respectively, the enzyme activity in extracts of these cell types increased, suggesting that virus infection induces an activity indistinguishable (at least in terms of cofactor requirement) from that present in tissues from animals. If such an enzyme is actually virus coded, then obvious problems in the interpretation of tissue culture studies could exist. More detailed studies on uninfected and virus-infected cells in culture were reported by Beard (1972). He observed that the ligase activities isolated from uninfected and polyoma virus-infected mouse embryo cells were identical with respect to pH and ionic strength optima, K_m for ATP, and molecular weight (about 200,000). Furthermore, in no case in which cells were infected with temperature-sensitive polyoma mutants did the ligase activity show any temperature dependence for activity. Beard (1972) concluded that it is most probable that the ligase of polyoma-infected cells is specified by the host genome.

Tsukada and Ichimura (1971) observed that the activity in extracts of rat liver had properties very similar to the rabbit enzyme, and that following partial hepatectomy the specific activity increased 305-fold. Spadari *et al.* (1971) observed the activity in extracts of a cultured human cell line (EUC). These authors noted that the partially purified enzyme had bimodal pH optima at 7.5 and 8.1 when assayed in either *tris*-HCl buffer or potassium phosphate buffer, and suggested the possibility of two enzymes in human tissues (Fig. 12). They also noted a requirement for crude boiled extract after the ligase had been purified by DEAE-cellulose chromatography, but later studies suggest that this is a rather nonspecific phenomenon (Pedrali

TABLE 6. Mammalian DNA Ligase Activities

Authors	Source	Molecular weight	pH Optima	Cofactors	Other characteristics
Söderhäll and Lindahl (1973a,b, 1975)	Assorted mammalian tissues	175,000 and	7.4–8.0	Mg^{2+}, ATP	Ligase I: $1000\times$ purified, high levels in rapidly dividing cells, 100% active after 5′ at 45°C, may be polymeric, freeze sensitive
		85,000	7.8	Mg^{2+}, ATP	Ligase II: insensitive to antibodies to ligase I, no activity after 5′ at 45°C, freeze sensitive
Beard (1972)	Mouse embryo	220,000	7.2–7.8	Mg^{2+}, ATP	$180\times$ purified K_m for ATP $= 1.5 \times 10^{-6}$ M
Pedrali Noy et al. (1973)	Human cells in culture	95,000 190,000	7.5 8.1	Mg^{2+}, ATP Mg^{2+}, ATP	$350\times$ purified, interconversion between the two forms by handling
Bertazzoni et al. (1972)	Calf thymus	—	7.8 8.2	Mg^{2+}, ATP	$390\times$ purified, Mn^{2+} as cofactor gives 60% of Mg^{2+} activity, freeze inactivated, K_m for ATP $= 1.5 \times 10^{-6}$ M
Tsukada and Ichimura (1971)	Rat liver	—	8.0	Mg^{2+}, ATP	Fourfold stimulation of activity in regenerating liver, PCMB sensitive
Sambrook and Shatkin (1969)	Human cells in culture	—	—	Mg^{2+}, ATP	Increased activity after viral infection (SV40), polyoma, vaccinia)

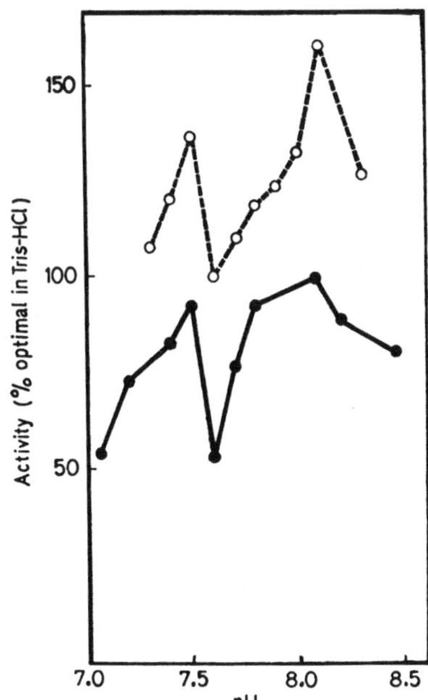

Fig. 12. Dependence of polynucleotide ligase activity on pH. The reaction was performed in 50 mM *tris*-HCl (●) or 50 mM potassium phosphate (O) buffers at the indicated pH, at room temperature. The activity is expressed as the percentage of that found at pH 8.1 with *tris*-HCl. From Spadari *et al.* (1971).

Noy *et al.*, 1973). Bertazzoni *et al.* (1972) also observed that DNA ligase activity obtained from calf thymus had two pH optima at 7.2 and 8.2, but were unable to observe any difference in K_m values for ATP at these two values. More specific evidence for the existence of two forms of DNA ligase activity in mammalian cells stems from the studies of Pedrali Noy *et al.* (1973) and of Söderhäll and Lindahl (1973*b*, 1975).

Pedrali Noy *et al.* (1973) characterized two distinct forms of the enzyme by gel filtration, with estimated molecular weights of about 95,000 and 190,000, respectively. They demonstrated that the enzyme from fresh crude extracts of EUC cells eluted off a Sephadex G100 column as a single peak of molecular weight 190,000. When this peak of activity was pooled and rechromatographed on Sephadex G100, about 40% of the activity eluted as the 95,000 molecular weight fraction. Furthermore, if fresh crude extract was allowed to stand at 0°C for 20 days, the first gel filtration yielded the two molecular weight forms. Pedrali Noy *et al.* (1973) concluded that both aging and handling of the enzyme can cause this conversion, and "it seems legitimate to conclude that we are dealing with a dimer structure going into a monomer." They did point out that the definition of "monomer" in this sense referred only to the smallest enzymatically active unit observed thus far, and did not imply a single polypeptide chain. They also pointed out that the apparent tendency for mammalian cell DNA ligase to aggregate and disaggregate may explain the different molecular weights observed by different authors.

In contrast, Söderhäll and Lindahl (1973*b*, 1975) provided evidence suggestive of two distinct polynucleotide ligase activities from calf thymus. They observed two species of enzyme activity separable by hydroxylapatite chromatography (Söderhäll and Lindahl, 1973*b*). Calculation of molecular weight based on sedimentation coefficients of the two activities yielded values of 175,000 and 85,000. Both activities required Mg^{2+} and ATP as cofactors; however, a number of other properties distinguished the two activities:

1.   DNA ligase I (high-molecular-weight fraction) was more stable to both heat and storage at 2°C than was DNA ligase II.

2.   DNA ligase I was observed to have a broad pH optimum between 7.4 and 8.0, while DNA ligase II was maximally active at pH 7.8.

3.   At pH 7.0, ligase I was more active in 2-(*N*-morpholino)ethanesulfonic acid–KOH buffer than in *tris*-HCl buffer, while the reverse was observed for ligase II.

In subsequent studies, Söderhäll and Lindahl (1975) purified both DNA ligases from calf thymus, and prepared an antiserum against DNA ligase I. The antiserum specifically inhibited this enzyme, but not DNA ligase II at equivalent protein concentrations. The antiserum was also specific when tested against a mixture of DNA ligases I and II. It was shown that, in addition to its inability to inhibit DNA ligase II activity, the DNA ligase I antiserum failed to bind to that enzyme, suggesting that the two ligases are antigenically unrelated. Examination of extracts of calf spleen and liver, rabbit spleen, human plaenta, and mouse ascites cells confirmed the presence of the two DNA ligase forms. In all cases, including calf thymus, DNA ligase I activity was present in about 10 times higher amount than ligase II.

Söderhäll and Lindahl (1975) reported that they too observed the dissociation of DNA ligase I to a lower-molecular-weight form, which they referred to as DNA ligase I monomer. This form of DNA ligase, however, was distinguished from DNA ligase II in that it had the same heat sensitivity and pH optimum as DNA ligase I, and was also inhibitable by DNA ligase I antiserum. Attempts to generate a DNA ligase II type of activity by proteolysis of DNA ligase I were unsuccessful. These authors concluded, therefore, that the DNA ligase I monomer of Pedrali Noy *et al.* (1973) and their DNA ligase II are two distinct proteins.

3.3. The Repair of Monoadduct Damage to DNA

The agents most frequently associated with single-base modifications are ionizing radiation, alkylating agents, and other chemicals that interact with DNA. These agents often produce rather complex and as yet incompletely characterized effects on DNA, resulting in the generation of multiple substrates, some of which are common to more than one type of damaging agent. Here, we will focus on those interactions that result in the generation of substrate sites with which specific enzyme interactions have been associated. The reader interested in a detailed description of the physicochemical effects of ionizing radiation and chemicals on DNA is referred to a number of excellent discussions on these topics (Ward, 1975; Lawley, 1966).

3.3.1. Alkylation Damage

Alkylating agents are electrophilic reagents that can combine with nucleophilic centers in DNA by S_N1- or S_N2-type substitution reactions. The products formed are numerous and varied, and differ both qualitatively and quantitatively for different alkylating agents (Table 7). Alkylation of all purines and pyrimidines has been demonstrated, and most bases react at several sites; however, damage to purines is quantitatively more significant than to pyrimidines. In addition to monoadduct base damage, alkylating agents can cause phosphotriester bond formation in DNA, and the bifunctional alkylating agents can produce interstrand cross-links between affected bases. The latter is appropriately considered as a form of diadduct base damage, and will not be discussed from the point of view of DNA repair in this section.

3.3.2. Ionizing Radiation Damage

Base damage from ionizing radiation primarily involves the pyrimidines (Scholes *et al.,* 1960). Hydroxyl radicals produced by the ionization of water interact with the 5,6 double bond, resulting in saturation through a peroxide intermediate (Hariharan and Cerutti, 1971, 1972). The best-characterized reaction products in this category are 5,6-dihydroxydihydrothymine and 5,6-dihydroxydihydrocytosine. Besides base damage, ionizing radiation is known to cause damage to sugar residues and to phosphodiester linkages (Ward, 1971, 1975; Daniels *et al.,*1956).

TABLE 7. Reported Sites of Attack by Electrophilic Reagents on DNA

Compound	Reaction Sites[a]					
	Adenine	Cytosine	Guanine	Thymine	Phosphate	Reaction type
N-Acetoxy-acetylaminofluorine (N-acetoxy-AAF)	C-8	—	C-8(100)	—	—	S_N1
7-Bromomethylbenz[α]anthracene (7-BMBA)	N^6 (6-amino)	N^4 (4-amino) (probable)	N^2 (2-amino) (major product)	—	—	S_N1
N-Methyl-N-nitrosourea (MNUA)	N-1(1.4) N-3(11.2) N-7(2.5)	N-3(0.6)	N-3(1.1) N-7(75.7) 0-6(7.3)	N-3(0.3)	(18)	S_N1
N-Methyl-N-nitro-N-nitrosoguanidine (MNNG)	N-1(1) N-3(12)	N-3(2)	N-3(2) N-7(67) 0-6(7)			S_N1
Ethylmethanesulfonate (EMS)					(15)	$S_N1 + S_N2$
Methylmethanesulfonate (MMS)	N-1(5) N-3(9)	N-3(1)	N-3(0.68) N-7(86) 0-6(0.34)		(1)	S_N2

[a] Values in parentheses give approximate relative amounts of reaction products. In some cases, these values are compiled from different publications and therefore do not add up to 100%. An entry means that the product has been reported. Lack of entry means only that the reference given does not record that product. From Strauss *et al.* (1975).

3.3.3. Deamination of Bases

Shapiro and Klein (1966) studied the pH dependency of cytosine deamination to uracil. They determined that the hydrolytic deamination of cytidine to uridine is a slow reaction, with 16 h required for half-reaction in 2 M citrate buffer at pH 4 and 95°C. Nonetheless, they suggested that the deamination of cytosine to uracil in DNA could provide the molecular basis for base transition mutations, and speculated that cytosine may be responsible in part for the mutations produced on heating *E. coli* in neutral solution.

Lindahl and Nyberg (1974) investigated the deamination of cytosine residues in DNA. They found that heat-induced deamination in single-stranded DNA, poly(dC), or dCMP occurred at readily detectable rates with rate constants of 2×10^{-7} s^{-1} at 95°C. The cytosine residues in native DNA were deaminated at rates of less than 1% of that observed with poly(dC).

Deamination of bases is the most widely recognized mode of action of nitrous acid (Shapiro and Yamaguchi, 1972). In addition to cytosine deamination to uracil, nitrous acid promotes conversion of guanine to xanthine, and deamination of adenine to hypoxanthine. Sodium bisulfite is also very efficient in causing deamination of cytosine in heat-denatured DNA, but not in duplex calf thymus DNA (Shapiro *et al.*, 1973).

3.3.4. Base Loss from DNA

3.3.4a. Depurination

The loss of purines from double-stranded DNA can occur both spontaneously (Lindahl and Karlström, 1973) and following the interaction of DNA with alkylating agents (Lawley and Brookes, 1973), nitrous acid, and ionizing radiation (Ducolomb *et al.*, 1974; Ljungquist *et al.*, 1974). With regard to spontaneous depurination, Greer and Zamenhof (1962) reported that purine bases were released in detectable quantities from DNA at neutral pH and temperatures near the T_m, and that the activation energy for depurination of denatured DNA in 5 mM phosphate buffer at pH 6.8 was about 28 kcal/mol. Subsequent studies by Lindahl and Nyberg (1972) investigated the rate of depurination of native DNA as a function of temperature, pH, and ionic strength. At 70°C, the rate of release of purines from *B. subtilis* proceeded at an initial rate of 4×10^{-9} s^{-1} in a Mg^{2+}-containing buffer. At lower ionic strength, the rate of depurination was higher. The temperature dependence of the rate of depurination of native DNA was associated with an activation energy of 31 ± 2 kcal/mol at both pH 6.0 and

5.0. Based on these experimentally determined data, extrapolation from Arrhenius plots provided an estimated value of 3×10^{-11} s^{-1} for the rate of depurination of native DNA at 37°C and pH 7.4. Assuming that the rate of depurination of DNA *in vitro* approximates that *in vivo*, Lindahl and Nyberg (1972) estimated that a diploid mammalian cell with a generation time of 20 h could lose 12,000 purines from its DNA per generation.

3.3.4b. Depyrimidination

Greer and Zamenhof (1962) did not observe heat-induced depyrimidination of DNA at neutral pH under conditions that clearly resulted in depurination. On reinvestigating this phenomenon, however, Lindahl and Karlström (1973) detected the loss of free pyrimidines from DNA in neutral aqueous buffers of physiological ionic strength heated to 80°C and 95°C. At 95°C and pH 7.4, the rate constant for the release of thymine from DNA was 1.8×10^{-8} s^{-1}. Thus depyrimidination of DNA was considerably slower than depurination at neutral pH. However, Lindahl and Karlström (1973) pointed out that a growing mammalian cell could well lose several hundred pyrimidines during an average generation period.

The rate of depurination of DNA is significantly increased by treatment with alkylating agents. Such treatment results in the formation of purine derivatives with very labile glycosidic bonds (Lawley and Brookes, 1963). At 37°C, for example, 7-methylguanine is released from DNA with a half-life of 3–6 days both *in vivo* and *in vitro*. Based on measurements of the extent of reaction of a number of alkylating agents with HeLa and HEp 2 cells, and on the rate constants estimated for spontaneous depurination quoted above, Strauss *et al.* (1975) calculated that depurination by loss of 7-methylguanine and 3-methyladenine would result in the generation of approximately one apurinic site per 2.5×10^5 base pairs during the first hour after treatment with a monofunctional alkylating agent.

Treatment of DNA with nitrous acid also causes depurination (Schuster, 1960). In this instance, as indicated by Lindahl and Ljungquist (1975), the mechanism is believed to involve deamination of guanine residues to xanthine, followed by hydrolysis of the labile xanthine–deoxyribose bond. Following γ-irradiation of mononucleotides, glycosidic bond cleavage is a major lesion affecting both purines and pyrimidines (Ducolomb *et al.*, 1974). Bond rupture occurs by attack of a hydroxyl radical at the C-1 of the sugar, leading to immediate cleavage of the glycosidic bond and release of an unaltered base. In addition, initial damage to the base residue leads to weakening of the glycosidic bond. On γ-irradiation of double-stranded DNA in neutral solution, lesions of both these types are introduced (Strauss *et al.*, 1975; Strauss and Hill, 1970).

This brief overview of monoadduct base damage is by no means comprehensive, but should serve to indicate that identical or structurally very similar lesions can be produced in DNA in a variety of ways. Thus the association of a given enzyme activity with a particular damaging agent (i.e., ionizing radiation or alkylating agents), rather than with a defined substrate produced by that agent, can be very confusing. Indeed, considerable confusion already exists with respect to the enzymology of the repair of monoadduct damage to DNA in prokaryotes. Since the classification of mammalian cell enzymes is frequently based on the results of earlier prokaryote studies, we feel obliged to offer a brief summary of the current status of the prokaryote field in the hope that the mammalian cell enzymology will be more clearly defined.

3.3.5. Endonuclease II and Apurinic Endonuclease of *E. coli*

Friedberg and Goldthwait (1969) and Friedberg *et al.* (1969) reported the identification of an endonuclease activity in extracts of *E. coli* that extensively degraded DNA alkylated with the monofunctional agent methylmethanesulfonate (MMS). This activity was designated as endonuclease II. In early studies with this enzyme it was demonstrated that intact alkylated DNA was attacked by the enzyme, and that it was not necessary to produce depurinated sites in the substrate in order to demonstrate enzyme activity (Friedberg *et al.*, 1969). Shortly thereafter, Verly and his associates (Verly and Rassart, 1975; Verly *et al.*, 1974; Verly and Paquette, 1972, 1973; Paquette *et al.*, 1972) reported the isolation of an endonuclease activity from *E. coli* that attacked depurinated DNA. This endonuclease was purified to apparent homogeneity, and was shown to have a molecular weight of approximately 32,000. Evidence has been presented that this enzyme is specific for apurinic sites and does not attack alkylated DNA unless the substrate has undergone depurination at the alkylation sites. In view of the known capacity of alkylated DNA to undergo depurination, the question arose as to whether or not the apurinic endonuclease described by Verly and his associates (Verly and Rassart, 1975; Verly *et al.*, 1974; Verly and Paquette, 1972, 1973; Paquette *et al.*, 1972) and endonuclease II are the same enzyme. Further studies from Goldthwait's laboratory showed that the preparation of an enzyme, designated as endonuclease II, attacks alkylated DNA by both *N*-glycosidic and endonucleolytic modes. The *N*-glycosidase activity was evidenced by the enzyme-catalyzed release of free O^6-methylguanine and 3-methyladenine (but not 7-methylguanine) from DNA treated with *N*-methyl-*N*-nitrosourea (Kirtikar and Goldthwait, 1974), and the release of N^6-(12)-methylbenz[*a*]anthracenyl-7-methyl)adenine and N^2-(12)-methylbenz[*a*]anthracenyl-7-methyl)guanine from DNA treated with 7-bromo-methyl-12-methylbenz[*a*]anthracene (Kirtikar *et al.*, 1975*a*).

In addition to the *N*-glycosidic removal of monoadduct base damage, endonuclease II catalyzes endonucleolytic cleavage of alkylated DNA. Kirtikar *et al.* (1976) have provided evidence that this phosphodiester hydrolase activity is distinct from the apurinic endonuclease described by Verly and Rassart (1975):

1.   Endonuclease II and the apurinic endonuclease have been separated by DEAE chromatography. The latter enzyme has been purified about 3000-fold and has been shown to possess minimal activity on alkylated DNA, presumably due to the presence of some depurinated sites. In agreement with the results of Verly and Rassart (1975), this enzyme has a molecular weight of about 33,000. The chromatographic fraction, designated as endonuclease II, degraded both alkylated and depurinated DNA, although the rate of degradation of the latter substrate was less than 10% that of the degradation of alkylated DNA.

2.   Mutants have been identified that are defective in either the apurinic endonuclease or endonuclease II. The latter activity was defective in a mutant of *E. coli* sensitive to MMS (Kirtikar *et al.*, 1976). The apurinic endonuclease activity was defective in a mutant isolated by Yajko and Weiss (1975) on the basis of decreased endonucleolytic activity on MMS-treated DNA.

3.   Phorbol myristate acetate, the active cocarcinogen in croton oil, inhibited both the hydrolysis of phosphodiester bonds and the *N*-glycosidase activity of endonuclease II. The hydrolysis of phosphodiester bonds in depurinated DNA by the apurinic endonuclease was not significantly inhibited by phorbol ester.

These lines of evidence make it unlikely that endonuclease II is an *N*-glycosidase acting in conjunction with the apurinic endonuclease, and suggest that the phosphodiester hydrolase activity associated with endonuclease II is a distinct enzyme. Assuming that the substrate sites for the phosphodiester hydrolase activity are depurinated sites produced by the activity of the *N*-glycosidase, it is not clear precisely how this activity differs from the apurinic endonuclease. Kirtikar *et al.* (1976) have shown that the rate of degradation of depurinated DNA by endonuclease II is always much slower than with alkylated DNA. These authors have advanced the possibility that endonuclease II is a single enzyme with two active sites (one for *N*-glycosidase activity and one for phosphodiesterhydrolase activity), and that the K_m of the enzyme for alkylated DNA is significantly lower than that for depurinated DNA. In addition to the substrate specificities designated above, endonuclease II of *E. coli* as defined by Kirtikar *et al.* (1976) attacks γ-irradiated DNA (Kirtikar *et al.*, 1975*b*), but the nature of the specific sites that are attacked has not yet been determined.

A number of other activities have been variously associated with endonuclease II of *E. coli*. Prior to the definitive separation of apurinic endonuclease from endonuclease II activity, the latter was found to have a very limited phosphodiesterhydrolase action on native DNA (Friedberg *et al.*, 1969), but approximately a thousandfold more enzyme was required for this hydrolysis than for the hydrolysis of alkylated DNA (Hadi *et al.*, 1973). At the time of writing, it is not clear whether this activity is associated with endonuclease II, the apurinic endonuclease, or neither. Finally, Yajko and Weiss (1975) reported the isolation of mutants of *E. coli* defective in either exonuclease III activity or an activity that degrades MMS-alkylated DNA (referred to by them as endonuclease II). In 12 independently derived mutants, deficiency of one enzyme was always accompanied by deficiency of the other. Kirtikar *et al.* (1976) have reported that in their hands endonuclease II could be clearly resolved by gel filtration from an exonuclease activity present in the later stages of purification. If this activity is exonuclease III, it is apparently not physically associated with endonuclease II. It is not clear from our review of the literature whether or not purified apurinic endonuclease has demonstrable exonuclease activity.

3.3.6. Other Enzymes That Recognize Monoadduct Base Damage

Aside from endonuclease II, other enzymes in *E. coli* have been identified that attack γ-irradiated DNA. Cerutti and co-workers (Hariharan and Cerutti, 1974*a,b*) have reported the selective excision of products of the 5,6-dihydroxydihydrothymine type from γ-irradiated, OsO_4-oxidized DNA or synthetic poly[d(A-T)] with crude extracts of *M. luteus*. The specific enzyme activities responsible for this excision reaction have not yet been identified. Striniste and Wallace (1975) have isolated an endonuclease activity from *E. coli* that attacks X-irradiated ϕX174 RF DNA. They have shown the presence of this activity in extracts of the MMS-sensitive strain AB3027 that is defective in endonuclease II activity, and suggest that their activity is distinct from endonuclease II.

The demonstration by Kirtikar and Goldthwait (1974) and Kirtikar *et al.* (1975*b*) of an *N*-glycosidase activity that can attack sites of monoadduct base damage is a highly significant observation. It indicates the existence of a new class of repair enzyme activity, which, instead of effecting base removal by direct phosphodiester bond hydrolysis, removes damaged bases by hydrolysis of the *N*-glycosidic bond, followed by endonucleolytic attack at the resultant apurinic or apyrimidinic site. In the light of this observation, we would stress the importance of specifically distinguishing between *N*-glycosidase and endonuclease activities when assaying crude extracts for excision repair of any type of base damage. These can be readily confused if, for instance, activity is measured by the release of radioactivity into an acid-soluble form from labeled DNA. Furthermore, since the sites of base loss left by *N*-glycosidase activity contain alkali-labile phosphodiester bonds, sedimentation velocity studies in alkaline gradients can generate results suggestive of an endonuclease activity.

It has recently been established that endonuclease II is not a unique example of an *N*-glycosidase activity. Another such activity that specifically removes uracil residues from DNA has been detected in prokaryotes. This activity was first reported in extracts of *E. coli* by Lindahl (1974), who used as a substrate *E. coli* DNA in which cytosine bases had been deaminated to uracil. A similar activity was discovered in *B. subtilis* by Friedberg *et al.* (1975), who used as a substrate the naturally occurring uracil-containing DNA from bacteriophage PBS2. Duncan *et al.* (1976) have shown that extracts of both *B. subtilis* and *E. coli* also contain a nuclease activity that attacks the depyrimidinated sites left in DNA following *N*-glycosidic removal of uracil. This nuclease is defective in extracts of a strain of *E. coli* (BW 2001) that has been shown by Kirtikar *et al.* (1976) to be defective in apurinic endonuclease activity. We therefore conclude that a single endonuclease can attack both apurinic and apyrimidinic sites in DNA. The designation "apurinic endonuclease" for this activity is perhaps confusing, since it implies a specificity for apurinic sites in DNA.

We would suggest that a clear distinction be made between excision repair involving the removal of damaged nucleotides from DNA (*nucleotide excision repair*) and that involving the removal of damaged bases from DNA (*base excision repair*). It is conceivable that the latter is a rather general type of excision repair mode based on the presence in cells of a number of *N*-glycosidases, each relatively specific for different types of monoadduct base damage.

Following *N*-glycosidic attack, a single endonuclease specific for sites of base loss in DNA (the so-called apurinic endonuclease) could hydrolyze a phosphodiester bond on either the 3′ or 5′ side of the site of base loss (Fig. 13). A second endonucleolytic incision on the opposite side would release a free phosphorylated deoxyribose residue, leaving a gap with a single nucleotide missing. Alternatiely, exonucleolytic degradation in either the 3′ → 5′ direction or the 5′ → 3′ direction (depending on the polarity of the endonucleolytic incision) would create a gap amenable to repair synthesis as in nucleotide excision repair (Fig. 13). If in *E. coli* exonucleolytic degradation occurs in the 3′ → 5′ direction, the purported association of exonuclease III with endonuclease II (Yajko and Weiss, 1975) may be of particular relevance.

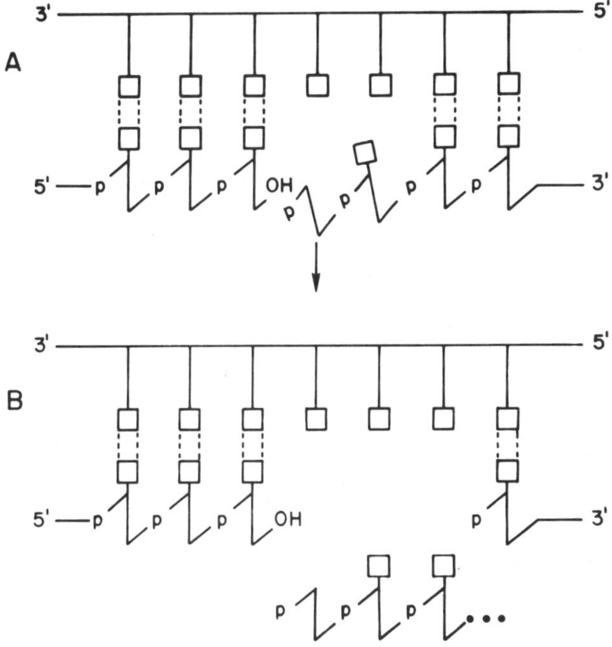

Fig. 13. Model for repair of sites of base loss in DNA. A: Endonucleolytic incision on the 5′ side of a site of base loss is shown leaving 3′ OH and 5′ P termini. B: Exonucleolytic degradation in the 5′ → 3′ direction removes the site of base loss as part of a larger oligonucleotide.

3.3.7. Enzymes from Mammalian Cells That Attack Sites of Monoadduct Base Damage

3.3.7a. *Apurinic Endonuclease*

Lindahl (1972) reported the presence of an activity in extracts of rabbit and calf thymus that selectively attacked heat-depurinated DNA. Subsequent reports indicate a similar activity present in extracts of rat liver (Verly and Paquette, 1973). The calf thymus apurinic endonuclease has been purified about 830-fold, and has been extensively characterized by Ljungquist *et al.* (1974) and Ljungquist and Lindahl (1974).

The purified enzyme had a pH optimum at 8.5, and enzyme activity was strongly stimulated by the presence of Mg^{2+} or Mn^{2+} in the reaction mixture, but not by Ca^{2+}. The enzyme did not have an absolute requirement for divalent cation, a significant residual activity being present in EDTA at

1 mM. The molecular weight was calculated to be about 32,000 based on gel filtration studies. The enzyme showed no activity on native PM2 DNA or on single-stranded DNA. However, heated DNA (from which purines were presumably lost) was efficiently attacked by the endonuclease, and DNA treated by other depurinating agents was also an effective substrate. Thus, when MMS-treated PM2 DNA containing two or three alkylated purines per molecule was exposed to the calf thymus endonuclease, very little chain cleavage was observed. But when the alkylated DNA was heated to depurinate the alkylated sites, there was essentially quantitative enzymatic cleavage. Single-stranded depurinated DNA was not a substrate for the enzyme. The enzyme was reported to create endonucleolytic incisions on the 3′ side of the apurinic sugar residues, with the formation of 5′ phosphate termini (Ljungquist and Lindahl, 1974).

Kirtikar *et al.* (1976) have purified an endonuclease specific for apurinic sites from calf liver approximately 900-fold. Interestingly, the enzyme has a number of properties that distinguish it from the apurinic endonuclease from calf thymus. The pH optimum of the calf liver enzyme was higher than that for the calf thymus enzyme, and it did not show the marked dependence on divalent metal shown by the thymus enzyme. While the liver enzyme was very sensitive to ionic strength (50% inhibition by 23 mM NaCl), the thymus enzyme was stimulated by low concentrations of salt. Additionally, the liver enzyme created both single- and double-strand breaks in depurinated, reduced DNA, while the thymus enzyme was reported to make only single-strand breaks.

3.3.7b. *Apurinic Endonuclease in Xeroderma Pigmentosum*

In a recent study, Kuhnlein *et al.* (1976) have investigated endonuclease activity on depurinated phage PM2 DNA in extracts of normal and xeroderma pigmentosum fibroblasts. They observed that the specific activity of apurinic endonuclease was lower in extracts of all XP cells, including the XP variant and an XP heterozygote. The decrease in activity was especially pronounced in extracts of cells from complementation group D, which had about one-sixth the specific activity of control cell extracts. Phosphatase activity measured in the same extracts showed no deficiency in the XP cells compared to the normal fibroblasts, although there was a considerable range of variation in the extracts examined. Mixing experiments showed no evidence of an inhibitor in the group D extracts. Examination of the pH optimum of apurinic endonuclease activity in extracts of normal and XP cells, excluding group D, revealed peaks at pH 7.5, 8.0, and 8.3, suggesting the presence of multiple activities in normal cells. This suggestion was supported by the observation that extracts of human placenta contain at least four different apurinic endonuclease activities separable by column chromatography (S. Linsleo, E. E. Penhoet, and S. Linn, unpublished observations). The extracts of group D cells showed a constant activity between pH 7.0 and 8.5.

Apparent K_m values for apurinic DNA were measured in the cell extracts. Kuhnlein *et al.* (1976) observed that the K_m values in extracts of group D cells were about sixfold higher than normal, and in complementation group A, the K_m was about ninefold higher. Thus, in group D cells, there may be both quantitative and qualitative alterations in apurinic endonuclease activity, and significant qualitative alterations in the activity in group A cells. All other XP cells tested (complementation groups B, C, and E) appeared to be quantitatively defective to some degree in apurinic endonuclease activity *in vitro*.

3.3.7c. Other Enzyme Activities

Enzyme activities in extracts of mammalian cells have been identified that attack alkylated DNA (Frei and Lawley, 1975; Maitra and Frei, 1975; Margison and O'Conner, 1973; Capps *et al.*, 1973; O'Conner *et al.*, 1973), γ-irradiated DNA (Brent, 1973; Mattern *et al.*, 1973; Hariharan and Cerutti, 1971), and DNA containing uracil residues (K. H. Cook and E. C. Friedberg, unpublished observation; Lindahl, 1976). In no case of which we are aware have these activities been purified or characterized with respect to their substrate specificity, and we believe that a detailed discussion of these enzyme activities is best deferred until further information is available.

3.3.7d. Enzymatic Deficiency in Ataxia Telangiectasia

Ataxia telangiectasia (AT) is an autosomal recessive defect in humans in which a number of clinical defects have been reported (McFarlin *et al.*, 1972). These include cerebellar ataxia, telangiectasia, and IgA deficiency. As in XP and FA, an increased frequency of malignancy and an enhanced level of spontaneous chromosomal aberrations have been documented (Harnden, 1974). Cunliffe *et al.* (1975), Morgan *et al.* (1968), and Gotoff *et al.* (1967) have reported an increased sensitivity of AT patients to ionizing radiation, while Rary *et al.* (1974) and Higurashi and Conen (1973) have reported increased frequency of chromosome aberrations in leucocyte cultures of AT patients after ionizing radiation.

Taylor *et al.* (1975) have studied fibroblasts in culture from AT patients. They observed a significant reduction in the survival of cells from three AT patients following exposure to ^{60}Co γ-rays (Fig. 14). In preliminary attempts to characterize the biochemical basis of the enhanced

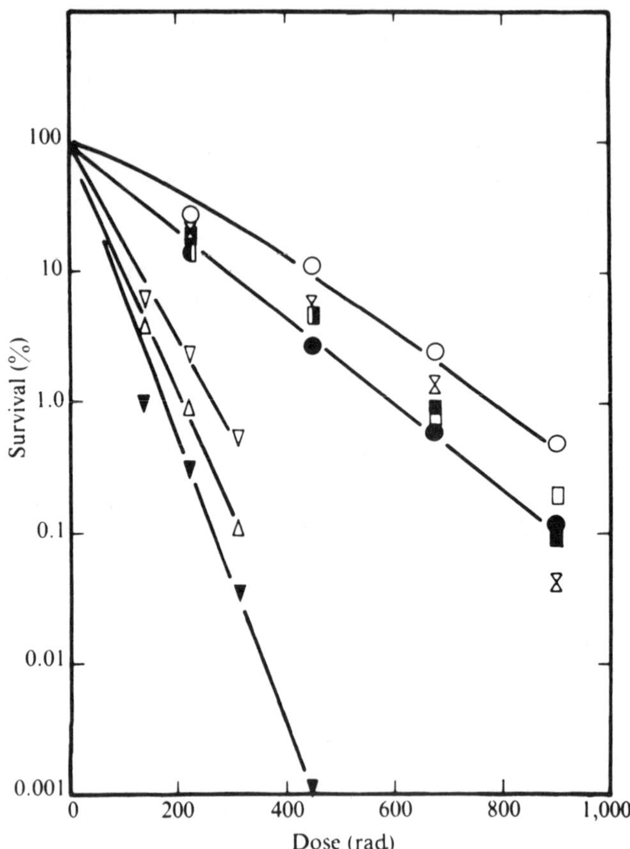

Fig. 14. γ-Ray response of cultured human fibroblasts. Lines fitted by eye. O, Normal 25-year-old male, seven experiments; ●, normal 25-year-old male, eight experiments; □, 38-year-old male with basal cell nevus syndrome, five experiments; ⊠, 74-year-old female with basal cell nevus syndrome, four experiments; ■, 14-year-old male with basal cell nevus syndrome, five experiments; △, 7-year-old male with AT, three experiments; ▽, 4-year-old male with AT, two experiments; ▼, 6-year-old male with AT, one experiment. From Taylor *et al.* (1975).

sensitivity, they compared the capacity of normal and AT cells to rejoin DNA strand breaks after ionizing radiation. The results of sedimentation velocity studies on DNA in both alkaline and neutral gradients showed no significant difference between normal and AT cells.

More recently, Paterson and Smith (1976) have reported a defect in the excision of damaged bases after γ-irradiation in AT cells, and have suggested that these patients may have a defect in an endonuclease required for excision repair of damage induced by ionizing radiation.

4. CONCLUSIONS

It is clear that since the observations of Rasmussen and Painter (1964) first suggested the existence of DNA repair in mammalian cells, great strides have been made in understanding the response of such cells to DNA damage. From the perspective of the enzymologist, the phenomenon of enzymatic photoreactivation, if fully substantiated, holds the promise of becoming a powerful probe in numerous types of photobiological experiments aimed at examining the role of pyrimidine dimers in pathological effects of UV radiation. Similarly, the establishment of cell-free systems for measuring the excision of damaged bases and nucleotides from both DNA and chromatin will hopefully facilitate our understanding of the enzymology of excision repair in mammalian cells. Such information should provide significant insights into the nature of the biochemical defects in diseases such as xeroderma pigmentosum, Fanconi's anemia, and ataxia telangiectasia.

What is sorely lacking in the field of mammalian cell DNA repair is the availability of mutant cells that are defective in different steps in the various DNA repair modes, so that the biochemist has a meaningful way of assessing the biological relevance of the numerous observations in cell-free systems being reported. It is to be hoped that the next dozen years will fill this need.

ACKNOWLEDGMENTS

The studies in the authors' laboratory were supported by Research Grants CA-12428 from the United States Public Health Service and NP 174 from the American Cancer Society as well as by Contract No. E (04-3)326 with the United States Energy Research and Development Administration. E. C. Friedberg is a research fellow of the Andrew W. Mellon Foundation and the recipient of Research Career Development Award CA 71005 from the National Cancer Institute, United States Public Health Service.

5. REFERENCES

Bacchetti, S., and Benne, R., 1975, Purification and characterization of an endonuclease from calf thymus acting on irradiated DNA, *Biochim. Biophys. Acta* **390**:285–297.

Bacchetti, S., Van der Plas, A., and Veldhuisen, G., 1972, A UV-specific endonucleolytic activity present in human cell extracts, *Biochem. Biophys. Res. Commun.* **48**:662–669.

Beard, P., 1972, Polynucleotide ligase in mouse cells infected by polyoma virus, *Biochim. Biophys. Acta* **269**:385–396.

Bertazzoni, U., Mathelet, M., and Campagnari, F., 1972, Purification and properties of a polynucleotide ligase from calf thymus glands, *Biochim. Biophys. Acta* **287**:404–414.

Bertazzoni, U., Stefanini, M., Pedrali Noy, E. G., Nuzzo, F., Falaschi, A., and Spadari, S., 1976, Variations of DNA polymerases α and β during prolonged stimulation of human lymphocytes, *Proc. Natl. Acad. Sci. USA* **73**:785–789.

Bloom, G. E., Warner, S., Gerald, P. S., and Diamond, L. K., 1966, Chromosome abnormalities in constitutional aplastic anemia, *N. Engl. J. Med.* **274**:8–14.

Boyce, R. P., and Howard-Flanders, P., 1964, Release of ultraviolet light-induced thymine dimers from DNA in *E. coli*, *Proc. Natl. Acad. Sci. USA* **51**:293–300.

Brent, T. P., 1972, Repair enzyme suggested by mammalian endonuclease activity specific for ultraviolet-irradiated DNA, *Nature (London) New Biol.* **239**:172–173.

Brent, T. P., 1973, A human endonuclease activity for gamma-irradiated DNA, *Biophys. J.* **13**:399–401.

Brent, T. P., 1975, Partial purification of endonuclease activity from human lymphoblasts. Separation of activities for depurinated DNA and DNA irradiated with ultraviolet light, *Biochim. Biophys. Acta* **407**:191–199.

Burt, D. H., and Brent, T. P., 1971, A deoxyribonuclease activity of HeLa cells specific for UV-irradiated DNA, *Biochem. Biophys. Res. Commun.* **43**:1382–1387.

Byrnes, J. J., Downey, K. M., and So, A. G., 1973, Bone marrow cytoplasmic deoxyribonucleic acid polymerase. Variation of pH and ionic environment as a possible control mechanism, *Biochemistry* **12**:4378–4384.

Capps, M. J., O'Connor, P. J., and Craig, A. W., 1973, The influence of liver regeneration on the stability of 7-methylguanine in rat liver DNA after treatment with *N,N*-dimethylnitrosamine, *Biochim. Biophys. Acta* **331**:33–40.

Cerutti, P., 1975, Repairable damage in DNA: Overview, in: *Molecular Mechanisms for Repair of DNA* (P. C. Hanawalt and R. B. Setlow, eds.), pp. 3–12, Plenum Press, New York.

Chang, L. M. S., and Bollum, F. J., 1972, Low molecular weight deoxyribonucleic acid polymerase from rabbit bone marrow, *Biochemistry* **11**:1264–1272.

Chang, L. M. S., and Bollum, F. J., 1973, A comparison of associated enzyme activities in various deoxyribonucleic acid polymerases, *J. Biol. Chem.* **248**:3398–3404.

Cleaver, J. E., 1968, Defective repair replication of DNA in xeroderma pigmentosum, *Nature (London)* **218**:652–656.

Cleaver, J. E., 1969, Xeroderma pigmentosum: A human disease in which an initial stage of DNA repair is defective, *Proc. Natl. Acad. Sci. USA* **63**:428–435.

Cleaver, J. E., 1971, Repair of alkylation damage in ultraviolet-sensitive (xeroderma pigmentosum) human cells, *Mutat. Res.* **12**:453–462.

Cleaver, J. E., 1973, DNA repair with purines and pyrimidines in radiation- and carcinogen-damaged normal and xeroderma pigmentosum cells, *Cancer Res.* **33**:362–369.

Cleaver, J. E., 1974*a*, Repair processes for photochemical damage in mammalian cells, *Adv. Radiat. Biol.* **4**:1–76.

Cleaver, J. E., 1974*b*, Sedimentation of DNA from human fibroblasts irradiated with ultraviolet light: Possible detection of excision breaks in normal and repair-deficient xeroderma pigmentosum cells, *Radiat. Res.* **57**:207–227.

Cleaver, J. E., and Bootsma, D., 1975, Xeroderma pigmentosum—biochemical and genetic characteristics, *Annu. Rev. Genet.* **9**:19–38.

Cleaver, J. E., and Trosko, J. E., 1970, Absence of excision of ultraviolet-induced cyclobutane dimers in xeroderma pigmentosum, *Photochem. Photobiol.* **11**:547–550.

Cleaver, J. E., Bootsma, D., and Friedberg, E. C., 1975, Human diseases with genetically altered DNA repair processes, *Genetics* **79**:215–225.

Cleaver, J. E., Paterson, M., and Friedberg, E. C., 1976, Absence of photoenzymatic monomer-isation of pyrimidine dimers in normal and xeroderma pigmentosum cells, *Biophys. J.* **16**:185a.

Cook, J. S., 1970, Photoreactivation in animal cells, *Photophysiology* **5**:191–233.

Cook, J. S., and McGrath, J. R., 1967, Photoreactivating enzyme activity in metazoa, *Proc. Natl. Acad. Sci. USA* **58**:1359–1365.

Cook, K. H., and Friedberg, E. C., 1976, Measurement of thymine dimers in DNA by thin-layer chromatography. II. The use of one-dimensional systems. *Anal. Biochem.* **73**:411–418.

Cook, K., Friedberg, E. C., Cleaver, J. E., and Slor, H., 1975, Excision of thymine dimers from specifically incised DNA by extracts of xeroderma pigmentosum cells, *Nature (London)* **256**:235–236.

Cunliffe, P. N., Mann, J. R., Cameron, A. H., Roberts, K. D., and Ward, H. W. C., 1975, Radiosensitivity in ataxia-telengiectasia, *Br. J. Radiol.* **48**:374–376.

Cunningham, L., and Laskowski, M., 1953, Presence of two different desoxyribonu-cleodepolymerases in veal kidney, *Biochim. Biophys. Acta* **11**:590–591.

Daniels, M., Scholes, G., and Weiss, J. J., 1956, Chemical action of ionizing radiations in solu-tions. Part XVI. Formation of labile phosphate esters from purine and pyrimidine ribonu-cleotides by irradiation with X-rays in aqueous solution, *J. Chem. Soc. (London)*, 3771–3779.

Day, R., III, 1975, The use of human adenovirus 2 in the study of the xeroderma DNA-repair defect, *Molecular Mechanisms for the Repair of DNA* (P. C. Hanawalt and R. B. Setlow, ed.), pp. 747–752, Plenum, New York.

Doniger, J., and Grossman, L., 1975, The characterization of an exonuclease purified from human placenta capable of pyrimidine dimer excision, *Biophys. J.* **15**:297a.

Ducolomb, R., Cadet, J., and Teoule, R., 1974, Irradiation γ de l'acide uridylique-rupture de la liaison *N*-glycosidique, *Int. J. Radiat. Biol.* **25**:139–149.

Duker, N. J., and Tabor, G. W., 1975, Different ultraviolet DNA endonuclease activity in human cells, *Nature* **255**:82–84.

Duncan, J., Slor, H., Cook, K., and Friedberg, E. C., 1975, Thymine dimer excision by extracts of human cells, in: *Molecular Mechanisms for Repair of DNA* (P. C. Hanawalt and R. B. Setlow, ed.), pp. 643–649, Plenum, New York.

Duncan, J. A., Hamilton, L. D., and Friedberg, E. C., 1976, The enzymatic degradation of uracil-containing DNA. II. Evidence for *N*-glycosidase and nuclease activities in unfrac-tionated extracts of *B. subtilis*, *J. Virology* (in press).

Epstein, W. L., Fukuyama, K., and Epstein, J., 1971, Ultraviolet light, DNA repair and skin carcinogenesis in man, *Fed. Proc.* **30**:1766–1771.

Eron, L. J., and McAuslan, B. R., 1966, The nature of pox-virus induced deoxyribonucleases, *Biochem. Biophys. Res. Commun.* **22**:518–523.

Fanconi, G., 1927, Familiarer infantile perniziosaartige anämie (perniziöses blutbild und konstitution), *Jahrb. Kinderheilkd.* **117**:257–280.

Fanconi, G., 1967, Familial constitutional panmyelocytopathy, Fanconi's anemia (FA). I. Clinical Aspects, *Semin. Hematol.* **4**:233–240.

Fansler, B. S., 1974, Eukaryotic DNA polymerases: Their association with the nucleus and relationship to DNA replication, *Int. Rev. Cytol. Supp.* **4**:363–415.

Frei, J. V., and Lawley, P. D., 1975, Methylation of DNA in various organs of C57B$_1$ mice by a carcinogenic dose of *N*-methyl-*N*-nitrosourea and stability of some methylation products up to 18 hours, *Chem. Biol. Interact.* **10**:413–427.

Fridlender, B., Fry, M., Bolden, A., and Weissbach, A., 1972, A new synthetic RNA-dependent DNA polymerase from human tissue culture cells, *Proc. Natl. Acad. Sci. USA* **69**:452–455.

Friedberg, E. C., 1975, DNA repair of ultraviolet-irradiated bacteriophage T4, *Photochem. Photobiol.* **21**:277–289.

Friedberg, E. C., and Clayton, D. A., 1972, Electron microscopic studies on substrate specificity of T4 excision repair endonuclease, *Nature (London)* **237**:99–103.

Friedberg, E. C., and Goldthwait, D. A., 1969, Endonuclease II of *E. coli.* I. Isolation and purification, *Proc. Natl. Acad. Sci. USA* **62**:934–940.

Friedberg, E. C., and Lehman, I. R., 1974, Excision of thymine dimers by proteolytic and *amber* fragments of *E. coli* DNA polymerase I, *Biochem. Biophys. Res. Commun.* **58**:132–139.

Friedberg, E. C., Hadi, S., and Goldthwait, D. A., 1969, Endonuclease II of *E. coli.* II. Enzyme properties and studies on the degradation of alkylated and native deoxyribonucleic acid, *J. Biol. Chem.* **244**:5879–5889.

Friedberg, E. C., Duncan, J., and Cleaver, J. E., 1974, Thymine dimer excision nuclease in extracts of human cells, *Radiat. Res.* **59**:98.

Friedberg, E. C., Ganesan, A. K., and Minton, K., 1975, *N*-Glycosidase activity in extracts of *Bacillus subtilis* and its inhibition after infection with bacteriophage PBS2, *J. Virol.* **16**:315–321.

Gellert, M., 1967, Formation of covalent circles of lambda DNA by *E. coli* extracts, *Proc. Natl. Acad. Sci. USA* **57**:148–155.

Georgatsos, J. G., and Symeonidis, A., 1965, A deoxyribonuclease from mammary tumors of C3H mice preferentially hydrolyzing heat-denatured DNA, *Nature (London)* **206**:1362–1363.

German, J., 1972, Genes which increase chromosomal instability in somatic cells and predispose to cancer, *Prog. Med. Genet.* **8**:61–101.

Gotoff, S. P., Amirmokri, E., and Liebner, E. J., 1967, Ataxia telangiectasia: Neoplasia, untoward response to X-irradiation, and tuberous sclerosis, *Am. J. Dis. Child.* **114**:617–625.

Greer, S., and Zamenhof, S., 1962, Studies on depurination of DNA by heat, *J. Mol. Biol.* **4**:123–141.

Grossman, L., 1974, Enzymes involved in the repair of DNA, *Adv. Radiat. Biol.* **4**:77–129.

Grossman, L., 1975, Excision repair of DNA, in: *DNA Synthesis and Its Regulation* (M. Goulian and P. C. Hanawalt, eds.), pp. 791–814, W. A. Benjamin, San Francisco.

Grossman, L., Kaplan, J. C., Kushner, S. R., and Mahler, I., 1968, Enzymes involved in the early stages of repair of ultraviolet-irradiated DNA, *Cold Spring Harbor Symp. Quant. Biol.* **33**:229–234.

Grossman, L., Braun, A., Feldberg, R., and Mahler, I., 1975, Enzymatic repair of DNA, *Annu. Rev. Biochem.* **44**:19–43.

Hadi, S., and Goldthwait, D. A., 1971, Endonuclease II of *Escherichia coli*: Degradation of partially depurinated deoxyribonucleic acid, *Biochemistry* **10**:4986–4994.

Hadi, S., Kirtikar, D. M., and Goldthwait, D. A., 1973, Endonuclease II of *Escherichia coli*: degradation of double and single-stranded deoxyribonucleic acid, *Biochemistry* **12**:2747–2754.

Hanawalt, P. C., 1968, Cellular recovery from photochemical damage, *Photophysiology* **4**:203–251.

Hariharan, P., and Cerutti, P., 1971, Repair of γ-ray-induced thymine damage in *Micrococcus radiodurans, Nature (London) New Biol.* **229**:247–249.

Hariharan, P., and Cerutti, P., 1972, Formation and repair of γ-ray-induced thymine damage in *Micrococcus radiodurans, J. Mol. Biol.* **66**:65–81.

Hariharan, P. V., and Cerutti, P. A., 1974*a*, Excision of damaged thymine residues from gamma-irradiated poly(dA-dT) by crude extracts of *Escherichia coli, Proc. Natl. Acad. Sci. USA* **71**:3532–3536.

Hariharan, P. V., and Cerutti, P. A., 1974*b*, The incision and strand rejoining step in the excision repair of 5,6-dihydroxy-dihydrothymine by crude *E. coli* extracts, *Biochem. Biophys. Res. Commun.* **61**:375–379.

Harm, H., 1974, Biological action of DNA photoreactivating enzyme from mammalian cells, *Abst. Am. Soc. Photobiol.*, p. 137.

Harnden, D. G., 1974, Ataxia telangiectasia syndrome: Cytogenetic and cancer aspects, *Chromosomes and Cancer* (J. German, ed.), pp. 619–636, Wiley, New York.

Hart, R. W., and Setlow, R. B., 1973, Evidence for a role of UV-induced pyrimidine dimers in malignant transformation, *Abst. Am. Soc. Photobiol.*, p. 120.

Hart, R. W., and Setlow, R. B., 1975, Direct evidence that pyrimidine dimers in DNA result in neoplastic transformation, in: *Molecular Mechanisms for Repair in DNA* (P. C. Hanawalt and R. B. Setlow, eds.), pp. 719–728, Plenum Press, New York.

Haynes, R. H., 1966, The interpretation of microbial inactivation and recovery phenomena, *Radiat. Res. Suppl.* **6**:1–29.

Higurashi, M., and Conen, P. E., 1973, *In vitro* chromosomal radiosensitivity in "chromosomal breakage syndromes," *Cancer* **32**:380–383.

Howard-Flanders, P., 1968, DNA repair, *Annu. Rev. Biochem.* **37**:175–200.

Howard-Flanders, P., and Boyce, R. P., 1966, DNA repair and genetic recombination: Studies on mutants of *Escherichia coli* defective in these processes, *Radiat. Res. Suppl.* **6**:156–184.

Huang, P. C., and Vincent, R., Jr., 1975, Repair deficiency and genetic complementarity of fibroblast cells in culture from six xeroderma pigmentosum patients, in: *Molecular Mechanisms for the Repair of DNA* (eds., P. C. Hanawalt and R. B. Setlow, eds.), pp. 729–733, Plenum Press, New York.

Jacobs, A. J., O'Brien, R. L., Parker, J. W., and Paolilli, P., 1972, Abnormal DNA repair of 4-nitroquinidine-1-oxide damage by lymphocytes in xeroderma pigmentosum, *Mutat. Res.* **16**:420–424.

Keir, H. M., and Smellie, R. M. S., 1962, Intracellular location of DNA nucleotidyl transferase, *Nature (London)* **196**:752–754.

Kirtikar, D. M., and Goldthwait, D. A., 1974, The enzymatic release of O^6-methylguanine and 3-methyladenine from DNA reacted with the carcinogen *N*-methyl-*N*-nitrosourea, *Proc. Natl. Acad. Sci. USA* **71**:2022–2026.

Kirtikar, D. M., Dipple, A., and Goldthwait, D. A., 1975*a*, Endonuclease II of *Escherichia coli*: DNA reacted with 7-bromomethyl-12-methylbenz[*a*]anthracene as a substrate, *Biochemistry* **14**:5548–5553.

Kirtikar, D. M., Slaughter, J., and Goldthwait, D. A., 1975*b*, Endonuclease II of *Escherichia coli*: Degradation of γ-irradiated DNA, *Biochemistry* **14**:1235–1244.

Kirtikar, D. M., Kuebler, J. P., Dipple, A., and Goldthwait, D. A., 1976, Enzymes involved in repair of DNA damaged by chemical carcinogens and γ-irradiation, in: *Cancer Enzymology—8th Miami Winter Symposium* (J. Schultz and F. Ahmad, eds.), Academic Press, New York.

Kleijer, W. J., Lehman, P. H. M., Mulder, M. P., and Bootsma, D., 1970, Repair of X-ray damage in DNA of cultivated cells from patients having xeroderma pigmentosum, *Mutat. Res.* **9**:517–523.

Kondo, S., and Jagger, J., 1966, Action spectra for photoreactivation of mutation to prototrophy in strains of *Escherichia coli* possessing and lacking photoreactivating-enzyme activity, *Photochem. Photobiol.* **5**:189–200.

Kornberg, A., 1974, *DNA Synthesis*, W. H. Freeman, San Francisco.

Kuhnlein, U., Penhoet, E. E., and Linn, S., 1976, An altered apurinic DNA endonuclease activity in xeroderma pigmentosum fibroblasts, *Proc. Natl. Acad. Sci. USA* **73**:1169–1173.

Lawley, P. D., 1966, Effects of some chemical mutagens and carcinogens on nucleic acids, *Prog. Nucleic Acid Res. Mol. Biol.* **5**:89–131.

Lawley, P. D., and Brookes, P., 1963, Further studies on the alkylation of nucleic acids and their constituent nucleotides, *Biochem. J.* **89**:127–138.

Lindahl, T., 1970, An exonuclease specific for double-stranded DNA: Deoxyribonuclease IV from rabbit tissues, in: *Methods in Enzymology,* Vol. 21 (L. Grossman and K. Moldave, eds.), pp. 148–153, Academic Press, New York.

Lindahl, T., 1971, Excision of pyrimidine dimers from ultraviolet-irradiated DNA by exonucleases from mammalian cells, *Eur. J. Biochem.* **18**:407–414.

Lindahl, T., 1972, Mammalian deoxyribonucleases acting on damaged DNA, in: *Molecular and Cellular Repair Processes* (R. F. Beers, R. M. Herriott, and R. C. Tilghman, eds.), pp. 3–13, Johns Hopkins University Press, Baltimore, Maryland.

Lindahl, T., 1974, An *N*-glycosidase from *Escherichia coli* that releases free uracil from DNA containing deaminated cytosine residues, *Proc. Natl. Acad. Sci. USA* **71**:3649–3653.

Lindahl, T., 1976, New class of enzymes acting on damaged DNA, *Nature (London)* **259**:64–66.

Lindahl, T., and Andersson, A., 1972, Rate of chain breakage at apurinic sites in double-stranded deoxyribonucleic acid, *Biochemistry* **11**:3618–3623.

Lindahl, T., and Edelman, G. M., 1968, Polynucleotide ligase from myeloid and lymphoid tissues, *Proc. Natl. Acad. Sci. USA* **61**:680–687.

Lindahl, T., and Karlström, O., 1973, Heat-induced depyrimidination of deoxyribonucleic acid in neutral solution, *Biochemistry* **12**:5151–5154.

Lindahl, T., and Ljungquist, S., 1975, Apurinic and apyrimidinic sites in DNA, in: *Molecular Mechanisms for Repair of DNA* (P. C. Hanawalt and R. B. Setlow, eds.), pp. 31–38, Plenum Press, New York.

Lindahl, T., and Nyberg, B., 1972, Rate of depurination of native deoxyribonucleic acid, *Biochemistry* **11**:3610–3618.

Lindahl, T., and Nyberg, B., 1974, Heat-induced deamination of cytosine residues in deoxyribonucleic acid, *Biochemistry* **13**:3405–3410.

Lindahl, T., Gally, J. A., and Edelman, G. M., 1969*a*, Deoxyribonuclease IV: A new exonuclease from mammalian tissues, *Proc. Natl. Acad. Sci. USA* **62**:597–603.

Lindahl, T., Gally, J. A., and Edelman, G. M., 1969*b*, Properties of deoxyribonuclease III from mammalian tissue, *J. Biol. Chem.* **244**:5014–5019.

Ljungquist, S., and Lindahl, T., 1974, A mammalian endonuclease specific for apurinic sites in double-stranded deoxyribonucleic acid. I. Purification and general properties, *J. Biol. Chem.* **249**:1530–1535.

Ljungquist, S., Andersson, A., and Lindahl, T., 1974, A mammalian endonuclease specific for apurinic sites in double-stranded deoxyribonucleic acid. II. Further studies on the substrate specificity, *J. Biol. Chem.* **249**:1536–1540.

Loeb, L. A., 1974, Eucaryotic DNA polymerases, in: *The Enzymes,* Vol. 10 (P. D. Boyer, ed.), pp. 173–209, Academic Press, New York.

Maher, V. M., Douville, D., Tomura, T., and Van Lancker, J. L., 1974, Mutagenecity of reactive derivatives of carcinogenic hydrocarbons: Evidence of DNA repair, *Mutat. Res.* **23**:113–123.

Maitra, S. C., and Frei, J. V., 1975, Organ-specific effects of DNA methylation by alkylating agents in the inbred Swiss mouse, *Chem. Biol. Interact.* **10**:285–293.

Margison, G. P., and O'Connor, P. J., 1973, Biological implications of the instability of the *N*-glycosidic bond of 3-methyldeoxyadenosine in DNA, *Biochim. Biophys. Acta.* **331**:349–356.

Mattern, M., Hariharan, P., Dunlop, B., and Cerutti, P., 1973, DNA degradation and excision repair in γ-irradiated Chinese hamster ovary cells, *Nature (London) New Biol.* **245**:230–232.

McFarlin, D. E., Strober, W., and Walmann, T. A., 1972, Ataxia telangiectasia, *Medicine* **51**:281–314.

Morgan, J. L., Holcomb, T. M., and Morrissey, R. W., 1968, Radiation reaction in ataxia telangectasia, *Am. J. Dis. Child.* **116**:557–558.

Morrison, J. M., and Keir, H. M., 1966, Heat sensitive deoxyribonuclease activity in cells infected with herpes simplex virus, *Biochim. J.* **98**:37c–39c.

Morse, L. S., and Pauling, C., 1975, Induction of error-prone repair as a consequence of DNA ligase deficiency in *Escherichia coli, Proc. Natl. Acad. Sci. USA* **72**:4645–4649.

Mortelmans, K., Friedberg, E. C., Slor, H., Thomas, G., and Cleaver, J. E., 1976, Evidence for a defect in thymine dimer excision in extracts of xeroderma pigmentosum cells, *Proc. Natl. Acad. Sci. USA* **73**:2757–2761.

Morton, H., 1970, A survey of commercially available tissue culture media, *In Vitro* **6**:89–108.

Nishioka, H., and Harm, W., 1972, Analysis of photoenzymatic repair of UV lesions in DNA by single light flashes. IX. Excess production of photoreactivating enzyme in *E. coli* B_{s-1}-160 under different growth conditions, and its suppression by adenine, *Mutat. Res.* **16**:121–131.

O'Connor, P. J., Capps, M. J., and Craig, A. W., 1973, Comparative studies of the hepatocarcinogen *N,N*-dimethyl nitrosamine *in vivo*: Reaction sites in rat liver DNA and the significance of their relative stabilities, *Br. J. Cancer* **27**:153.

Olivera, B. M., and Lehman, I. R., 1967, Linkage of polynucleotides through phosphodiester bonds by an enzyme from *Escherichia coli, Proc. Natl. Acad. Sci. USA* **57**:1426–1433.

Painter, R. B., 1970, The action of ultraviolet light on mammalian cells, *Photophysiology* **5**:169–189.

Paquette, Y., Crine, P., and Verly, W. G., 1972, Properties of the endonuclease for depurinated DNA from *Escherichia coli, Can. J. Biochem.* **50**:1199–1209.

Paterson, M. C., and Setlow, R. B., 1972, Endonucleolytic activity from *Micrococcus luteus* that acts on γ-ray-induced damage in plasmid DNA of *Escherichia coli* minicells, *Proc. Natl. Acad. Sci. USA* **69**:2927–2931.

Paterson, M. C., and Smith, B. P., 1976, Defective excision repair of γ-ray-damaged DNA in human (ataxia telangiectasia) fibroblasts, *Biophys. J.* **16**:183a.

Paterson, M. C., Lohman, P. H. M., and Sluyter, M. L., 1973, Use of a UV endonuclease from *Micrococcus luteus* to monitor the progress of DNA repair in UV-irradiated human cells, *Mutat. Res.* **19**:235–256.

Paterson, M. C., Lohman, P. H. M., De Weerd-Kastelein, E. A., and Westerveld, A., 1974, Photoreactivation and excision repair of ultraviolet radiation-injured DNA in primary embryonic chick cells, *Biophys. J.* **14**:454–466.

Pedrali Noy, G. C. F., Spadari, S., Ciarrocchi, G., Pedrini, A. M., and Falaschi, A., 1973, Two forms of the DNA ligase of human cells, *Eur. J. Biochem.* **39**:343–351.

Pettijohn, D., and Hanawalt, P. C., 1964, Evidence for repair-replication of ultraviolet damaged DNA in bacteria, *J. Mol. Biol.* **9**:395–410.

Poon, P. K., O'Brien, R. L., and Parker, J. W., 1974, Defective DNA repair in Fanconi's anemia, *Nature (London)* **250**:223–225.

Rary, J. M., Bender, M. A., and Kelly, T. E., 1974, Cytogenetic studies of ataxia telangiectasia, *Am. J. Hum. Genet.* **26**:70a.

Rasmussen, R. E., and Painter, R. B., 1964, Evidence for repair of ultraviolet damaged deoxyribonucleic acid in cultured mammalian cells, *Nature (London)* **203**:1360–1362.

Rasmussen, R. E., and Painter, R. B., 1966, Radiation-stimulated DNA synthesis in cultured mammalian cells, *J. Cell Biol.* **29**:11–19.

Regan, J. D., and Setlow, R. B., 1974, Two forms of repair in the DNA of human cells damaged by chemical carcinogens and mutagens, *Cancer Res.* **34**:3318–3325.

Regan, J. D., Setlow, R. B., Carrier, W. L., and Lee, W. H., 1970, Molecular events following the ultraviolet irradiation of human cells from ultraviolet-sensitive individuals, *Adv. Radiat. Res. Biol. Med.* **1**:119.

Riklis, E., 1965, Studies on mechanism of repair of ultraviolet-irradiated viral and bacterial DNA *in vivo* and *in vitro, Can. J. Biochem.* **43**:1207–1219.

Robbins, J. H., and Burk, P. G., 1973, Relationship of DNA repair to carcinogenesis in xeroderma pigmentosum, *Cancer Res.* **33**:929–935.

Robbins, J. H., Kraemer, K. H., Lutzner, M. A., Festoff, B. W., and Coon, H. G., 1974, Xeroderma pigmentosum, *Ann. Intern. Med.* **80**:221–248.

Rupert, C. S., 1975, Enzymatic photoreactivation: Overview, in: *Molecular Mechanisms for Repair of DNA* (P. C. Hanawalt and R. B. Setlow, eds.), pp. 73–87, Plenum Press, New York.

Sambrook, J., and Shatkin, A. S., 1969, Polynucleotide ligase activity in cells infected with simian virus 40, polyoma virus, or vaccinia virus, *J. Virol.* **4**:719–726.

Sasaki, M. S., and Tonomura, A., 1973, A high susceptibility of Fanconi's anemia to chromosome breakage by DNA cross-linking agents, *Cancer Res.* **33**:1829–1836.

Scholes, G., Ward, J. F., and Weiss, J. J., 1960, Mechanism of the radiation-induced degradation of nucleic acids, *J. Mol. Biol.* **2**:379–391.

Schroeder, T. M., Anshütz, F., and Knopp, A., 1964, Spontane Chromosomenaberrationen bei familiäer Panmyelopathie, *Humangenetik* **1**:194–196.

Schuler, D., Kiss, A., and Fabian, F., 1969, Chromosomal peculiarities and "in vitro" examinations in Fanconi's anemia, *Humangenetik* **7**:314–322.

Schuster, H., 1960, Die Reaktionsweise der Desoxyribonucleinaure mit salpetriger Säure, *Z. Naturforsch.* **15b**:298–304.

Setlow, R. B., and Carrier, W. L., 1964, The disappearance of thymine dimers from DNA: An error-correcting mechanism, *Proc. Natl. Acad. Sci. USA* **51**:226–231.

Setlow, R. B., and Regan, J. D., 1972, Defective repair of N-acetoxy-2-acetyl-aminofluorene-induced lesions in the DNA of xeroderma pigmentosum cells, *Biochem. Biophys. Res. Commun.* **46**:1019–1024.

Setlow, R. B., and Setlow, J. R., 1972, Effects of radiation on polynucleotides, *Annu. Rev. Biophys. Bioeng.* **1**:293–346.

Setlow, R. B., Regan, J. D., German, J., and Carrier, W. L., 1969, Evidence that xeroderma pigmentosum cells do not perform the first step in the repair of ultraviolet damage to their DNA, *Proc. Natl. Acad. Sci. USA* **64**:1035–1041.

Shapiro, R., and Klein, R. S., 1966, The deamination of cytidine and cytosine by acidic buffer solutions: Mutagenic implications, *Biochemistry* **5**:2358–2362.

Shapiro, R., and Yamaguchi, H., 1972, Nucleic acid reactivity and conformation. I. Deamination of cytosine by nitrous acid, *Biochim. Biophys. Acta* **281**:501–506.

Shapiro, R., Braverman, B., Louis, J. B., and Servis, R. E., 1973, Nucleic acid reactivity and conformation. II. Reaction of cytosine and uracil with sodium bisulfite, *J. Biol. Chem.* **248**:4060–4064.

Slor, H., 1973, Induction of unscheduled DNA synthesis by the carcinogen 7-bromomethylbenz (A) anthracene and its removal from the DNA of normal and xeroderma pigmentosum lymphocytes, *Mutat. Res.* **19**:231–235.

Slor, H., and Lev, T., 1973, Ultraviolet-induced changes in DNA: Possible confusion of repair and degradation enzymes, *Biochim. Biophys. Acta* **312**:637–644.

Smith, K. C., 1971, The roles of genetic recombination and DNA polymerase in the repair of damaged DNA, *Photophysiology* **6**:209–278.

Söderhäll, S., and Lindahl, T., 1973*a,* Mammalian deoxyribonucleic acid ligase, *J. Biol. Chem.* **248:**672–675.

Söderhäll, S., and Lindahl, T., 1973*b,* Two DNA ligase activities from calf thymus, *Biochem. Biophys. Res. Commun.* **53:**910–916.

Söderhäll, S., and Lindahl, T., 1975, Mammalian DNA ligases, *J. Biol. Chem.* **250:**8438–8444.

Spadari, S., and Weissbach, A., 1974, HeLa cell R-deoxyribonucleic acid polymerases, *J. Biol. Chem.* **249:**5809–5815.

Spadari, S., Ciarrocchi, G., and Falaschi, A., 1971, Purification and properties of a polynucleotide ligase from human cell cultures, *Eur. J. Biochem.* **22:**75–78.

Stich, J. F., 1975, Response of homozygous and heterozygous xeroderma pigmentosum cells to several chemical and viral carcinogens, in: *Molecular Mechanisms for Repair of DNA* (P. C. Hanawalt and R. B. Setlow, eds.), pp. 773–784, Plenum Press, New York.

Strauss, B. S., 1968, DNA repair mechanisms and their relation to mutation and recombination, *Curr. Top. Microbiol. Immunol.* **44:**1–85.

Strauss, B. S., 1974, Repair of DNA in mammalian cells, *Life Sci.* **15:**1685–1693.

Strauss, B. S., and Hill, T., 1970, The intermediate in the degradation of DNA alkylated with a monofunctional alkylating agent, *Biochim. Biophys. Acta* **213:**14–25.

Strauss, B., Scudiero, D., and Henderson, E., 1975, The nature of the alkylation lesion in mammalian cells, in: *Molecular Mechanisms for Repair of DNA* (P. C. Hanawalt and R. B. Setlow, eds.), pp. 13–24, Plenum Press, New York.

Striniste, G. F., and Wallace, S. S., 1975, An *Escherichia coli* endonuclease which acts on X-irradiated DNA, in: *Molecular Mechanisms for Repair of DNA* (P. C. Hanawalt and R. B. Setlow, eds.), pp. 201–204, Plenum Press, New York.

Sugimoto, K., Okazaki, T., and Okazaki, R., 1968, Mechanism of DNA chain growth. II. Accumulation of newly synthesized short chains in *E. coli* infected with ligase-defective T4 phages, *Proc. Natl. Acad. Sci. USA* **60:**1356–1362.

Sutherland, B. M., 1974, Photoreactivating enzyme from human leukocytes, *Nature (London)* **248:**109–112.

Sutherland, B. M., and Oliver, R., 1975, Low levels of photoreactivating enzyme in xeroderma pigmentosum variants, *Nature (London)* **257:**132–134.

Sutherland, B. M., and Oliver, R., 1976, Culture conditions affect photoreactivating enzyme levels in human fibroblasts, *Biochim. Biophys. Acta* **442:**358–367.

Sutherland, B. M., Runge, P., and Sutherland, J. C., 1974, DNA photoreactivating enzyme from placental mammals. Origin and characteristics, *Biochemistry* **13:**4710–4715.

Sutherland, B. M., Rice, M., and Wagner, E. K., 1975, Xeroderma pigmentosum cells contain low levels of photoreactivating enzyme, *Proc. Natl. Acad. Sci. USA* **72:**103–107.

Sutherland, B. M., Oliver, R., Fuselier, C. O., and Sutherland, J. C., 1976, Photoreactivation of pyrimidine dimers in the DNA of normal and xeroderma pigmentosum cells, *Biochemistry* **15:**402–406.

Sutherland, J. C., and Sutherland, B. M., 1975, Human photoreactivating enzyme. Action spectrum and safelight conditions, *Biophys. J.* **15:**435–440.

Tanaka, K., Sekiguchi, M., and Okada, Y., 1975, Restoration of ultraviolet-induced unscheduled DNA synthesis of xeroderma pigmentosum cells by the concomitant treatment with T4 endonuclease V and HVJ (Sendai virus), *Proc. Natl. Acad. Sci. USA* **72:**4071–4075.

Taylor, A. M. R., Harnden, D. G., Arlett, C. F., Harcourt, S. A., Lehmann, A. R., Stevens, S., and Bridges, B., 1975, Ataxia telangiectasia: A human mutation with abnormal radiation sensitivity. *Nature (London)* **258:**427–429.

Todaro, G. J., Green, H., and Swift, M. R., 1966, Susceptibility of human diploid fibroblast strains to transformation by SV40 virus, *Science* **153:**1252–1254.

Tomura, T., and Van Lancker, J. L., 1975, The effect of a mammalian repair endonuclease on X-irradiated DNA, *Biochim. Biophys. Acta* **402**:343–350.

Town, E. D., Smith, K. C.,and Kaplan, H. S., 1973, Repair of X-ray damage to bacterial DNA, *Curr. Top. Radiat. Res. Q.* **8**:351–399.

Tsukada, K., and Ichimura, M., 1971, Polynucleotide ligase from rat liver after partial hepatectomy, *Biochem. Biophys. Res. Commun.* **42**:1156–1161.

Van Lancker, J. L., and Tomura, T., 1974, Purification and some properties of a mammalian repair endonuclease, *Biochim. Biophys. Acta* **353**:99–114.

Verly, W. G., and Paquette, Y., 1972, An endonuclease for depurinated DNA in *Escherichia coli* B, *Can. J. Biochem.* **50**:217–224.

Verly, W. G., and Paquette, Y., 1973, An endonuclease for depurinated DNA in rat liver, *Can. J. Biochem.* **51**:1003–1009.

Verly, W. G., and Rassart, E., 1975, Purification of *Escherichia coli* endonuclease specific for apurinic sites in DNA, *J. Biol. Chem.* **250**:8214–8219.

Verly, W. G., Gossard, F., and Crine, P., 1974, *In vitro* repair of apurinic sites in DNA, *Proc. Natl. Acad. Sci. USA* **71**:2273–2275.

Wagner, E. K., Rice, M., and Sutherand, B. M., 1975, Photoreactivation of herpes simplex virus in human fibroblasts, *Nature (London)* **254**:627–628.

Wang, T. Y., 1967, The isolation and purification of mammalian cell nuclei, in: *Methods in Enzymology,* Vol. 22A (L. Grossman and K. Moldave, eds.), pp. 417–421, Academic Press, New York.

Wang, T. S. F., Sedwick, W. D., and Korn, D., 1975, Nuclear deoxyribonucleic acid polymerase, *J. Biol. Chem.* **250**:7040–7044.

Ward, J. F., 1971, Deoxynucleotides models for studying mechanisms of strand breakage in DNA. I. Protection by sulfhydryl compounds, *Int. J. Radiat. Phys. Chem.* **3**:239–249.

Ward, J. F., 1975, Molecular mechanisms of radiation-induced damage to nucleic acids, *Adv. Radiat. Biol.* **5**:181–239.

Weiss, B., and Richardson, C. C., 1967, Enzymatic breakage and joining of deoxyribonucleic acid. I. Repair of single strand breaks in DNA by an enzyme system from *Escherichia coli* infected with T4 bacteriophage, *Proc. Natl. Acad. Sci. USA* **57**:1021–1028.

Weissbach, A., 1975, Vertebrate DNA polymerases, *Cell* **5**:101–108.

Wilkins, R. J., 1973, DNA repair: A simple enzymatic assay for human cells, *Int. J. Radiat. Biol.* **24**:609–613.

Witkin, E. M., 1964, Photoreversal and "dark repair" of mutations to prototrophy induced by ultraviolet light in photoreactivable and non-photoreactivable strains of *Escherichia coli,* *Mutat. Res.* **1**:22–64.

Witkin, E. M., 1969, Ultraviolet-induced mutation and DNA repair, *Annu. Rev. Microbiol.* **23**:487–514.

Yajko, D. M., and Weiss, B., 1975, Mutations simultaneously affecting endonuclease II and exonuclease III in *Escherichia coli, Proc. Natl. Acad. Sci. USA* **72**:688–692.

Index